SUSTAINABLE MOUNTAIN AGRICULTURE

Farmers' Strategies and Innovative Approaches

VOLUME 2

Editors

N.S. JODHA, M. BANSKOTA and TEJ PARTAP

International Centre for Integrated
Mountain Development, Kathmandu

INTERMEDIATE TECHNOLOGY PUBLICATIONS 1992

Intermediate Technology Publications
103/105 Southampton Row, London WC1B 4HH, UK

Set ISBN 1 85339 114 X
Vol. 1 ISBN 1 85339 130 1
Vol. 2 ISBN 1 85339 131 X

Originally published in India by Mohan Primlani for Oxford & IBH Publishing Co. Pvt. Ltd., 66 Janpath, New Delhi 110 001. Phototypeset by Laserwords, Madras, processed and printed in India at Pauls Press, E44/11 Okhla Industrial Area, Phase II, New Delhi 110 020.

CONTENTS

VOLUME 1

FOREWORD

This document has been produced as part of the work programme of the Mountain Farming Systems Division of ICIMOD. Mountain Farming Systems is one of ICIMOD's four thematic research programmes.

The programme has now completed the first phase of work on Strategies for Sustainable Mountain Agriculture. To complete this phase, problem-oriented research work, in selected areas of the Hindu Kush-Himalayan Region, was conducted. The areas of the Region focussed upon were West Sichuan (China), Himachal Pradesh (India), the North West Frontier Province (NWFP) (Pakistan), and the Middle Hills of Nepal. The work involved thematic reviews of selected development policies and programmes, site-specific studies covering crop, livestock, and horticulture-dominated farming systems as well as other reviews covering specific topics such as underexploited plant genetic resources, mountain beekeeping, and innovative and successful initiatives in the field of agricultural technology and rural institutions.

Work was conducted in collaboration with national institutions and experts from the regional countries. During the course of the studies, different national workshops were held (in China, India, Nepal, and Pakistan) to discuss the findings in the countries where research was being conducted. The experience gained from these national workshops became the input for the International Symposium for Strategies on Sustainable Mountain Agriculture, held at Kathmandu from September 10–14, 1990.

In addition to papers based on the HKH Region, papers were presented on the Andean Region. This volume includes these papers as well as a number of papers selected from background material, produced through the auspices of ICIMOD, in order to give the reader a broader picture of the problems and concerns of Sustainable Mountain Agriculture.

In publishing this volume, ICIMOD is fulfilling a part of its mandate to facilitate the economically and environmentally sustainable development of mountain areas through the dissemination of relevant information. Since agriculture remains the principal occupation of the majority of the inhabitants of the Hindu Kush-Himalaya, it is understood that Mountain Farming Systems merit a great deal of study, especially within the context of long-term sustainability. ICIMOD hopes that this volume will serve to add to the understanding of these systems and to the long-term benefit of mountain peoples.

Several organisations have supported ICIMOD's endeavours in this particular study, and I would like to acknowledge the support of the Asian Development Bank (ADB) and the Ford Foundation in helping us to undertake this work. In addition, the UNDP, the Rockefeller Foundation, and the Swiss Development Cooperation facilitated the participation of various experts in the workshops held. His Majesty's Government of Nepal and a number of other organisations in Nepal have been generous in extending their logistical support. Finally, I would like to thank all those institutions—such as the Chinese Academy of Sciences, Agro-economic Research Centre, the University of Himachal Pradesh, the Agricultural Projects Services' Centre, Nepal, the Aga Khan Rural Support Project Pakistan, and the Planning Commission and Ministry of Agriculture of the Royal Government of Bhutan—which have given this search for Strategies for Sustainable Agriculture their valuable support.

DR. E.F. TACKE
Director, ICIMOD

PREFACE

Economic development has been a long, slow process for much of the developing world. This is especially so for the peoples who inhabit the mountainous region of Asia. In the two decades after World War II development strategies emphasised the importance of growth in per capita income and food production, in the expectation that the benefits of such growth would be universal, ''trickling down'' to the poorest communities. Growth was achieved, but there was little evidence of trickle down. Rapidly increasing food production was brought about by the Green Revolution but the benefits were mostly confined to the flat, well irrigated lands of Asia. Few of the new miracle varieties of rice and wheat were suitable for mountain environments and the technologies that went with them were inappropriate, if not damaging. Development thinking in the late 1960s and 1970s shifted to an emphasis on growth with redistribution and on satisfying certain ''basic needs''—nutrition, health, water supply, shelter, sanitation and education. The new strategies accepted that poverty could not be eliminated solely by economic growth and required a direct attack, mobilising government and aid spending rather than relying on market forces. This has undoubtedly brought considerable benefits. Mountain people are now better served than at any time in the past by schools and primary health care centers, potable water supplies and veterinary clinics as well as by improved roads and access to markets. Yet, at the same time, the resource base on which these people depend has deteriorated at an accelerating rate. The roads and markets have been a mixed blessing, encouraging unbridled exploitation. Forests have been felled, the vegetation cover removed and steep slopes have become severely eroded. Together with exploitation, the impact of population pressure and inappropriate technologies have severely degraded the mountain environment.

Such resource degradation is not unique to the mountains although is perhaps more dramatically obvious there. Elsewhere in the world has come a growing realisation of the perils of the damage we are causing to our natural resources and environment. In 1987 the issues were given worldwide attention by the publication of the World Commission on Environment and Development report ''*Our Common Future*''. In addition to illustrating the problems we face the Brundtland Report, as it is known, called for a new strategy of *sustainable development* as a basis for economic growth in the coming decades.

Following the publication of the report the concept of sustainable development has caught the imagination of policy makers, economic, social and environmental advisers and leaders of the non government community. Policy and programme proposals, project documents and plans are invariably replete with references to the need for sustainability. Yet few use the term precisely. It has come to mean all things to all men and women.

In the Brundtland Report sustainable development is ''development that meets the needs of the present without compromising the ability of future generations to meet their own needs.'' Such a definition resonates with our ideals and our humanity but is not readily translated into a plan of practical action.

Today we are in urgent need of more practical, down to earth interpretations of sustainability. We need to be in a position to ask and answer, with precision, such questions as ''Is this cropping system sustainable?'' ''Is this forest sustainable under the present pattern of exploitation?'' ''Is this watershed sustainable under the present regime?'' ''Is the current lifestyle of these people sustainable?'' And, if the answers are ''no'', how can the sustainability be improved? We also need to be aware that sustainability—the ability of the field or village or watershed to survive may involve costs. There may well be trade-offs in development between sustainability, economic growth and equity for instance. Designing a sustainable, but essentially static, resource management regime may be relatively easy. It will be considerably more difficult to design one that is sustainable yet also provides growing incomes that are equitably shared.

These are the formidable challenges of the 1990s and the 21st century. They are faced by all peoples of the world, but perhaps most acutely by mountain people. The concept of sustainable development faces its severest test in mountain environments. At the same time, the acute and dramatic nature of resource degradation in the mountains brings the issues into sharp relief. Success may well carry lessons for other environments elsewhere in the world.

This book addresses many of the central issues of sustainable agricultural development in the mountains. It provides innovative concepts, detailed case studies and thoughtful analysis. Readers will find a wealth of information, ideas and examples of practical experiences culled from a wide range of countries and mountain environments.

PROFESSOR GORDON R. CONWAY

Representative, India, Sri Lanka, Nepal
and Professor, Centre for Environmental Technology
Imperial College of Science, Technology and Medicine, London

EDITORS' NOTE

The work presented in this book covers various dimensions and issues relating to mountain agriculture in the overall context of development strategies, experiences, problems and prospects in the mountain areas with special focus on the Hindu Kush-Himalayan region. ICIMOD's work on the subject during the last three years supplemented by contributions from other institutions and experts in the International Symposium on Strategies for Sustainable Mountain Agriculture in September 1990, determined the overall scope of the book. Besides adding to the depth and variety of issues, this also increased the size of the book. For the reasons of comprehensive reporting on the subject and protecting linkages between different aspects covered by conceptual and empirical work extending to different geographical areas and relating to a range of themetic contexts, it was difficult to exclude some material to reduce the size of the book.

Concerned with the possible operational problems and convenience of readers, the whole book is divided into two volumes, published simultaneously. The thematic coverage of the two volumes is governed by closer linkages between the issues and evidence covered by different papers. Accordingly, Volume 1 focusses on agricultural development perspectives, approaches, and strategies in the HKH region. The long-term sustainability issues and use of agro-ecological zonation as a tool for development planning in mountain areas are other important aspects covered by Volume 1. Volume 1 consists of 15 chapters.

Volume 2 covers relatively more operational dimensions of the subject where micro-level or project-level realities, experiences and implications are reported and analysed. Chapters dealing with farmers' strategies and some innovative project initiatives are covered by Volume 2. Volume 2 consists of 18 chapters.

N.S. JODHA, M. BANSKOTA and TEJ PARTAP (editors)

ACKNOWLEDGEMENTS

Acuteness of environmental degradation and poverty in mountain regions of the developing countries is now well recognised. The problems are sharply manifested by the stagnation and decline of Agriculture the major activity of mountain people. Understanding of the factors and processes contributing to this decline and identification and adoption of possible solutions is a herculean task. This is more so due to specific conditions characterising mountains and general disregard of mountain regions in the past by researchers and policy makers alike.

To address the above problem ICIMOD attempted to mobilise available knowledge and evidence on different aspects of the problem through an International Symposium on Strategies for Sustainable Mountain Agriculture and subsequently present the relevant material through the two volumes before all those associated with development of mountain areas. We take this opportunity to thankfully acknowledge the help of all those who have contributed in this task.

Our thanks are due to the authors who not only managed to prepare and revise their contributions under tight time schedules but many of them presented their evidence and argument in the overall context of mountain-perspective sustainability framework suggested by us.

We would also like to thank Mrs. Greta Rana and Mrs. Archana Karki for their assistance in language editing. This made the job easier for the publishers. The secretarial assistance from Mrs. Sami Joshi and Reeta Rana deserves praise specially for the hardwork they did in typing the manuscript time and again for several months.

Finally we thank ICIMOD Director and all our professional staff colleagues. Their cooperation and involvement in programme activities from time to time saw successful completion of the project and production of this book.

N.S. JODHA, M. BANSKOTA and TEJ PARTAP (editors)

THE CONTRIBUTORS

Bajracharya B.B.
Centre for Economic Development and Administration (CEDA)
Tribhuvan University
P. O. Box 797, Kirtipur
Nepal

Bajracharya D.
Division Head
Mountain Population and Employment Division
ICIMOD, G.P.O. Box 3226, Kathmandu, Nepal

Banskota M.
Chief Programme Coordinator and Programme Head,
Area Development Planning and Implementation
ICIMOD, G.P.O. Box 3226, Kathmandu, Nepal

Bhati J.P.
Department of Economics
Himachal Pradesh University
Shimla, India

Burathoki K.
Lumle Agriculture Centre (LAC)
P. O. Box 1
Pokhara, Kaski, Gandaki Zone
Nepal

Camino A.
Director
Fundacion Peruana
Para la Conservacion de la Naturaleza (FPCN)
Chinchon 858/A
San Isidro, Aptdo.18–1393
Lima, Peru

Carson, B.
Project Leader
Master Plan for Horticulture, HMG,
East Consultants Co. Ottawa,
Canada

Chand S.P.
Agronomist
Pakhribas Agriculture Centre (PAC)
c/o BTCO, P. O. Box 106
Kathmandu, Nepal

Dafu Y.
Chengdu Institute of Mountain Disaster and Environment
Chengdu, P.O. Box 417
Sichuan, China

Denholm J.
Mountain Farming Systems Division
International Centre for Integrated Mountain Development (ICIMOD), G.P.O. Box 3226,
Kathmandu, Nepal

Fei W.
Chengdu Institute of Mountain Disaster and Environment
Chengdu, P. O. Box 417
Sichuan, China

Gupta A.K.
Centre for Management in Agriculture,
Indian Inst. of Management
Vastrapur, Ahmedabad
India

Hongbin G.
Leading Group Office for the Economic Development of Poor Areas
Beijing
China

Husain T.
Managing Director
Development Research and Management Services
P.O. Box 2389
Islamabad, Pakistan

Jodha N.S.
Division Head
Mountain Farming Systems Division
International Centre for Integrated Mountain Development
(ICIMOD), G.P.O. Box 3226, Kathmandu, Nepal

Joshi B.R.
Lumle Agricultural Centre (LAC)
P. O. Box 1
Pokhara, Kaski, Gandaki Zone,
Nepal

Katwal B.
Agriculture Projects and Services Centre (APROSC)
G. P. O. Box 1440
Ramshah Path
Kathmandu, Nepal

Karim Z.
Mountain Environment and Natural Resources Information Services (MENRIS)
ICIMOD, G.P.O. Box 3226, Kathmandu, Nepal

Keatinge J.D.H.
International Centre for Agricultural Research in the Dry Areas (ICARDA)
P.O. Box 5466, Aleppo, Syria

Khan R.
Director
Arid Zone Research Institute
P.O. Box 362
Quetta, Pakistan

Kotta P.K.
Mountain Environment and Natural Resources Information Services (MENRIS)
ICIMOD, G.P.O. Box 3226, Kathmandu, Nepal

Lundberg, P.A.
Chief Technical Advisor to Planning Commission
HMG Nepal on Regional Development Planning
United Nations Development Programme (UNDP)
Pulchowk
Kathmandu, Nepal

Mr. Mulk M. Ul
Aga Khan Rural Support Programme
District Programme Office, (AKRSP)
NWFP, Pakistan

Mya S.
Mountain Environment and Natural Resources Information Services (MENRIS)
ICIMOD, G.P.O. Box 3226, Kathmandu, Nepal

Mr. Nakarmi G.B.
Geologist
Integrated Survey Section Topographic Survey Branch
P. O. Box 1611, Min Bhawan, Kathmandu, Nepal

Ning W.
Chengdu Institute of Biology
Academia Sinica
P.O. Box 416, 610015
Chengdu, Sichuan, China

Partap T.
Mountain Farming Systems Division
ICIMOD, G.P.O. Box 3226, Kathmandu, Nepal

Pound B.
Director
Lumle Agricultural Centre
c/o BTCO, P. O. Box 106
Kathmandu, Nepal

Pradhan P.
Mountain Environment and Natural Resources Information Services (MENRIS)
ICIMOD, G.P.O. Box 3226, Kathmandu, Nepal

Rathore M.S.
Senior Research Fellow
Institute of Development Studies
B-124, Mangal Marg
Bapunagar, Jaipur
India

Rhoades R.E.
Professor and Head

Department of Anthropology
University of Georgia Athens
Georgia, 30602
USA

Sharma K.S.
Integrated Development Service
New Baneswor
Kathmandu, Nepal

Sharma L.R.
Agroeconomic Research Centre
Himachal Pradesh University
Shimla, India

Sharma P.
Mountain Population and Employment Division
ICIMOD, G.P.O. Box 3226, Kathmandu, Nepal

Sharma S.
Mountain Infrastructure and Technology Division
ICIMOD, G.P.O. Box 3226, Kathmandu, Nepal

Shrestha S.
Mountain Farming Systems Division
ICIMOD, G.P.O. Box 3226, Kathmandu, Nepal

Tapia M.E.
Proyecto de Investigacion
de Sistemas Agropecurios Andenos (PISA)
INIAA, Box 110697
Lima 11, Peru

Thapa B.
Pakhribas Agricultural Centre
Dhankuta District
Koshi Zone, Nepal
c/o BTCO, P.O. Box 106
Kathmandu, Nepal

Ura K.
Planning Officer
Planning Commission
Royal Government of Bhutan
Thimpu, Bhutan

Velde E.J.V.
Professor
International Irrigation Management Institute (IIMI)
Country Office, Islamabad, Pakistan

Verma L.R.
Mountain Farming Systems
ICIMOD, G.P.O. Box 3226, Kathmandu, Nepal

Xingqing Y.
Leading Group Office for the Economic Development of Poor Areas
Beijing
China

Yanhau Liu
Institute of Geography
Chinese Academy of Sciences (CAS)
Beijing, 100012
China

Yadav Y.
Agricultural Projects and Services Centre (APROSC)
Post Box 1440,
Kathmandu, Nepal

Zhaoguang L.
Director
Chengdu Institute of Biology
Academia Sinica
P. O. Box 416, 610015
Chengdu, Sichuan, China

Zimmerer K.
Assistant Professor
Department of Geography
University of Wisconsin, Madison
WI 53706, USA

PART 4

Mountain Farmers' Strategies and Sustainability Implications

16

FARMERS' INNOVATIONS AND AGRICULTURAL TECHNOLOGIES

A.K. Gupta

Anil K. Gupta is a professor at the Centre for Management in Agriculture, the Indian Institute of Management, Ahmedabad, India

CONTENTS

INTRODUCTION

Mountain regions are characterized by low population densities, weak market infrastructures, high levels of emigration (particularly of males), the predominance of female-headed or female-managed households, money order economies,[1] poor linkages between the formal and informal sectors, and low levels of social and political articulation.

The value of traditional skills has gone down over time to such an extent that in many parts of the world these regions are seen as a reservoir of 'unskilled labour'. It is not without significance that the majority of emigrating labour occupies a low status employment 'niche' in urban or rural labour markets. Such a low rating of their skills by society leads to a decline in the pride of mountain peoples in their indigenous methods of resource management. Often, their low self-esteem is reinforced by public policies which fail to even recognize (much less to appreciate) the strength of indigenous institutions and traditional technological innovations. Jodha (Chapter 2) and Sanwal (1989) have rightly argued that despite a great deal of talk about integrated mountain development, the approaches are often segmented, sectoralised, and, in some cases, even dehumanizing. Official documents on the development of backward areas in India have, in fact, cautioned against an attempt to stem the migration of people from such regions elsewhere lest the supply of 'cheap' labour for infrastructural development projects be cut off!

It is in this context that we have to look at the strength of innovations developed by mountain people all over the world. Given the high ecological heterogeneity, it is not conceivable that technological transformation through diffusion of standardized technologies can take place. Organizational rewards for developing technologies that can diffuse only in small, localized 'niche' have been seen to be generally poor (Gupta 1985). Thus, eliminating the mismatch between design of R&D organizations and the expectations of the people is the central issue for recasting development strategies for sustainable mountain agriculture. We recognize that not all technological alternatives can be either anticipated or demanded by the farmers.[2] Therefore, we do not believe that future options will increase without simultaneously increasing the responsibility of the supply system to widen the decision-making horizon of the farmers. At the same time, we do recognize that the study of farmers' innovations, both technological and institutional, can broaden the vision of the scientists themselves (Gupta 1987b, 1987c).

This paper presents an analytical framework to look at the eco-institutional aspects of the choice of technology, and then briefly reviews some of the innovative technologies and institutions developed by the farmers and the issues involved in their search and scrutiny.

---

1 The implications of this for the nature of cash flows and attendant patterns of demand have to be carefully worked out. If surplus exists in even a few hands, without suitable opportunities for investment, it can establish wrong role models. The increasing influence of liquor consumption in many of the hill areas receiving inward remittances is a case in point.

2 I strongly disagree with those who argue that farmers can always demand what they need. It has to be appreciated that our capacity to demand what we need depends upon our prior experience with the supply side, exposure to various alternatives, the skill to convert a need into a demandable output, and an understanding of the language in which the supply side understands our need as being distinct from a demand.

An Eco-institutional Framework for Analysing Choice of Technology

Human choices in a given eco-sociological configuration are circumscribed by the historical evolution of institutional structures. Institutions provide a framework of rules, sanctions, and meanings that is commonly understood by people grouped within a common boundary. In a way, institutional behaviour relies more and more upon internal commands rather than upon external demands. However, a combination of both moral and material sanctions provides legitimacy to an institution. In the present context, we are drawing upon another feature of institutions which, in the context of farming systems research in mountain regions, is extremely vital; that is the assurance provided by the institutions—formal and informal—to individuals and groups about various uncertainties faced over time and space.

We deal with mainly two types of assurances—horizontal and vertical. The former type includes the assurances that provide guarantees about others' behaviour vis-a-vis one's own. Thus, if I sow my crop early will others also? Or if I do not graze my animals on common land will others also not do so? Vertical assurances refer to the future returns from present investments. If I plant trees on common or private land will I be allowed to harvest them? Or if I apply organic fertilizer to a particular plot of land taken on lease will I be allowed access to it next year also (in view of the slower release of nutrients from the organic fertilizer)?

Assurances by themselves, however, are not sufficient. If I have assurances of better prices or better returns for certain kinds of collective behaviour but I do not have *access* to the given resource, or I do not have the skill or ability to convert a resource into an investment, or both, then assurances are of little use.

Assurances help in generating cooperative behaviour when we deal with common properties (Sen 1967, Runge 1986, Gupta 1985). In the case of private resources, assurances may stimulate demand for better access or technical skills or both. Likewise, if we have an institution in which people have access to resources and also have assurances, but do not have the skills or abilities, the investments will not follow.

All the three vectors of choice, that is, *access*, *assurances*, and *abilities*, must be synchronized to generate appropriate attitudes for change or maintenance of a resource use system. Thus, within a specific spatial, sectoral, and seasonal configuration, portfolios may vary within a given range because of changes in access, assurances, and abilities.

As we note in Fig. 16.1, the access to natural resources, to assurances from the institutions, to ability in terms of technology, and to attitudes in terms of culture, collectively influence the household portfolios. This framework also helps in designing interventions. Thus, if we want to introduce technologies that presuppose the existence of certain skills, access modes, or institutional structure, but some or all of these vectors are missing, we should not fault people for not using the given opportunity. It may be useful, therefore, to recognize that this framework can be used as a tool or as a filter to assess available information and generate further choices. If we know the given complexity of access and the abilities of the people in a given system, we should be able to anticipate what kind of assurances will generate or respond to the given attitudes. Attitudes here are both an outcome of historical experiences and inputs into future choices. The culture, I must add, does become modified over a period of time.

Ecology	Traditional	Modern				
Space	Season	Sector	Formal	Informal	Technology	Culture
Public	Private	Communal				
Access						
Assurance						
Ability						
Attitude						

Figure 16.1: Eco-institutional framework for analysing choice of technology
Source: Author's compilation

The same framework can be used to analyse the supply side that constitutes the response of scientists to various types of problems or social situations. For instance, if scientists do not have (1) assurance of peer approval (collective choice or horizontal assurances) or (2) career rewards (vertical assurances), but have (3) access to the facilities for on-farm research and also have (4) skills for performing experiments, we should not be surprised if they develop attitudes that are conservative or non-enterprising. In the same manner, changes in different parameters may help us to identify the corresponding changes required in other parameters.

The ecological dimension can be further looked at in terms of space, season, and sectoral interactions. Institutions could be traditional or modern, formal or informal, public, or communal. The technology could be based on local inputs or external inputs and culture may reinforce compliance (because of a feudal past), or trigger innovations, or both. For the sustainable development of mountain regions, we have to appraise every intervention through the matrices given in Figs. 16.1 and 16.2. We can, as a consequence, anticipate the likely changes that will come about in various subsystems if interventions are made to modify access, assurance, ability, or attitudes. An important point of departure in this framework is that sustainability is being defined primarily in its institutional context.

The socio-ecological paradigm (Gupta 1984, 1985, 1989, 1990a) provides the basis for understanding the choice of technology through the interaction of ecological, technological, and institutional variables. The eco-institutional framework provides additional linkages to the cultural core of the mountain society or other high-risk environments and the attitudinal basis for their current behaviour. It is accepted that inappropriate policies in the recent past have changed attitudes significantly in several parts of the world.

The 4-S Model

Several studies on farmers' adjustments to risk have shown a multi-market, multi-enterprise, and multi-institutional approach to survival (Jodha 1975, 1978, Jodha and Mascarenhas 1985, Gupta 1981, 1984, 1988, 1990a, Ostrom et al. 1989). The multi-market approach refers to the farmers' attempts to adjust to risks through simultaneous operations in different factor and product markets. The factor markets include land, labour, capital, and even information. The product markets include crops, livestock, and trees as well as various technologies of land and water use. The higher the risk in the environment, the greater the dependence between the decisions made in one resource market and those made in others. These links are important in well-developed regions also but, in these regions, many imperfections in respective markets can be offset through market mechanisms themselves over time and space. In high-risk environments, the cost at which these errors may be corrected will be far higher, and thus there is greater dependence on inter-market adjustments.

The multi-enterprise approach implies that farmers' adjustments to risks or the evolution of portfolios cannot be understood by concentrating on any one enterprise such as crops, livestock, labour, or trees. The 4-S model helps in understanding these linkages at the macro-level.

The multi-institutional perspective is helpful because various resources or enterprises, as mentioned earlier, may be governed by various kinds of property right regimes, in combination or separately. Livestock, for instance, may be managed by some households through biomass derived from private lands only. In other cases, it may be derived from private as well as common and/or open access lands. Thus, various institutional arrangements, whether or not regulated by the State, market, or both, further influence the choices at micro-level. Any framework that ignores the multi-market, multi-enterprise, and multi-institutional dimensions of household portfolios will generate only a partial understanding of the survival logic of the people. The innovative technologies or institutional arrangements are a part of dealing with these complexities. Innovations for survival sometimes may follow rules that are different from innovations for accumulation.

To use the 4-S model we use a three-dimensional matrix as shown in Fig. 16.2. Each dimension can be dichotomized for the purpose of creating ideal types. The basic principle of logic that we use here is 'compare and contrast'. If we want to understand a phenomenon it is useful to begin by comparing and contrasting the extreme values of its distribution. For instance, 'space' can be dichotomized in terms of high or low land, undulated or plain topography, higher slope or lower slope in the mountain regions. Likewise, 'sector' can be dichotomized as agriculture or industry, public or private, specialized or diversified, and single crop or diversified crop combinations, cash crop or food crop dominated. 'Season' can also be divided into uni- or bi-modal rainfall regimes, arid or humid, low or high rainfall, low or high diurnal temperature variations, or low or high seasonal fluctuations. (This is essentially the dimension or time with which the uncertainty is associated.)

Given any two parameters we can speculate about the third. For instance, in a region with low population density and high seasonality (low rainfall and high diurnal temperature variations in the arid plains and low diurnal temperature variations at high altitudes) the sectoral characteristics may be highly diversified. Instead of a single crop, farmers may prefer mixed or intercropping in several plots, if not all. Households may simultaneously pursue many activities such as crops, crafts, and livestock rather than being dependent on any one of these. The social exchange relations in such regions will be quite different from those in the regions with high population density, low seasonality, and specialized sectoral activities or diversification for accumulation rather than survival.

Some of the characteristic ways of social exchange relations may include the following: the predominance of kinship and external family networks over the nuclear family system to hedge risks; the preponderance of non-monetary exchanges and the informal mechanisms of pooling bullocks, implements, and inputs;[3] dominance of generalized reciprocities over specific ones;[4] and choice of a much longer time frame to settle accounts.[5]

The communication systems in these regions are more metaphorical or *analogical* than digital. The strategies of technology transfer in on-farm research and extension systems would obviously have to be tailored to the typologies that can emanate from the simple matrix given above. One can make it more complex and generate richer insights but parsimony always has a price. I must acknowledge that the nature of institutions and market interventions can modify the initial conditions that may be predicted by the configuration of spatial, sectoral, and social variables. It might appear that some of the social relations are defined by the ecological variables in a deterministic manner. We have seen that the relationship between pastoral and cultivating communities in the Swiss Alps (Netting 1972), Northern Pakistan (Buzdar 1988), Bhutan, and some other Himalayan mountain regions (Gupta and Ura 1990) have striking similarities, although specific parameters may vary due to cultural and religious differences. Over time, however, formal

---

3 Anthropologists have provided rich insights into the pooling mechanisms in various societies living in hill areas, arid regions, or forests. The pooling of bullocks in Maharashtra, for instance, is called 'irjik'. As many as 10 to 12 pairs of bullocks can be seen ploughing the land in a particular catchment area across the fields on a specific gradient. Since the moisture recedes faster near the ridge line, the plots along the contour towards the ridge have to be ploughed first. It is possible that some people may contribute one bullock pair though they may have only half an acre or even no land in that areas, while others may have a much larger tract of land in this area and yet contribute the same pair of bullocks. The obligations for feeding the cattle and the ploughman are also worked out in many diverse ways. What is important to understand is that uncertainties over time and space may generate reciprocities that may be settled over a longer period of time and thus generate rationality of choice in the short term.

4 The generalized reciprocities refer to exchange of labour for thatching huts with labour contributed for ploughing the land. It is very difficult to work out the equivalence between such related activities. How critical thatching is before the rains, only a poor family living in such a hut can realize. Likewise, the criticality of draft power in receding moisture conditions in light soil regions can be understood by someone who may miss the entire season in the event of failure to sow the crop at the right time. Traditional economic theories are of limited help because equivalence is not just the value of labour as assessed in the market place. Sometimes help provided in such a context may generate an IOU that can be redeemed much later. The specific reciprocities on the other hand refer to exchange of the same goods or services. I have paid for your tea today, you should pay for me tomorrow or I have given you five kilogramme of wheat seeds and you return the same amount of the same crop later. Commercialized societies often would have a dominance of specific reciprocities. Decision-making with constrained resources cannot be analysed without looking at these reciprocities.

5 Studies have shown that IOUs are settled in the regions described here over far longer time periods extending sometimes to several generations. A good or bad turn may invoke a return gesture not necessarily on the same day or in the same month or even the same year. Even the nature of factional leadership remains divided at village level for longer periods than at the State or national level where loyalties can shift quickly without generating problems of legitimacy or social acceptance.

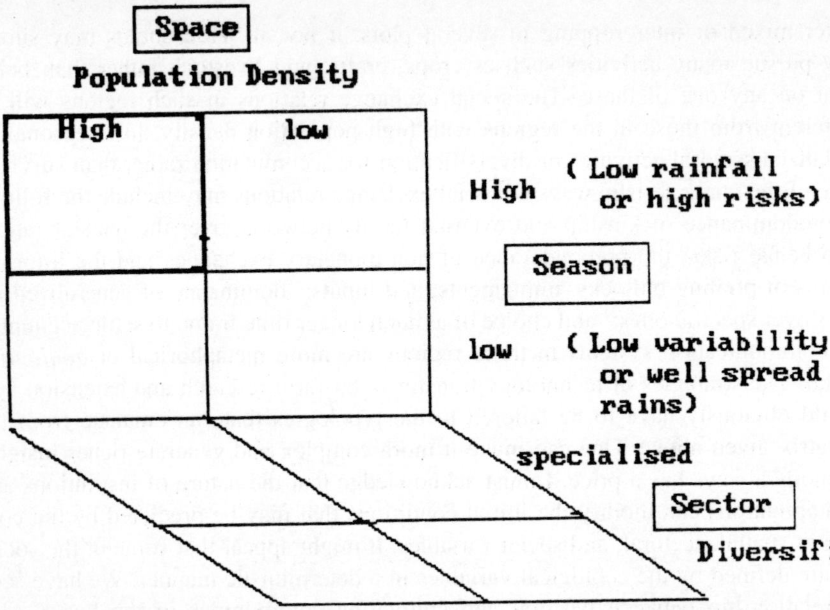

Figure 16.2: Three-dimensional matrix

institutional inroads and market developments do modify these strategies. The availability of walkie talkies means that Swiss pastoralists do not have to develop specific whistling styles as observed in the Andean mountains or in the Himalayan Mountain regions. However, the need for surviving collectively is felt in almost all such socio-ecological conditions.

Socio-ecological Framework

I make two assumptions: (1) ecological conditions define the range of economic choices that can be sustained in a given region and (2) the scale on which different enterprises are selected, however, is a function of the access to factor and product markets; kinship networks, public, private, and common institutions; and historical resource reserves. Instead of calling the framework socio-ecological, as I have so far, I shall now call it eco-sociological because of the dominance of the ecological dimensions of the socio-economic processes.

Earlier it was assumed that in any given ecological 'niche' only certain economic enterprises were feasible at the given level of technological and institutional infrastructure. However, I modify this condition to suggest that the ecological endowments of the proximal environment of a particular social community need not be the major determinant of portfolio. Distant environments where the community has customary or traditional rights through migration, or other mechanisms, have also to be taken into account.

Thus, once a mix of enterprises or a portfolio is selected, drawing upon resources from private, public, and common properties, the nature of risk inherent in this portfolio can be analysed through a matrix of mean or average return and variance in returns. The high mean, low variance portfolios would obviously have different implications for individual and collective behaviour than portfolios with low mean and high variance.

Given an initial portfolio and its mean variance or risk return characteristics, households may respond to a given risk in the environment and any of the following alternative means: household-level risk adjustments; public and market risk-reducing mechanisms; and communal and common property risk adjustments.

Household risk adjustments can be further analysed at intra-household levels and inter-household levels. The intra-household risk adjustments include measures that the household employs by negotiations within the household. For instance, asset disposal, migration, and reduction or modification of family consumption. The inter-household risk adjustment strategies include tenancy, borrowing, labour contracts, and group ploughing.

Public risk-adjustment mechanisms imply the availability of drought and flood relief insurance mechanisms, public employment programmes, etc. The market-based risk adjustment option includes forward trading, the interlocking of factor and product markets, insurance cover, and so on.

Communal risk-adjustment strategies refer to the group-based measures that require collective decision-making either to use or to preserve private or common property resources. The pooling of resources, such as bullocks or implements, is also a part of communal risk-adjustment strategies.

Once the range of risk-adjustment options is known the households may modify either their perceptions or their response or both by changing the discount rate or time frame used for appraising returns from each investment. Thus, while discount rates capture the control the household has in a given resource market, the time frame may capture the certainty with which the household views a particular resource stream. In fact, either of the two can be used to derive risk preference. The shorter the time factor in which households (or scientists) appraise their choices, the less likely it is for technology to be sustainable. Development, I have argued, is nothing but widening the decision-making horizon and extending the time frame of disadvantaged households (Gupta 1981). It is obvious that not everybody's choices can be widened at the same time and in the same proportion given the limitations of resources in a developing society. It is at this stage that an eco-sociological framework has to become an eco-political framework. Constraining the choices of some while widening those of others is an institutional issue which is discussed elsewhere (Gupta 1990b).

The uncertainty of an outcome may vary differently for different households depending upon previous experience with a particular enterprise or crop; immediate past experience; successive losses or gains; accumulated defects or surpluses in the household cashflow; future expectations of returns; and complementarity between other assets or enterprises and the proposed investment.

The cashflows of the households resulting from a given portfolio, modified by various risk-adjustment options, may be in surplus, deficit, or subsistence. In addition, the variability in these cashflows may be evened out over space, season, sector, and social networks. The stakes of different social groups in the management of ecological systems will vary in each resource market.

The trick is to develop a calculus in which unequal stakes of different groups in various resource systems or regimes generate a set of expectations that are equitable or appear equitable (given differences in cultural and social ways of perceiving returns) at the portfolio level of the households. The group-level estimation of the aggregated effects of individual portfolios may generate rules that modify the conditions for use of resources, technology, and institutions. Under extraordinary circumstances, the cultural norms are also modified to accommodate ecological and sociological imperatives.[6]

The household budget influences the choices differently than would be the case if the budget was even, that is, sufficient for subsistence, or if it was in surplus, that is, more than subsistence. A number of researchers have mistakenly grouped deficit budget groups with surplus ones. A sustained deficit may shift the portfolio in favour of low mean, low risk assets and in some cases low mean, high risk assets, provided the risk is not co-variant. In some cases, low mean, high risk assets can be accommodated in the portfolio also, because much of the cost is transferred on to the open access or common property resources. Sheep herding is one such example. Stall-feeding is rarely practized and sheep herds are characteristically maintained by some of the poorest households.

At aggregate level, shifts in the portfolio can be seen by differential growth rates of various species and varieties of crops, trees, and livestock. Public policy at the macro- and micro-level influences the portfolios through changes in the access modes, assurances (through various risk-adjustment strategies), and abilities. The attitudes are also modified by the expectation of future changes in the various sources of subsistence in future.

Changes in the individual stakes in various resource systems feed back into ecological conditions. Once the ecological conditions are modified, the changes in the enterprise mix become inevitable. It may be necessary to note here that I am not underplaying the importance of changes in the institutional conditions or the technological choices as already mentioned earlier. However, a multi-stage or a multi-plane analysis requires that we do not mix the assumptions necessary for analysis on one plane with the assumptions relevant on another plane. These frameworks can be used to understand the context in which people survive at altitude.

The innovations in hill areas were necessary because the complexity of the environment would not permit any one set of strategies to sustain livelihood indefinitely into the future. Since technology cannot be understood without looking at its ecological, cultural, and institutional context, we will try to isolate the patterns in the innovations that hill people have tried to develop. It is obvious that the systematic taxonomy of innovative strategies in different parts of the world will require a much wiser collaborative effort. This paper only illustrates the legitimacy and the feasibility of such an intellectual adventure. Various dimensions identified by Jodha (Chapter 2), as a part of the mountain perspective, provide a very rich basis to begin the synthesis of the analytical framework, which should be responsive to the strengths of people's own technologies, institutions,

---

6 Aggarwal (personal communication) provides an interesting example of a village in which the punishment for poaching on a common property was to offer grains to the birds while standing barefoot under the sun. Such logic cannot be analysed in the classical tradition of institutional analysis. Such sanctions cannot be justified on economic grounds at all. The reciprocities here extend to claimants of resources that do not have a vote, that is, the birds. But in the process, suffering in public by standing barefoot in the sun generates a collective responsibility. It is recognized that the moral appeal may have a longer-lasting effect than an economic tax or fine. The public display of the punishment may also generate guilt.

The cultural norms for individual and group behaviour thus do modify the perception and response to the risks and resources. While the fuelwood crisis may generate a tendency to poach, the sanctions generated by institutions may safeguard, to some extent, the scarce and depletable common property resource. In Southern Bhutan we came across a case in which a group of villagers had restricted the introduction of a male bull of exotic breed lest the local breed become adversely affected (Gupta and Ura 1990). Even today, many villages in South Asia follow a similar practice.

and culture. Such a framework will have to explicitly reject the possibility of any person in a high risk environment being thought of as totally 'unskilled'.

FARMERS' INNOVATIONS: SEARCH AND SCRUTINY

Several researchers have identified the barriers to scientific perception (Chambers 1983, Chambers et al. 1989) and curiosity (Gupta 1985, 1987a, 1987b, 1988a, 1988b, and Richards 1989). The assumption that the low level of literacy is responsible for the backwardness of such regions is criticized by Richards (1989). He argues that such an approach discounts heavily the *oral* and *practical* skills that have been developed over hundreds of years. The success of technological change in high-growth regions should not blind us to the richness of the ethnic basis of local knowledge systems. Studies have shown that the more diverse the environment and the lower the population density, the greater the need for social networking. Box (1988) argues that knowledge networks provide a platform for farmers to satisfy their curiosity about the different innovations being tried by different people, not always successfully. It is important to note that communications among people about innovations very often are not purposive. Often, the search for innovations is a set of continuous events rather than discrete events, accidents, or milestones.

The process of innovation involves constant experimentation, improvisation, adaptation, and simultaneous rejection of certain results either partly or completely depending upon individual or collective feedback. Many times, while searching for innovations, people have drawn negative inferences about the innovative potential of peasants. They were either looking for the wrong things, or looking through inappropriate prisms, or asking the wrong questions in the wrong places.

The interest in indigenous innovations in the Indian subcontinent and China has been there for several centuries, although the intensity of the interests has varied (ICAR 1964). Munshi (1952), in his lecture 'The Gospel of Dirty Hand' to agricultural scientists, highlighted the relationship between soil and soul. He emphasized the need for close cooperation between farmers and scientists. More recently, Verma and Singh (1969) and Verma (1967) provided a rich account of the indigenous innovations that animal husbandry farmers had developed in the hill areas of Himachal Pradesh as well as the then Punjab. Verma (1967), Dharampal (1983), and several others have done considerable work in India on local innovations. It is a pity that scholars in the third world have often ignored the paths opened by local studies and have tried to follow a trail only when it emerged from the West. Even so, nothing much would be lost if that were to happen in a proper manner. The damage is really done when, in place of culturally rooted concepts and terms, we try to analyse or catalogue the innovative genius of local peasants in alien concepts or categories. Such an approach often results in the indigenous knowledge of the people becoming inaccessible to us.

The lack of interest among scientists (biological or social) concerning innovations could be a function of the evaluative criteria. The scales by which they evaluate local innovations may be calibrated by the cultural biases inherent in western philosophy. Tillman (1988), Warren (1986), and Dharampal (1983) highlighted the scientific and technological strengths in the Indian subcontinent which were acknowledged to be superior to some of the western technologies available in the 18th century. Part of the reason for the decline of some of the traditional, more sustainable technologies could be that innovations of a collective nature (Osti 1988), concerned with survival through collective action or sharing (Gupta 1988), were generally less well integrated into the formal networks of institutions. We know much less about what worked and much more about what did not.

Prain (Rhoades and Bebbington 1988) described the case of a farmer in Chicche village in Montoro Valley who developed the hypothesis that varieties expressing apical dominance would yield fewer but larger potato tubers. The market price for such tubers was also higher. There are several other examples which Rhoades gives about the curiosity experiments pursued by farmers, sometimes alone but sometimes collectively. He also illustrates the problem-solving and adaptation experiments pursued by farmers. The former included an attempt by farmers to drive away the Andean weevil by sunning the potatoes (as noted by Prain). The fact that aphids were attracted to green sprouts and not to red sprouts was another observation. The diffused light method of storing potatoes is too well known to warrant a repetition as an illustration of adaptive experiments.

An important conceptual contribution made by Rhoades and Bebbington (1988) is that wherever a transition between two major vegetative communities or biomes exists, the probability of experimentation by farmers is much greater. This has an operational implication for studies on innovation. (Fig. 16.3) Innovations could possibly come about through 'comparing and contrasting' the opposites, i.e., farmers with extremely divergent practices (possibly using ecological maps).

What Rhoades suggests can be linked to the 'compare and contrast' idea. It would thus be useful to pursue such explorations from the transitional boundaries towards the centre of the niche. This has a sociological implication also. The social communities that inhabit the transitional zones are also often new settlers or people from outside. Therefore, experimentation is a necessity for survival. Scientists desirous of looking for innovative practices may be disappointed if they try to concentrate only on the clustered settlements or densely cultivated 'niche'. Verma (1967) described several practices with regard to navel cutting, disposal of the placenta, and therapeutic measures against ingestion of the placenta.

Recognizing that even the availability of human health services is scarce in various developing countries, the possibility of providing animal husbandry services in mountain regions would definitely be an even more distant goal. In a way there may be a virtue in this vice. Indigenous innovations that evolved to manage pasturelands over large stretches are equally fascinating. Our accompanying paper (Chapter 23) provides an example about the way yak and cattle herds coordinate their movements so that the two herds do not meet (because some disease may be transferred from the cattle to the yak). The paper also discusses imaginative institutional rules that nomadic peoples or communities have evolved in consultation with settled communities to manage the access to pastures and the exchange of livestock goods. Equally rich illustrations are available regarding cropping systems, forestry systems, and the interactions among crops, livestock, trees, and tools.

Disadvantaged households should never be called resource-poor, as long as we believe that knowledge is a resource. Formal institutions may not price this knowledge properly or build upon it adequately.

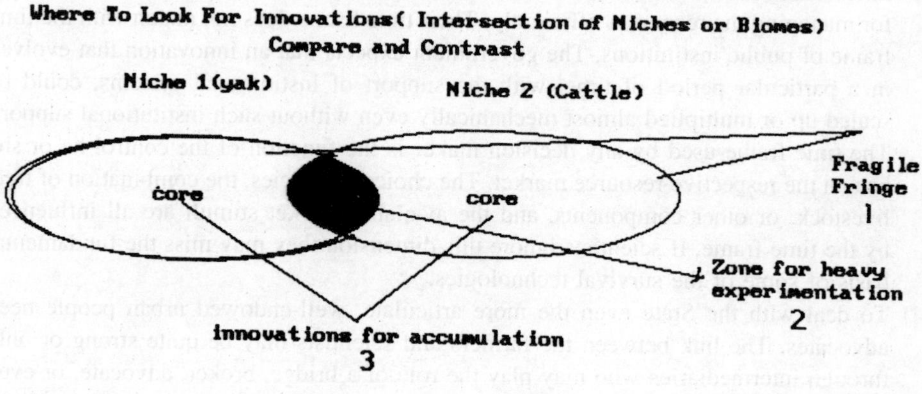

Figure 16.3: Where does one look for innovations?

Interaction between Scientists and Innovating Farmers

Recently we analysed the linkages between demand and supply through the following interfaces (Gupta et al. 1988, Gupta 1987a, 1987b, 1987d, 1989).

(1) The needs of the well-endowed areas and sections of society often merged with the objectives of the research system because market forces were there to steer. Input selling agencies cannot sell many inputs in hill areas, or in flood or drought-prone regions. With all the uncertainties of transportion and insurance costs, why should these institutions locate distribution points in the mountains? In the absence of distribution points for various inputs, located nearby, the real price of using many technologies becomes very high and farmers rationally avoid trying otherwise useful technologies. There is an essential need to modify the policies of public finance for this purpose, for example, with freight equalization policies aimed at reducing the transportation costs or incentives for smaller packages (of fertilizers, for example) through rebate in excise and sales tax.

The markets acted as monitors for only certain types of links. For example, suggestions from the association of industries or the chambers of commerce are often heard by finance ministers before they finalize their budget proposals. Such a consultation is unlikely to take place with disadvantaged households inhabiting mountain regions. As if to compensate for the isolation from the market and the public institutions, people evolve knowledge networks embedded in very strong kinship networks. Policy makers will have to appreciate the dynamics of these knowledge networks if strategies for sustainable technological change in mountain areas are to evolve.

(2) Given the poor articulation of their needs, farmers do not even demand a different type of technology when the opportunity for such dialogues emerges. Scientists can then 'safely' believe that since there was no demand or complaint nothing much needed to be done. However, with the increasing concern for the environment, migration, and sustainability of even the valley production systems, the attitude of the supply-side agencies is slowly changing.

(3) The link between the skills of disadvantaged households and their resources is sometimes weak because of the time frame in which they want to use the skills available for managing the resources efficiently. Their time frame does not match with the time frame of public institutions. The government expects that an innovation that evolved in a particular period of time, with the support of institutional systems, could be scaled up or multiplied almost mechanically even without such institutional support. The time frame used by any decision maker is the function of the control he or she has on the respective resource market. The choice of species, the combination of tree, livestock, or other components, and the available market stimuli are all influenced by the time frame. If scientists ignore this dimension they may miss the fundamental basis of some of the survival technologies.

(4) To deal with the State even the more articulate, well-endowed urban people need advocates. The link between the farmers and scientists may be quite strong or only through intermediaries who may play the role of a bridge, broker, advocate, or even *bania* (moneylender). If new links have to be forged, some of the debilitating links will have to be broken or weakened. The role of extension machinery is quite different in mountain regions. In the early years, they will have less to transfer from the lab to the land and more from the land to the lab or from the tiller to the technologist. The training of disadvantaged people needs careful conceptualisation. The training and visit system is singularly unsuitable for mountain regions, notwithstanding the millions of dollars that the World Bank invests in it.

(5) The link between the individual and collective rationality of the scientists and the farmers often poses the biggest challenge. The ability of scientists to deal with the farmers in groups rather than individually has not been built up over the years because of the individual-oriented approach of technology transfer. Given the fact that the common property resources and even open access resources, managed through collectives, play a pivotal role in the survival mix of the poor people, we have to learn to deal with collective choice problems (Gupta 1989).

The household's portfolio of resources, skills, and opportunities has to be carefully understood and analysed before embarking upon technological interventions. The more diverse the environment, the greater the linkage between different subsystems and the need for scientists to talk across disciplines.

IMPLICATIONS FOR THE FORMAL R&D ORGANIZATIONS

The learning at individual and organizational level can be strengthened if the emphasis is on monitoring the *context* in which the scientists work rather than the *content*. The barriers to learning in any developmental organisation (whether in a mountain region or otherwise) could include the following propositions (Gupta 1984, 1987c, and 1987d).

(1) My learning is not enough, others must also learn,
(2) The benefits assumed from learning are not sure and sufficient,
(3) The cost of my not learning is borne by others, why does it matter if I do not learn?
(4) Learning takes time, one is always in a hurry while planning, who has the time to review past experiences and learn from previous mistakes?
(5) Learning from below and outside (i.e., from juniors, farmers, and extension workers

for researchers and vice versa) requires the capacity to acknowledge the lack of correlation between status and skills,

(6) Replicating success rather than the process of discovering the rules or the grammar of success is most admired in bureaucracies. Allowance for learning the process may mean providing room for decentralized designing. Who will take the risk of having diversity in programme content? Will it not increase the burden of monitoring?

Learning implies being accountable both horizontally (towards the clients) and vertically as well as taking care to monitor client satisfaction or creativity at lower levels. In the mountain regions, the excessive emphasis on budget exhaustion as an indicator of project success will invariably lead the researchers, administrators, and extension workers to concentrate their efforts on the valley regions. Thus, as I have argued earlier (Gupta 1987c), a change not monitored is a change not desired.

To overcome various barriers, several strategies have to be tried depending upon the institutional and political economic context.

(1) Creation of *demand groups* of the 'farmers on the fringe' by the scientists may help to counteract the demands made by already well-endowed and articulate farmer groups. Let us recognize that such dispersed, disadvantaged, and inarticulate farmers cannot be expected to demand different types of technologies from scientists in the short run.

(2) There are several other pressures that scientists have to face, including pressure from parliament, media, donors, and public administrators. The strengthening of a research management system cannot take place unless we study these pressures and ways of coping with them. Too much emphasis on the technical aspect, disregarding management and organization, might meet the same fate as many five year plans have met in India in terms of the goal of poverty alleviation or balanced regional development.

(3) When resources are scarce, the need for networking is higher. However, which scientists' group will network with whom will often depend upon the way the top leaders of the R&D system monitor performance. If the purpose is to reorient forestry or watershed development, appropriate arrangements for networking and inter-organizational coordination will have to accompany the technology development and transfer. For far too long, the institutional issues have been taken as constraints for which adaptations have to be made. If technologies and institutional arrangements for managing the natural streams or '*kools*' have to be modified then the strategic linkages will be of a qualitatively different type than would be the case if scientists were responsible for most of the functions over a given spatial unit (as is the case in on-station research). The need for on-farm research is higher when ecological diversity is higher and technologies developed at one location cannot be replicated at another location even a short distance away.

(4) The links among farmers and scientists have to be placed in the ecological context by use of a mean-variance matrix (Fig. 16.4). The eco-institutional perspective requires scientists to take care of the vertical assurances and the horizontal assurances while providing either new resources or new skills or both. In other words, even if scientists are trying to strengthen an already existing indigenous innovation they should make the boundaries of their role very clear. Undue expectations can lead to mutual distrust and disrespect.

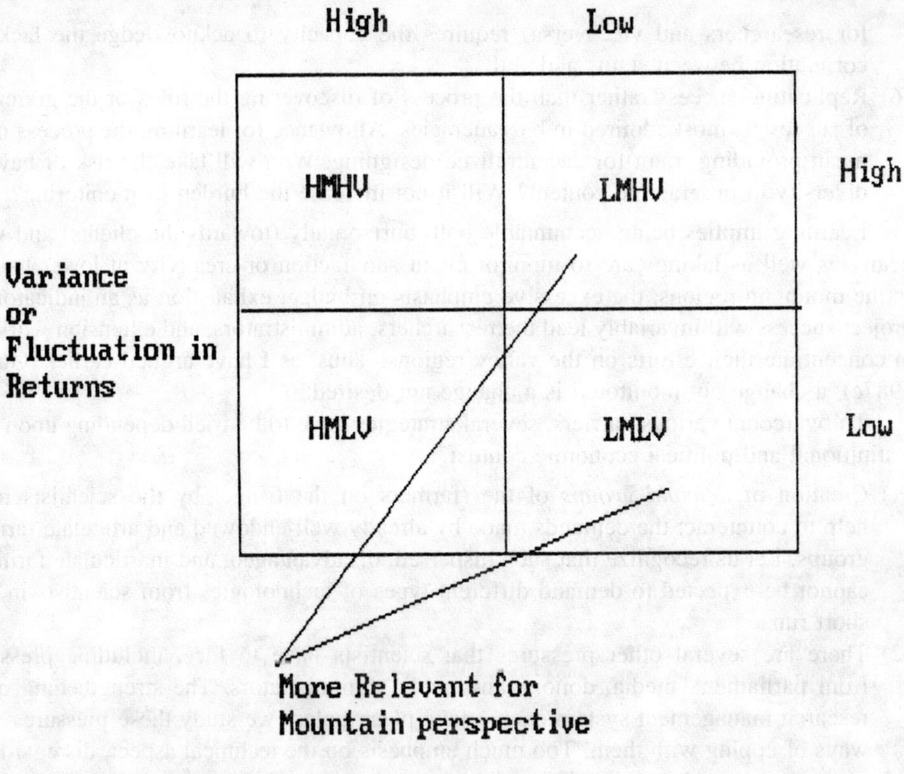

Figure 16.4: Average or mean return
Source: Author's compilation

We can understand how different types of linkages will need to be forged in different types of regions having a varying combination or portfolio of endowments. For example, if the household portfolios of various enterprises comprising their farming system generate a *high average or mean income with high variance*, one can anticipate the availability of market channels. Public interventions can be restricted to regulation. The extension system could be commodity-based rather than household or region-based. In such a context there may not be any need for village-based extension workers but there would be a need for focussing on the issues of sustainability. Farmers may try in the short run to reduce their risks by excessive use of chemical inputs. The households with such portfolios may perhaps be served through contractual services rather than through a mandatory public supply system.

Farmers having *high mean and low variance* portfolios would really be the best-endowed farmers whose goals would be to move towards value-adding technologies. In such cases, farmers' groups can even hire the scientists rather than relying on public or private channels. The government should provide tax concessions to promote such organizations and reduce the size of the public bureaucracy. In the case of households with *low mean and low variance* or *low mean and high variance* portfolios the role of scientists has to be much more evident. Given the high vulnerability in the case of high variance portfolios around low mean returns, the risk-bearing systems are necessary for technological trials on farmers' fields. We have argued that there will be limited use of the transferral of ready-made technologies in such a context. One has to transfer scientific principles so that farmers can develop their own technology. However, there would be areas such as biological pest control or the management of synchronized sowing or other farm operations which need institution building support.

(5) Horizontal links among farmers and scientists cannot be built without weakening the vertical links among junior and senior scientists. The accountability to peers, including farmers, at local level can be strengthened only if top-level scientists recognize the need to be accountable towards the lower levels in the organisations. This principle is valid even in other cases, but in mountain and other risk regions it is critically important. Decentralized experimentation in collaboration with farmers will not take place if there is tolerance of such a process at a higher level. Given the communication system, there is no way innovations can be general in a tightly coupled or linked system.

(6) Reinforcing the pride of mountain people is important. If the pride of the people is a major casualty of wrongly designed policies in the recent past, then restoration of that pride should be the first priority now. Awards for innovative systems and technologies developed by farmers and pastoralists must be given. Scientists should give due acknowledgement to farmers in their publications when their ideas provide the precursor for research insights or experimental design.

(7) Recognition for developing technologies for limited diffusion is essential. If sustainability of technology in fragile regions depends upon compatibility with diverse ecological systems, then the possibility of developing technologies that diffuse widely is limited. Organizational rewards for work that cannot be measured in terms of numbers of farmers, or acreages under cultivation of a new technology, may be necessary.

(8) There is a need for a larger number of experimental sites and higher budgets. The higher the risks and the greater the variability in the production environment the greater will be the number of observations required to verify any experimental research. The challenge before scientists is to develop a network of experiments that are sufficiently broad-based to produce good results.

Ethical and Moral Issues in Knowledge Transfer

It is unfortunate that, while looking for alternatives for the sustainable development of high-risk environments, we often ignore the ethical and moral issues involved. For instance, if the major resources of hill regions are herbal medicines, honey, and other such products, how do we ensure that the interest of the corporate world (national and international) will not lead to reduction of biodiversity and the extinction of certain species?

How do we apportion responsibility for taking undue risks at the farmer's cost? Do we educate the farmer about the various implications of a particular experiment? Do we do it individually or in groups? How do we bear the cost of our mistake, e.g., recommending a variety that has not been tested adequately or bringing seeds without proper processing so that new weeds or diseases come into the region with disastrous long-term consequences.

When we do learn from farmers' innovations we expect rent, royalty, or profit for documenting or sharing their knowledge. How do we share part of the rent with the provider of the knowledge? What are the institutional mechanisms to monitor their sharing? Should judgments on the above issues be left to individual choices? And if so how do we evaluate the morality of such freedom? How do we judge the ethics of the assumption that farmers can always guide the direction of the research that scientists should take? Can farmers demand when they do not even know what scientists can deliver? How do we incorporate the innovating farmers as a part of the educational system in which they also teach and we also learn? How can post-graduate curricula be modified so that future resource managers in high-risk environments develop sensitivity to the above concerns?

In this paper we have discussed the dilemma that we face when we try to develop a system of *lateral learning* and *mutual monitoring* among farmers and scientists. There is no doubt that there is a tremendously rich reserve of innovations available with the farmers which can guide or influence the direction of research. There may be an equally rich reserve of ideas available with the scientists who crave different sets of rewards than the ones available within the organizations.

The challenge is to generate institutional innovations which can link such scientists with innovative farmers in marginal regions. We have no doubt that these links cannot be forged unless the legitimacy of such linkages is established. There is no escape from recognizing that the sustainable development of mountain regions requires that we rethink the very basis of the ethics and politics of resource management. The experimental ethics of mountain peoples need to be nurtured and their skills need to be properly priced. A viable strategy will require that we not only improve their access to resources and upgrade their skills but also provide assurances to them that their restraint in resource use will be valued. Why should they conserve genetic diversity, technological skills, and ecological balance if the benefits accrue only to the people in the plains?

REFERENCES

Box, L. *Networks and Nitpricks, Socio-ecological Perspective on Agri-knowledge*. Wageningen: Agricultural University, 1988.

Buzdar, N. *Property Rights, Social Organisation and Resource Management in Northern Pakistan*. Honolulu: Environment and Policy Institute East West Centre, 1988.

Chambers, R. *Rural Development; Putting the Last First*. London: Longman, 1983.

Chambers, R., A. Pacey and L. Thrupp (eds.). *Farmer First: Farmer Innovation and Agricultural Research*. London: Intermediate Technology Publications, 218 pp., 1989.

Gupta, A.K. 'Viable projects for unviable farmers.' Paper presented at a Symposium on Rural Development in South Asia, Amsterdam, Apr. 1981.

Gupta, A.K. 'Impoverishment in Drought Prone Regions: A View from Within.' Joint field study of SDC/NABARD/IIMI-A CMA, Indian Institute of Management, Ahmedabad, p. 573, 1983.

Gupta, A.K. Small Farmer Household Economy in Semi-Arid Regions (Mimeo). Ahmedabad: CMA, IIM, 1984.

Gupta, A.K. 'Sociology of Stress: Why Do Common Property Resource Management

Projects Fail?' In Proceedings of the Conference on Common Property Resource Management. Washington, D.C.: BOSTID & National Research Council, 1985.

Gupta, A.K. Role of Women in Risk Adjustment in Drought Prone Regions with Special Reference to Credit Problems. Working paper No. 704, Ahmedabad: Indian Institute of Management, 1987a.

Gupta, A.K. 'Generating Ecology and Class Specific Research Priorities: Socio-Ecological Perspective on FSR.' Paper presented at International Conference on Farming Systems Research. University of Kansas, Kansas, Oct. 5–7, 1987b.

Gupta, A.K. Technology for Dry Farming: How the Scientists, Students and Farmers View the Challenge. Working Paper No. 708. Ahmedabad: IIM, 1987c.

Gupta, A.K. 'Organising the Poor Client Responsive Research System: Can Tail Wag the Dog?' Paper presented at a Workshop on Farmers and Agricultural Research: Complementary Methods, Institute of Development Studies at the University of Sussex, U.K. July 26–31, 1987d.

Gupta, A.K. 'Survival Under Stress: Socio Ecological Perspective on Farmers' Innovation and Risk Adjustment.' p. 16. Paper presented at International Congress on Plant Physiology, New Delhi, Feb. 1988a.

Gupta, A.K. 'On the Concept of Knowledge: The Conundrum of Criticism (ILEIA).' Control and Commitment in Peasant Science, presented as keynote paper a Conference on Farmer Participatory Research Institute for Low External Input Agriculture, Leusden, Apr. 11–12, 1988b.

Gupta, A.K. 'Ecology of Market Forces and Design of Resource Delivery Organisations.' Paper prepared for International Conference on Organizational and Behavioural Perspective for Social Development, Dec. 29 to Jan. 2, 1986. *International Studies in Management and Organisation*, 18 (4): 64–82, 1989.

Gupta, A.K. 'Eco-Sociology of Household Risk Adjustment and Commons: Performance in an Uncertain World.' Paper presented at International Conference on Designing Sustainability on the Commons, International Association for the Study of Common Property, Duke University, USA, 1990a.

Gupta, A.K. Politics of Articulation, Mediating Structures and Voluntarism: From 'Chauraha' to 'Chaupal.' Working paper No. 894. Ahmedabad: IIM, 1990b.

Gupta, A.K., N.T. Patel and R. Shah. Matching Farmers' Concerns with Technologies and Objectives in Dry Regions: An Exploratory Study of Scientific Goal Setting (Mimeo.) Ahmedabad: IIM, CMA, p. 103. 1988.

Gupta, A.K. and K. Ura. 'Blending Cultural Values, Indigenous Technology and Environment: The Experience of Bhutan.' Chapter 23, Vol. 2, this publication.

ICAR. *Agriculture in Ancient India*. New Delhi: ICAR 1964.

Jodha, N.S. 'Famine and Famine Policies: Some Empirical Evidence.' *Economic and Political Weekly*, 10 (41): 1609–1624, 1975.

Jodha, N.S. 'Effectiveness of Farmers' Adjustment to Risk.' *Economic and Political Weekly*, 18 (1): A38–A48, Bombay, 1978.

Jodha, N.S. and A.C. Mascarnehas. 'Adjustment to Climatic Variability is Self Provisioning Societies: Some Evidence from India and Tanzania.' ICRISAT, Economic Programme Report No. 48, March, 1983.

Jodha, N.S. 'Mountain Perspective and Sustainability: A Framework for Development Strategies.' Chapter 2, Vol. 1, this publication.

McCorckle, C.M. 'Veterinary Anthropology.' *Human Organisation*, 48(2): 156–162, 1989.

Munshi, K.M. *Land Transformation. A Philosophy and a Faith.* New Delhi: Ministry of Food and Agriculture, 1952.

Netting, R.M.C. 'Of Men and Meadows: Strategies of Alpine Land Use.' *Anthropology Quarterly*, 45: 132–144, 1972.

Pal, D. *Indian Science and Technology in the Eighteenth Century.* Hyderabad: Academy of Gandhian Studies, pp. 229–256. 1983.

Osti, G. 1988. 'Innovations in Disadvantaged Areas: The Case of the Eastern Alps.' Paper presented at the 7th Section on Science, Diffusion of Technological Knowledge and System of Local Knowledge, 7th World Congress of Rural Sociology, Bologna, 1988.

Ostrom, V., D. Fenny and Picht. *Rethinking Institutional Analysis and Development Issues, Alternatives and Choices.* Ilya: International Centre for Economic Growth, 1989.

Rhoades, R. and A. Bebbington, 'Farmers and Experiment: An Untapped Resource For Agricultural Research and Development.' Paper presented at a Session on Farmers' Knowledge and Risk Adjustment, International Conference on Plant Physiology, New Delhi, Feb. 1988.

Runge, C.F. 'Common Property and Collective Action in Economic Development.' *World Development*, 14 (5): 623–635, 1986.

Richards, P. 'Agriculture as a Performance.' In R. Chambers, A. Pacey and A. Thrupp, (eds), *Farmer First* London: Intermediate Technology, Publications, pp. 39–43, 1989.

Sanwal, M. 'What We Know about Mountain Development: Common Property, Investment Priorities and Institutional Arrangements.' *Mountain Development and Research*, 9(1), 1989.

Sen, A. 'Isolation, Assurance, and the Social Rate of Discount.' *Quarterly Journal of Economics*, 81: 112–124, 1967.

Tillman, H.J. 'Western Science and Andean Technology.' Paper presented at the International Rural Sociological Conference, Bologna, 1988.

Verma. M.R. 'Dairy Husbandry of Nomadic Gujjars in Six South-east Himachal Forest Ranges. A Study in Pastoral Animal Husbandry.' M.Sc. thesis, Hisar Punjab Agricultural University, 1967.

Verma, M.R. and Y. P. Singh. 'A Plea for Studies in Traditional Animal Husbandry.' *Farmers*, 63(2): 93–98, 1969.

Warren, D.M. 1986. 'Linking Scientific and Indigenous Agricultural Systems.' In J. L. Compton (ed.), *Transformation of International Agricultural Research and Development: Some US Perspectives*. Boulder: Westview Press, 1986.

17

LAND-USE MODIFICATION AND LABOUR SHORTAGE IMPACTS ON THE LOSS OF NATIVE CROP DIVERSITY IN THE ANDEAN HIGHLANDS

K. Zimmerer

Karl Zimmerer is an Assistant Professor, Department of Geography, University of Wisconsin-Madison, USA

CONTENTS

INTRODUCTION

Biological Diversity in Mountain Agriculture

The agriculture of peasants and indigenous people in the mountainous regions of the tropics and sub-tropics has long been recognized as containing an extraordinary diversity of crops (Vavilov 1951, Sauer 1952). It also supports concentrated numbers of the intra-specific variants (cultivars or landraces) of these species. Most major crops as well as dozens of secondary ones underwent the majority of their evolutionary divergence in mountain environments. Three characteristics of mountain agriculture especially promoted the diversification of crops: (1) its relative antiquity, having comprised many of the earliest sites of plant and animal domestication in South and Middle America, Africa, China, and the Southwest, South, and Southeast portions of Asia; (2) biophysical diversity of the environment; and (3) the ethnic diversity of cultivators.

The maintenance and expanded use of biologically diverse native crops have been identified as the major focusses necessary for sustainable agricultural development in mountain regions (Camino, Chapter 22, Partap, Chapter 15, Tapia, Chapter 4). Successful development along this path, however, cannot be guided solely by the presumed persistence of mountain traditions but rather must examine the contingent conditions that will allow the continued integration of native crops into peasant production strategies.

The present paper focusses on one especially important set of contingent conditions necessary for maintaining the diversity of native mountain crops, namely the recruitment of labour by peasant and indigenous farmers (Zimmerer n.d.a). It argues that in many mountain regions the cultivators of diverse crops are increasingly beset by an acute seasonal shortage of labour that leads to the modification and transformation of their agriculture. Such change can lead to the loss of native crop cultivars, a process referred to as 'cultivar loss'.

Crop Diversity and the Strategies of Peasant Farmers

The intra-specific diversity of native crops has been found to play various roles in the farming strategies of mountain peasants. Production purposes, consumption rationales, and cultural reasons indirectly related to subsistence constitute three primary areas of usefulness. Diverse cultivars of native crops are often maintained by mountain peasants for production purposes, especially as a means of maximizing the certainty of yield. Falling under the general category of risk aversion strategies, the practice of planting multiple cultivars tends to assure a modicum of yield under the highly variable biophysical conditions characteristic of mountain environments. Within native crops, certain cultivars are most tolerant of drought while others best resist frost and yet others yield well despite waterlogged soils (Clawson 1985).

The marginality of mountain environments, especially their climatic and edaphic variations, favours the planting of multiple native cultivars. Mountain peasants continue to plant cultivar mixes also because of the uncertain availability of agrochemical inputs. Here the economic marginality of mountain people (Jodha, Chapter 2) is seen to encourage the planting of diverse cultivars for production purposes rather than the adoption of 'improved' varieties.

Consumption rationales are often a primary consideration in the maintenance of diverse native cultivars by mountain peasants. In both major and secondary crops, distinct sets of cultivars are used for separate culinary purposes. The maintenance of the diversity of maize cultivars in the Peruvian Andes, for instance, depends in part on the variety of preparations used by peasant farmers (Grobman et al. 1961). Maize preparations involving parching, boiling, and the making of beer each draw on a divergent, albeit overlapping, group of cultivars. The variety of culinary preparations enabled by the continued planting of native cultivars provides nutritional benefits as well as much-needed diversity in the diets of mountain peasants.

The importance of diverse native crops extends beyond production and consumption rationales to include symbolic usefulness. Mountain peasants often view native cultivars as symbols that are central to cultural and ethnic identity. Again, taking an example from Quechua peasant farmers in the Peruvian Andes, it has been shown that the maintenance of diverse native potato and maize cultivars forms an important part of 'being Quechua' (Zimmerer n.d.b). The symbolic importance of native cultivars highlights an often overlooked way in which biological diversity serves as a cultural resource for mountain peoples.

CULTIVAR LOSS AND AGRICULTURAL CHANGE: THE ROLE OF LABOUR

Cultivar Loss and Genetic Erosion

'Genetic Erosion' is a broad term that refers to the decline of biological diversity in crop species and their wild relatives resulting from various human-induced processes, including habitat destruction, plant breeding practices, germplasm storage techniques, and the disappearance of native crop cultivars from peasant and indigenous agriculture (Frankel and Bennett 1970, Frankel and Hawkes 1975). The last of these, in particular (cultivar loss), threatens the maintenance of diverse cultivars in mountain agricultural systems and hence undermines the feasibility of sustainable agricultural development based on biologically diverse crops.

Existing explanations of cultivar loss, which agree that the problem is due to agricultural change in peasant and indigenous societies, focus on a shortage of land as the primary cause. In the scenario posited by the land deficiency explanation, introduced 'improved' varieties directly displace native cultivars. The present paper contends that, in the marginal environments characteristic of mountain agriculture, a seasonally accentuated deficit of labour is as important as or more important than the shortage of land as a cause of cultivar loss.

The Shortage of Labour in Peasant Societies

The shortage of labour in mountain agricultural system needs to be viewed as part of a 'double reproduction squeeze' (Bernstein 1982). In essence, the 'double reproduction squeeze' refers to the condition whereby agricultural households have increased consumption requirements while, at the same time, they receive a decreased return per unit of their fixed inputs, especially land. Mountain societies, like other groups of peasants, are pressured due to combination of demographic, social, and political-economic changes characteristic of their marginal context (Jodha, Chapter 2).

Increased consumption needs, for instance, often arise from a combination of population growth and expectations concerning access to basic durable consumer goods such as cooking ware and bicycles. A decreased return to agriculture per unit of land frequently comes about as the result of unfavourable terms of trade (a political-economic condition reflecting the weakness of the agricultural sector and the general urban bias of development policies) and the diminished capacity for production that results from environmental degradation.

Peasant households in many mountain regions confront a growing deficit of labour as they must expand their work efforts to survive under the unfavourable conditions surveyed above. A major economic practice manifesting the household labour shortage is temporary labour migration (a commercialization strategy indicating the weakness of agricultural product markets, a condition due in part to the 'inaccessibility' referred to by Jodha in Chapter 2). Undertaken by a member of one-half or more of all households in regions of the Andes as well as the Nepal Himalaya, the decision to migrate leaves agricultural households with significantly less labour available for farming tasks.

In many regions again, including much of the Andes and Nepal, the increased prevalence of temporary labour migration has contributed to the 'feminization of agriculture'. Women, young children, and elders are left to carry out agricultural chores while young men typically migrate for short or medium periods to work in activities such as construction, commercial agriculture (usually in the lowlands), and extractive industries such as logging.

The contracting supply of labour available to the peasant household for land use activities and agriculture threaten the maintenance of native crops due to several features of diverse cultivar production (Zimmerer n.d.a). In particular, the staggered production calendar that underlies diverse-cultivar agriculture often conflicts with the highly seasonal bottlenecks in the household's capacity to recruit workers. Although peasant farmers are sometimes able to rearrange the production calendar without abandoning native crops, biological constraints ultimately limit the flexibility available for temporal changes.

Native crops in mountain environments, especially, contain cultivars possessing a wide range of growing seasons, a genetically controlled trait assumed to have been selected over millennia in order to stagger the supply of foodstuffs as well as the demand for labour. Yet today the same array of growing seasons, while still highly adapted to the diversity and marginality of mountain environments, is increasingly found to conflict with the need of peasant households to allocate labour to other economic activities, not only off-farm, such as temporary labour migration, but also intensified production on the farm.

Environmental resources, other than diverse native crops, can also be seen to be threatened by the diminishing capacity of peasant households to allocate labour for time-demanding sustainable practices. In mountainous regions, peasant and indigenous farmers have mastered a staggering repertoire of techniques and local knowledge used in the management of soil resources (Denevan 1980). Terracing, tillage techniques, and field boundaries are commonly managed to maintain high productivity on sloping land. Yet the capacity of contemporary cultivators to continue such practices is jeopardized by the shortage of household labour brought by the double reproduction squeeze. During certain periods of the year (which usually are defined by seasonality), environmentally sound practices must be modified, frequently in the form of 'short-cutting', or even abandoned.

CASE STUDY OF CULTIVAR LOSS IN ANDEAN MAIZE

The Region

A case study of cultivar loss in the Peruvian Andes can be used to illustrate the impact of labour shortages, within the peasant household economy, on the cultivation of diverse native crops and, by extension, on the possible maintenance of biologically diverse crops in sustainable development. The study area chosen is the highland Paucartambo region of Southern Peru (Fig. 17.1). Located in the eastern Andes between the urban centre of Cusco and the lowlands of the Amazon Basin, Paucartambo is characterized by high transportation costs '(inaccessibility') that limit the movement of goods from the economically more developed regions of the coast.

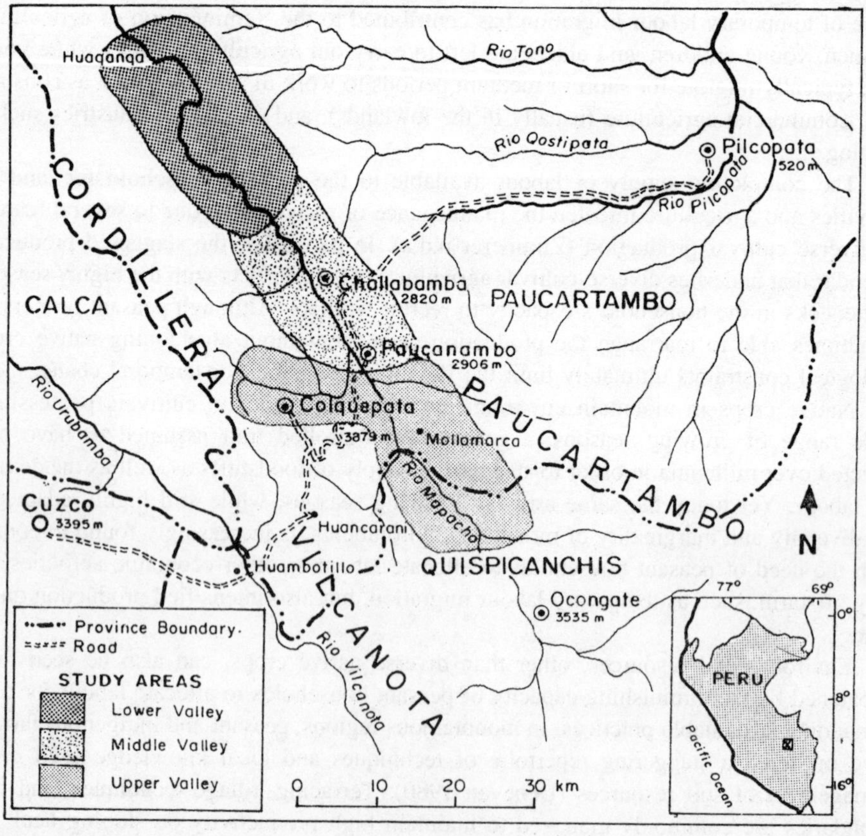

Figure 17.1: The Paucartambo region of Southern Peru

It also is marked by a diverse array of biophysical environments (marginal environments) poorly suited to mechanized agriculture and the reproduction of improved varieties. Moreover, the vast majority of the inhabitants of Paucartambo are Quechua-speaking agriculturalists and herders who make up an economically impoverished peasantry (economic marginality) that at the same time possesses little political power (political marginality).

Maize Diversity in the Region

The peasant farmers of highland Paucartambo produce an extremely diverse array of crops, as well as cultivars within these species. Among the native crops, the major ones are maize (domesticated in Middle America but introduced into South America at least 3,000 years ago) and potatoes, while secondary species include *quinoa* (*Chenopodium quinoa*), lisas (*Ullucus tuberosus*), anu (*Tropaelum tuberosum*), and oca (*Oxalis tuberosa*). Together, the native crops of highland Paucartambo contain hundreds of distinct cultivator types. The study region, like several others located in the easterly range of the Peruvian Andes, is distinguished by exceptionally high levels of biological diversity in its native crops due to the extreme variation of biophysical environments, the presence of the wild relatives (and presumed ancestors) of the major crop species, and the likelihood of early agricultural development.

The native maize crop of Paucartambo was found to represent at least 11 of the 33 major maize races found in Peru. (In maize taxonomy, 'race' is a supra-cultivator designation that is the most useful—and presently the only practical—systematic rank for comparing regional diversity.) Given the extraordinarily high morphological diversity of Peruvian maize, which has been estimated to be the most diverse in the world (Brandolini 1970, Grobman et al. 1961), the agricultural systems of highland Paucartambo are seen to encompass an impressive array of intra-specific variants within the crop.

Peasant farmers in the northern portion of Paucartambo produce the greatest diversity of native maize types. Individual households maintain extremely diverse sets of cultivars by organizing native maize production into a complex calendar made up of three staggered plantings referred to as 'big seed' (*hatun muhu*, planted in September and harvested in June), 'middle seed' (*chaupi muhu*, planted in October and harvested in either May or June), and 'small seed' (*uch'uy muhu*, planted in November and harvested in May). Each of the planting categories consists of a distinct set of native cultivars.

The length of the cultivar's growing season corresponds to its belonging to a certain planting. Slow-maturing cultivars, for instance, constitute the big-seed planting, i.e., the calendric 'niche' with the longest growing season. The ecological 'niche' concept is useful here in highlighting the close match between the diversity of planting schedules (temporal niche) and cultivar types. The maintenance of calendrically staggered maize plantings, along with the provisioning of a variety of maize cultivars useful for several culinary purposes, however, is not assured a permanent place in the mountain landscape of Paucartambo.

Temporary Labour Migration and Cultivar Loss

The peasant farmers of highland Paucartambo, especially in the region's northern portion, have become increasingly involved in temporary labour migration. Pressured (or squeezed) by the unfavourable terms of trade in highland agriculture, whereby the price of agricultural inputs has risen much more rapidly than the selling price of agricultural production (Alvarez 1983), many households periodically send one or more of their members to regions of expanding economies.

The lowlands adjacent to Paucartambo (the Pilcopata region) in particular have served as the destination for most temporary migrants from the study region. The production of rice, pineapples, and cassava on large commercial farms in Pilcopata draws labour from Paucartambo during the periods when lowland producers need workers in planting and harvesting (the months of August and September and those of June and early July, respectively).

Increased temporary labour migration in the northern portion of highland Paucartambo has reduced notably the calendric complexity of the native maize crop and threatens to eliminate completely the production of certain cultivars. Incipient cultivar loss is occurring as a consequence of local households being unable to marshall sufficient agricultural labour during certain periods, especially August-September and June-July. The seasonally specific bottleneck in the supply of labour is leading many peasant farmers to forego production of the big-seed planting of slow-maturing cultivars.

Maize cultivars belonging to the big-seed planting, which belong to the racial categories *Huancavelicano, Paro*, and *Ancashino*, are thus in jeopardy due to the impending disappearance of their temporal 'niche'. It is worth noting that, due to the close match between cultivar type and planting (the so-called temporal 'niche'), big-seed maize types cannot be transferred to either the middle-seed or the small-seed grouping.

A notable social cleavage has accompanied the growing disappearance of slow-maturing maize cultivars belonging to the big-seed planting. Unexpectedly, perhaps, the households that continue production rank among the wealthiest in the northern portion of highland Paucartambo. The majority of agriculturalists still producing big-seed maize are considered 'rich peasants', one of the designations used by local inhabitants in ranking the wealth of their fellow peasants. As a result of opportunities for earning income from large and relatively productive field areas as well as cattle herds, wealthier peasant households in the region rarely send migrants to the lowlands. Moreover, if short of labour for field production, the wealthy households can contract help for the production of slow-maturing big-seed cultivars from their neighbours in the region.

Socially and economically powerful, the relatively rich peasants in Paucartambo (like their counterparts in other mountain peasant societies) recruit labour through the use of a wide array of social relationships including wage labour, reciprocal labour, exchange, and payment in kind. Those peasant households possessing less social and economic power have less capacity to allocate labour and must therefore forego certain tasks, as in the case of the extended production calendar supporting biologically diverse native maize.

CONCLUSIONS AND IMPLICATIONS FOR PUBLIC INTERVENTION

The loss of native maize cultivars in the highland Paucartambo region of Southern Peru is seen to be arising from a seasonal shortage of labour in agricultural households. Slow-maturing maize types, occupying a distinct temporal 'niche', (the big-seed planting) are in jeopardy due to the conflict that peasant farmers face in continuing to allocate labour for maize production or instead channelling it to temporary off-farm migration, especially to the nearby lowlands. A so-called double reproduction squeeze characteristic of socially and sectorally disarticulated peasant economies such as those that mark mountain regions in the Third World is found to be forcing agricultural households to expend ever-larger amounts of labour in order to survive.

In the Peruvian Andes, the double reproduction squeeze results primarily from declining terms of trade and the chronic over-production of traditional crops as well as a stagnant or shrinking land base. The increased frequency of temporary labour migration, not only in the Andes but also in other economically marginal mountain regions (Jodha, Chapter 2), is a prominent manifestation of the growing shortage of household labour.

Results of the present study controvert the view that the loss of biologically diverse native crops is brought about solely by the displacement of native cultivars and subsistence production by commercial agriculture and improved varieties. The disappearance of slow-maturing maize cultivars is not associated with the adoption of improved varieties. It also is found not to signal the conversion of the peasant farmer into a completely commercial (capitalist) agriculturist.

In light of the present findings, public intervention promoting sustainable agricultural development needs to recognize that agricultural change is taking place in the context of a persistent peasantry. In mountain regions, the notable persistence of the peasant economy is linked to a growing shortage of labour within local farming households, a key condition demanding consideration by intervention efforts aimed at sustainable development.

The *in situ* conservation of native crops, which has been recognized as a key component of sustainable development in mountain regions (Camino, Chapter 22, Partap, Chapter 15, Tapia, Chapter 4), must be viewed as depending upon the peasant household's capacity to allocate land, capital, and labour. The latter, in particular, increasingly poses a constraint due to the increased frequency of temporary labour migration and the ensuing exacerbation of competition for labour that previously was devoted to the cultivation of diverse crops.

Public intervention in sustainable agricultural development therefore needs to address the causes and consequences of labour shortages in mountain regions. At perhaps the most immediate level, those designing intervention efforts should identify and address pronounced seasonality of labour bottlenecks that result in environmental degradation, including the loss of native crop diversity. If environmentally costly, acute labour shortages and their alleviation are compelling targets for intervention on behalf of sustainable agriculture.

REFERENCES

Alvarez, E. *Politica Economica Agricultura en el Peru, 1969–1979*. Lima: Instituto de Estudios Peruanos, 1983.

Bernstein, H. 'Notes on Capital and Peasantry.' In J. Harriss (ed.), *Rural Development: Theories of Peasant Economy and Agrarian Change*. London: Hutchinson, 1982.

Brandolini, A. 'Maize.' In O.H. Frankel, and E. Bennett (eds.), *Genetic Resources in Plants—Their Exploration and Conservation*. Philadelphia: F.A. Davis, pp. 273–310, 1970.

Camino, A. 'Andean Farming Systems: Farmers' Strategies and Responses.' Chapter 22, Vol. 2, this publication.

Clawson, D.L. 'Harvest Security and Intraspecific Diversity in Traditional Agriculture.' *Economic Botany*, 39(1):56–67, 1985.

Denevan, W.M. 'Latin America'. In G.A. Klee (ed.), *World Systems of Traditional Resource Management*. New York: Halstead Press, pp. 217–244, 1980.

Frankel, O.H. and E. Bennett. 'Genetic Resources—Introduction.' In O.H. Frankel, and E. Bennett (eds.), *Genetic Resources in Plants—Their Exploration and Conservation* Philadelphia: F.A. Davis, pp. 7–17, 1970.

Frankel, O.H. and J.G. Hawkes. 'Genetic Resources—The Past Ten Years and the Next.' In O.H. Frankel, and J.G. Hawkes (eds.), *Crop Genetic Resources for Today and Tomorrow*. Cambridge: Cambridge University Press, pp. 1–11 1975.

Grobman, A., W. Salhuana and R. Sevilla. *Races of Maize in Peru*. Washington, D.C.: National Academy of Sciences, National Research Council, 1961.

Jodha, N.S. 'Mountain Perspective and Sustainability. A Framework for Development Strategies.' Chapter 2, Vol. 1, this publication.

Partap, T. 'Biological Diversity and Genetic Resources: Some Issues for Sustainable Mountain Agriculture.' Chapter 15, Vol. 1, this publication.

Sauer, C.O. *Agricultural Origins and Dispersals*. Washington, D.C.: American Geographical Society, 1952.

Tapia, M. 'Mountain Agricultural Development Strategies: The Andean Perspective.' Chapter 4, Vol. 1, this publication.

Vavilov, N.L. 'The Origin, Variation, Immunity, and Breeding of Cultivated Plants.' *Chronica Botanica*, 13:1–366. (Translated from Russian by K. Stan Chester), 1951.

Zimmerer, K.S. Seeds of Peasant Subsistence: Agrarian Structure, Crop Ecology and Quechua Agricultural Knowledge with Special Reference to the Loss of Biological Diversity in the Southern Peruvian Andes. Ph.D. Dissertation, University of California, Berkeley, 1988.

Zimmerer, K.S. n.d.a. 'Labor Availability and the Decline of Crop Diversity in the Southern Peruvian Sierra.' *The Geographical Review*, 81(2), April 1991, forthcoming.

Zimmerer, K.S. n.d.b. 'Maintaining Diversity in Potato and Maize Fields of the Peruvian Andes'. *Journal of Ethnobiology*, 10 (2), June 1991, forthcoming.

18

FARMERS' STRATEGIES IN THE MOUNTAIN AREAS OF WEST SICHUAN: CHINA

Liu Yanhua, Wang Fei and Yu Dafu

Liu Yanhua is a geographer with the Chinese Academy of Sciences and Wang Fei and Yu Dafu are from the Institute of Mountain Disasters and Environment of the Chinese Academy of Sciences

CONTENTS

INTRODUCTION

Experience generally suggests that the regional economy of mountain areas, as well as people's livelihood, were founded upon agriculture. This indicates that agriculture is the principal form of production. A review of agricultural development in mountain areas further shows that its patterns and levels differ from place to place. These differences can be partly attributed to the diversity of farming systems characterizing mountain areas.

The development of mountain agriculture depends largely upon the scope and intensity of resource use and management, in relation to the physical and socioeconomic environment, for ensuring both people's basic needs and increased demands. By examining agricultural development approaches vis-à-vis mountain-specific conditions, Banskota and Jodha (Chapter 3) report the uneven progress of agriculture in different mountain areas. Rapid population growth, market forces, and the side-effects of public interventions have accentuated resource extraction in the mountains giving rise to indicators of unsustainability (Jodha, Chapter 2). For understanding a farming system and the role of farmers' responses to the constraints and potentialities of the production environment, in the following discussion we focus on the constraints in the production environment and people's adaptations to those constraints.

However, whatever the number and nature of forces indirectly influencing agricultural resource use and its sustainability, the farmer should be considered to be the central actor whose decisions and actions directly influence the pattern and intensity of resource use. These decisions and actions are also the manifestations of farmers' strategies to bring about adjustments to the biophysical environment. Farmers' strategies in the selected areas of West Sichuan were examined (Dafu et al. 1990) as a part of the studies on farming systems in different countries of the Hindu Kush-Himalayan region.

Although, owing to institutional changes in China following the Revolution, the sphere of individual farmers' decision-making and the choice of specific practices were reduced, in mountain areas some traditional practices survived. Following the reforms of 1978, in spite of some collective obligations, the farmer is able to make his own individual decisions. Our ensuing discussions are preceded by a brief introduction to West Sichuan and to agro-development in the area. Figure 18.1 sketches the overall framework in which the farmer operates.

BACKGROUND AND AGRO-DEVELOPMENT IN WEST SICHUAN

The Mountain Areas of West Sichuan

Located in the western part of Sichuan Province, the mountain region of West Sichuan (Fig. 18.2) embraces an area of 0.315 million sq km. It had a population of 6.18 million in 1986, accounting for 55.3 per cent and 6 per cent of the population of Sichuan Province. Administratively, West Sichuan includes three autonomous prefectures (Ganzi, Aba, and Liangshan), one municipality (Dukou), and a part of Ya-an District (Hanyuan County and Shimian County).

The West Sichuan region is mainly characterized by a complex topography with a high plateau and mountains. It forms a major section of the Hengduan Mountains in the eastern part of the Tibetan Plateau and belongs to the highest topographical realm

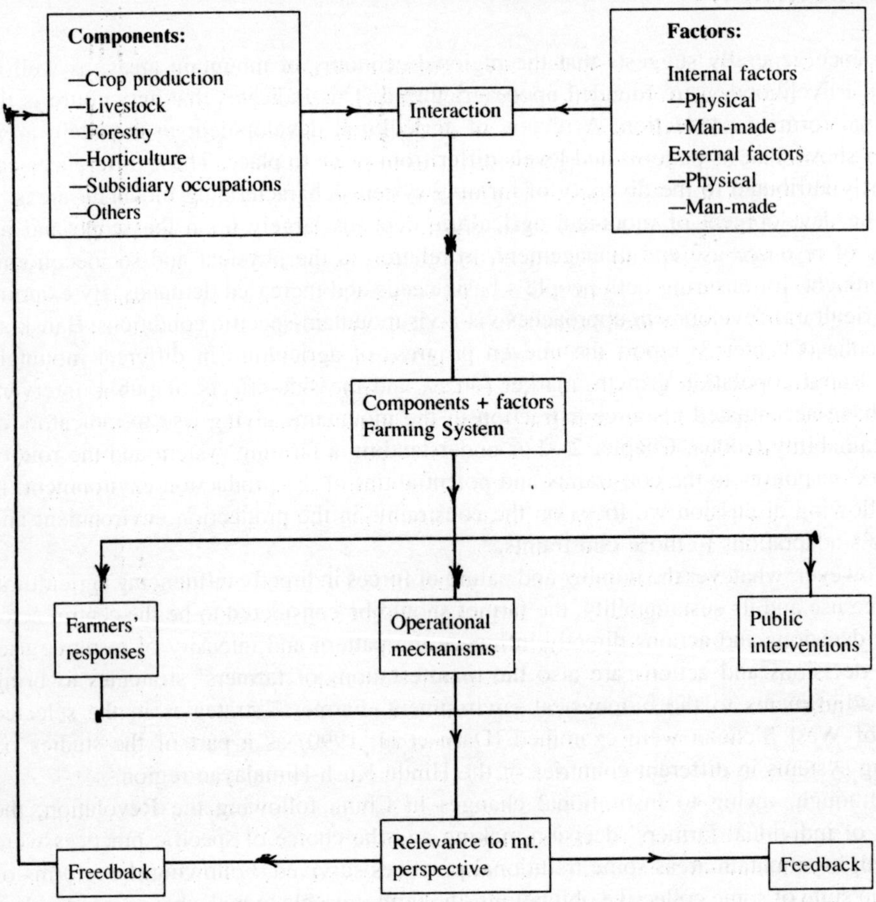

Figure 18.1: General scenario of interacting factors and mechanism within mountain farming systems

of the three realms or zones of China (plains, hills, and high plateaux/mountains). The elevation of West Sichuan, on average, is about 3,000 masl and declines from north-west to southeast. Three large rivers, namely the Jinsha, the Yalong, and the Dadou, and their tributaries dissect the plateau surface and form deep gorges and river valleys. A distinctive, vertical variation in relief dominates the general features of the region. In the northern part, the elevation is higher than 3,400 masl with comparatively level and wider landscapes. Southwards, lands are deeply dissected with height differences of 1,000–3,000 m between the valleys and the mountains. In contrast, the southern part of West Sichuan is relatively lower, averaging 2,000 masl with gently sloped mountains, broad intermontane plains, and wide river valleys. Out of the total land area, the plateau accounts for 50 per cent, the mountains 45.8 per cent, and the intermontane basin and broad river valleys only 2.2 per cent. The latter area is quite populous and is the most intensively cultivated.

Climatically, West Sichuan is a transitional zone between the sub-tropical climates

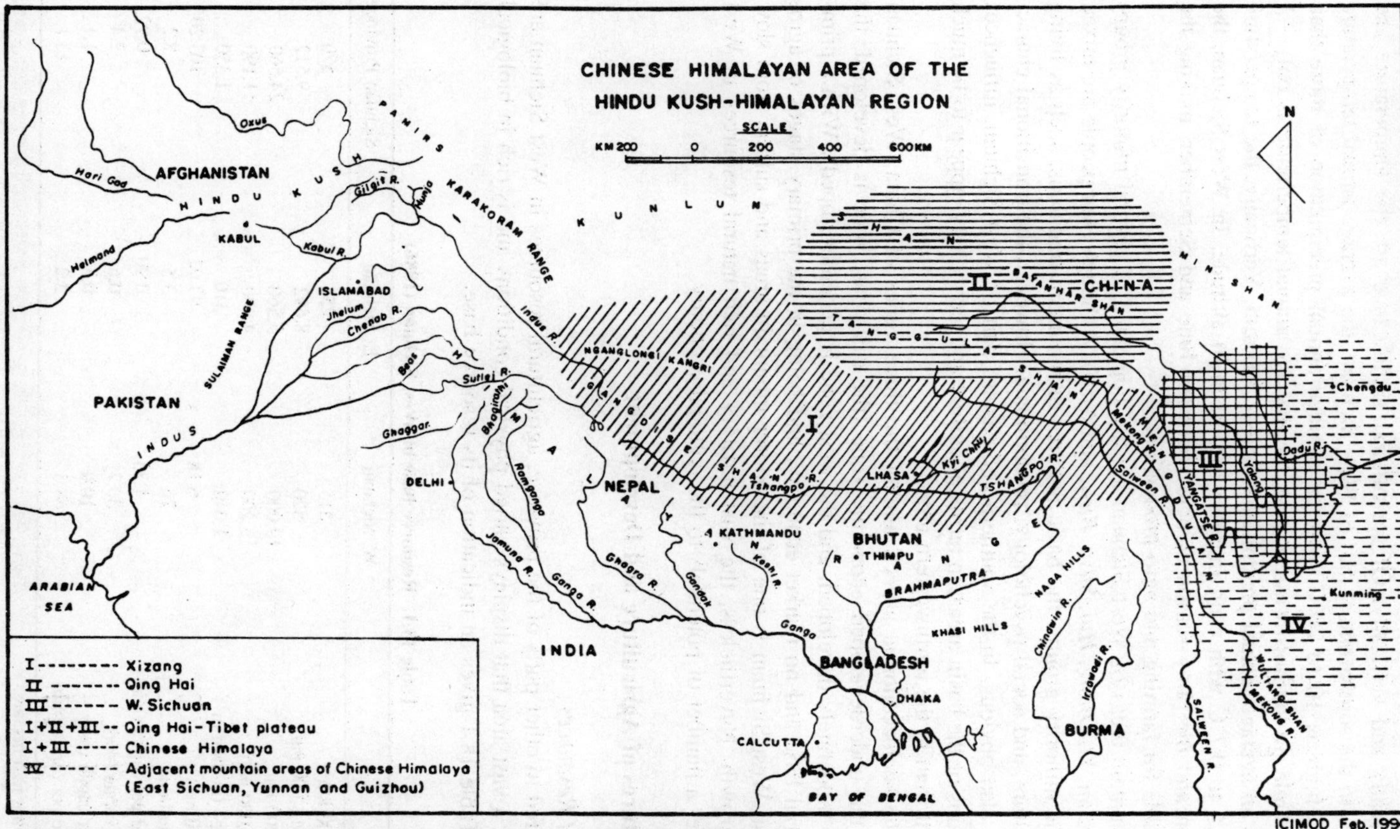

This map has been prepared by ICIMOD exclusively for the convenience of the readers.
The denomination used and the boundaries shown in this map do not imply, on the part of ICIMOD, any judgment on the legal status of any territory endorsement or acceptance of such boundaries.

Figure 18.2: Chinese Himalayan area of the Hindu Kush-Himalayan region

in the south and the highland, frigid climates in the northwest. Climatic conditions vary both horizontally and vertically. Of agricultural significance are the temperature and precipitation. In the southernmost part of West Sichuan, the average annual temperature can reach as high as 10 to 20°C, with an average annual precipitation of more than 1,000 mm, while, in the northwestern part, the average annual temperature is only −1 to 4°C with an average annual precipitation below 500 mm. Vertically, the temperature decreases 0.5 to 0.7°C with each increase of 100 m in altitude. In West Sichuan, the warm season and the rainy season coincide between June and September, making the area favourable for farming and agro-production.

The mountain areas of West Sichuan are the homeland of several minority groups (mainly Tibetan, *Yi, Qiang, Hui, Susu, Buyi*, and *Naxi*). Although *Han* people are mixed in among the minority groups, the basic features of mountain traditions, such as living customs, culture, and social psychology, predominate. However, the transitional characteristics are also obvious. In the southern and eastern part of West Sichuan, influences from the plains or the basin areas, in terms of culture and technology, appear to be much more evident than in the northwestern areas.

From the socioeconomic aspect, among the three zones of China, West Sichuan belongs to the underdeveloped economic zone (the east-coast zone is developed, the mid-zone is moderately developed, and the west zone is underdeveloped). With a primitive industrial base and no market economy, and being in the primary stages of agro-development, West Sichuan is one of the poorer mountain areas and suffers from slow economic growth. Nevertheless, the vast territory and rich natural resources in West Sichuan offer a number of potentials to the regional economy.

Major Features of Agriculture and Development

Agricultural Resources

Compared to other parts of the province, agricultural resources in West Sichuan are spread over a vast area that displays varied physical conditions and is rich in biological resources. Table 18.1 gives an indication of its resource base.

Table 18.1. Resource base of West Sichuan (1986)

Item	W. Sichuan	E. Sichuan	Sichuan Province
Land area ('000 km^2)	315	255	570
Cultivated land ('000 *mu*)[a]	820	8,692	9,512
Grassland (0,000 *mu*)	19,000	5,560	24,560
Forest land (0,000 *mu*)	6,200	4,990	11,190
Timber shortage (million m^3)	1,040	310	1,350
Population (million)	6.18	97.02	103.20
Land per capita (*mu*)	76	3.5	8.2
Cultivated land per capita (*mu*)	1.33	0.90	0.92
Grassland per capita (*mu*)	30.7	0.6	2.4
Forest land per capita (*mu*)	10.0	0.5	1.1
Timber storage per capita (*mu*)	168.3	3.2	13.1

[a]There are 15 *mu* in a hectare.

Source: Agricultural Statistics of Sichuan Province, 1987.

The varied topography and a number of micro-climates has led to the development of three-dimensional agriculture in these mountain areas. A wide range of resources characterizes mountain agriculture. For example, so far, more than 20 species of grain crop, 100 species of fruit trees, 600 species of animals, and 6,000 species of valuable wild plants (including about 2,000 species of medicinal plants) have been identified in West Sichuan. These diversified resources make it a promising area for agricultural development.

However, there are some weaknesses in resource conditions also. First, the mountain topography, which is dominated by slopy lands, makes the use and management of resources difficult; consequently, agricultural inputs such as irrigation and transportation are relatively high. Second, the sparse distribution of resources does not favour scaled production and mountain agriculture is restricted by small-scale operations and dispersed distribution. Third, the mountain environment is relatively fragile; soil erosion and natural hazards frequently occur. In West Sichuan, the main causes of unsustainable agriculture are landslides, drought, floods, and frost.

Features of the Agro-economy

Agriculture is the predominant sector in the regional economy. It is dependent largely upon natural resources and extensive management. In 1986, the gross output value of agriculture and industry (GOVAI) in the area was 2,550 million RMB yuan (according to the statistics for Ganzi, Aba and Liangshan, the gross output value from agriculture accounts for 57.63% of the GOVAI, and this is much higher than the provincial average, 34.11%).

The agricultural system is dominated by crop farming and animal husbandry. The allocation of agro-production in terms of output value is shown in Table 18.2.

Table 18.2. Gross output value of different agricultural sectors (1986)

	Sichuan	Liangshan	Aba	Ganzi
GOV of Ag. '000 yuan[a]	2,809,400	86,278	31,899	28,784
Animal husbandry (%)	26.90	22.67	46.46	59.06
Horticulture (%)	18.20	14.69	9.68	2.16
Crop (%)	42.88	43.76	23.76	19.09
Forestry (%)	5.42	7.94	8.14	4.01
Subsidiary occupations (%)	5.73	10.63	10.78	15.63
Fisheries (%)	0.87	0.30	0.03	0.05

[a]Calculated at the fixed price in 1980. In 1986, there were 3.20 years to one U.S. dollar.

Source: Agricultural Statistics of Sichuan Province, 1987.

Animal husbandry occupies a prominent position, because of the extensive rangelands that constitute one of the largest pastoral areas in China. Here 50 per cent of the large stock, 95 per cent of the sheep, and 80 per cent of the goats of Sichuan Province are raised, making West Sichuan the major source of livestock products for the whole province. In 1986, livestock products amounted to 131,000 tons of meat, 172,000 tons of milk, and 4,230 tons of wool.

Cropping is next in importance after animal husbandry. In the southern mountain areas, where there are a number of broad valleys, crop production is more developed than in the northern areas. Although the output from cropping in West Sichuan was only 24.4 per cent of the total for the whole province in 1986, it is an essential component of the mountain farmer's livelihood. In 1986, 1.75 million tons of crops (Liangshan, Aba, Ganzi, and Dukou) averaging 304 kg per capita for the rural population (less than 80% of the provincial average) were produced.

Although the agro-productivity is quite low—e.g., the grain yield was 219 kg/*mu* in 1986 (53% of the provincial average), the GOV of agriculture per unit of land was 4,886 yuan RMB/km^2 (9.9% of the provincial average). Thus, agriculture does have a meaningful contribution to make to the regional economy. Out of the total population in West Sichuan, 85 per cent are living in the rural areas and are engaged in agriculture.

Process of Agricultural Development

Based on recorded history, agriculture has been carried out for more than 2,000 years in West Sichuan. But, owing to the complicated physical and cultural conditions, the situation has not changed essentially throughout that period.

Up to 1950, the region remained feudal with numerous small tribes or even some primitive communes. Because of the mountains, West Sichuan was isolated from the mainstream of development and the local economy was characterized by very primitive practices. In the northern part of the region, nomadic pastoralism predominated, whereas, in the southern areas, two kinds of farming system, valley farming and hill farming, were practised.

By the end of 1949, a political change took place and West Sichuan entered a completely new development stage. The minority peoples from different social systems, no matter how uneven their development, all took part in the socialist transformation process. Through several stages of land reform, i.e. land redistribution, multi-aid teams, agricultural cooperatives, and people's communes, agriculture developed rapidly. By 1978, the gross agricultural output value (calculated at fixed prices) increased three times compared to that of 1950, a network of highways, linking each county of the region, had been constructed, and a research and extension system for agriculture had been established and consolidated. Most importantly, the agricultural production in West Sichuan broke through its isolation and began to develop in line with national standards.

Since 1978, a wave of economic reforms has swept the countryside in West Sichuan, particularly in the context of the implementation of the 'household responsibility system', which has opened up new prospects in agricultural production. The responsibility system allows farmers to run their farmlands on a contract basis and gives the right of decision-making in agro-production to the farmers. This new policy and its implementation has given the incentive to peasant farmers to adapt diversified agro-practices to their needs. To a great extent, this system suits the conditions of mountain areas where there are specific 'niche'. Accordingly, agriculture has progressed a great deal.

Agro-zones in West Sichuan

On the basis of broad similarities among physical conditions, socioeconomic situation, background of agricultural production, and prospects of area development, West Sichuan can be classified into three basic agro-zones (Fig. 18.2). The major features of the three agro-zones are described below.

The cropping-forestry-animal husbandry zone comprises the mid-mountain, wide valley areas of the Southwest. Situated in the transitional area of the Tibetan Plateau, Yun-Gui Plateau, and Sichuan Basin, this zone is characterized by mountains and wide valleys/basins, with altitudes ranging from 1,000 to 3,000 masl. The climatic conditions are sub-tropical and semi-humid. The valley areas, which have an average annual temperature of 17–20°C and an average annual precipitation of 800–1,000 mm, are extremely suitable for agriculture.

So far, cropping has been the predominant sector. Cultivated lands account for less than 6 per cent of the total area and are mostly concentrated in broad valleys. Crop production accounts for 60 per cent of the total output value of agriculture. The staple grain crops are paddy, maize, wheat, and potatoes. In terms of cash crops, sugarcane and tobacco are significant.

The area has 28 per cent forest coverage, out of which air-seeding forests account for a large proportion. Since the 1950s, the air seeding of Yunnan pine has been very successful, especially in the mid-mountain areas. Forestry development has a lot of potential, since it is estimated that 48 per cent of the total area of the zone, mainly on mountain wasteland, can be reforested.

The forestry-cropping-animal husbandry zone is situated in a very rugged area of high mountains and deep valleys. The relative difference in elevation is normally 1,000–3,000 m, while the highest altitude is 6,000 m. The steep slopes and narrow valleys constrain, to a great extent, the exploitation of resource and agricultural development. Influenced by the relief, climatic conditions are subject to significant variations vertically. In moving from the low land to the high mountain ridges, the average annual temperature ranges from 15°C to below 0°C and the accumulated temperature (> 10°C) from over 3,000°C to below 300°C. Deep valley areas are mostly dry and mid-mountain areas are relatively humid.

Forests, the most important resource in the zone, are mostly distributed throughout the mid-mountain areas where they are comparatively inaccessible. The area covered by forests and the reserves of standing timber account for 28.5 per cent and 44.4 per cent of the total of Sichuan Province. The major tree species are fir, spruce, birch, and alpine oak. Among the cash trees, apples and pears are the best known.

Although cultivated land is quite limited, crop cultivation is the activity on which people depend for their livelihood. Crop cultivation in the zone is characterized by small-sized and sparsely distributed farmland, rainfed agriculture, and low productivity. The predominant crops are maize, wheat, barley, and potatoes.

The animal husbandry zone, in the high plateau in the northwest, is the highest zone in Sichuan Province, with an average altitude of more than 3,500 masl. Its topography features a vast flat plateau surface, open valleys, and mountains with gentle slopes. Its climatic condition ranges from frigid to subhumid. The average annual temperatures (> 10°C) are from 400 to 1,500°C, and the area is free of frost for less than 100 days. The low temperature condition is the major constraint to agriculture. There are vast areas of pastoral land, accounting for 68 per cent of the total area. The vegetation is mainly alpine and sub-alpine meadow, belonging to the herbaceous, nutgrass, and leguminous species.

This zone has the largest pastoral area, and caters for 76 per cent of the yaks, 57 per cent of the horses, and 53 per cent of the sheep in Sichuan Province. Beef and wool production account for 33 per cent and 50 per cent of the provincial total. Because of poor physical conditions, poor management, a low level of commercialism, and poor manufacturing capabilities, animal husbandry is developed.

Forests, covering 7 per cent of the area, are distributed along the upper reaches of the Jinsha, Yalong, and Dadu rivers. These forests play a significant role in water resource conservation. Crop cultivation is limited in this zone, with highland barley and spring wheat as the main food crops and rapeseed as the major cash crop. The crop productivity is low (about 120 kg/*mu*), and grain crop production is far from sufficient for local consumption.

FARMERS' STRATEGIES FOR MOUNTAIN FARMING SYSTEMS IN WEST SICHUAN

Farmers' Strategies Defined

Historically, farming systems in mountain areas have evolved in response to the ecological and agro-climatic attributes of the mountain regions. These systems contained a lot of features in terms of choice of land-based activities and their management practices which helped to both protect the resource base and maintain the level of productivity.

Farmers' strategies include the response of farmers towards mountain environmental conditions in the context of the use of natural resources and management of socioeconomic circumstances. The adaptation of farmers' strategies is affected by both individual behaviour, in relation to changing economic variables, and the nature of policies pursued by the public sector. Farmers' strategies, to a great extent, reflect the scope for harnessing land use in a diversified or integrated manner to ensure the basic needs of the people and a degree of sustainable development for mountain farming systems. The study of farmers' strategies was involved in identification of elements (such as linkages between different land-based activities, indigenous methods of resource management, and an inventory of traditional technologies) that helped in the sustainability of mountain agriculture and was involved in the examination of the feasibility and viability of these elements in the context of recent changes.

Based on the case study (Dafu et al. 1990), several constraints and potentials of mountain farming systems in West Sichuan can be identified. Constraints refer to the restrictions or negative processes that deteriorate the production environment of farming systems in the long run and also adversely affect its current performance. Potentials imply the possibilities that could direct the mountain farming system towards sustainable development. This paper presents the constraint aspects of agriculture in West Sichuan and discusses the farmers' responses to them.

The basic constraints common to West Sichuan are summarized in Table 18.3. The table lists the constraints as they are commonly identified and understood both by development planners and knowledgeable members of the community. Column 2 of the table lists the relevant mountain specificities (Jodha, Chapter 2) with which these constraints are associated. Operational implications of the mountain specificities and farmers' responses to them, as observed in West Sichuan, are presented below. The constraints as both restrictions and negative processes are elaborated upon. The negative processes could also be viewed as emerging indicators of unsustainability as elaborated upon by Jodha (Chapter 2).

Table 18.3. Constraints of mountain farming systems

Constraints	Relevant Mt. Specificity associated with the constraint	Implication	Farmers' response
Environmental degradation	Fragility	* vulnerable to natural hazards * over use of resources * deforestation	* terracing * afforestation * fallow land * change land-use pattern
	Marginality	* inferior management * illiteracy/ignorance * poor resource position	* tendency to look for short-term gains * small production units * multiplication of options by diversification of activities
	Inaccessibility	* poor mobility * information gaps	* dependence on local resources
Limited land available per capita	Fragility	* population pressure * resource degradation * no intensive use * limited options for production	* out-migration * off-farm employment * increase input on unit of land * change land-use pattern
	Inaccessibility	* isolation from mainstream * low mobility * poor market access	* non-land-based activity
	Marginality	* focus on subsidy * high illiteracy * passive response to family planning	
	Adaptation mechanism	* local resource management system * land-use pattern	
Shortage of energy, water, funds, etc.	Inaccessibility	* long distance/high transport cost * poor market access * low mobility	* save the use * devote time to access * public regulation * increase prices
	Fragility	* low productivity * long cycle of production * population pressure * seasonality of resources	* change land-use pattern * collective management
	Marginality	* poverty * not integrated with other areas	
	Adaptation mechanism	* local conditions disregarded under restriction * local resource centered management	

Contd.

Table 18.3. Contd.

Constraints	Relevant Mt. Specificity associated with the constraint	Implication	Farmers' response
Low capability of small production unit	Fragility	* vulnerable to disasters * limited production options * instability of production	* more labour input * simplifying tools * collective management * reduce cash input
	Marginality	* poor exposure/knowledge	* specialized households
	Adaptation mechanism	* tendency to look for short-term gains * practising choice * local resource management system * lack of incentives by government	* specialized households * exchange labour
Raw material of integrated production	Inaccessibility	* isolation, poor mobility, * information/communication gap	* subsidiary activities * diversified production to ensure the basic needs
	Fragility	* limited production options	* local consumption or integrated management.
	Fragility Marginality	* instability of production * focus on subsidy * inferior quality of labour	* high value cash crop
	Adaptation mechanism	* incentive to the farmers * activity choice	
Frequent change of policies	Fragility	* instability of production * limited production options	* non-production or integrated management * cash allocation * quick return
	Marginality	* farmers' enthusiasm in production * incentive mechanisms	* house construction * yard cultivation
	Adaptation mechanism	* tendency to look for short-term gains * local conditions disregarded under commune system * activity choice * passive response in production * risk adaptation	
Insufficient supply of input	Inaccessibility	* high transport cost * unfavourable terms of exchange * slow transportation development * poor market access * information gaps	* change cropping pattern * change the allocation of available input * elimination of species needing high input

	Fragility	* vulnerable to disasters * limited production options * instability of production	* adjustment of land use * local production techniques * collective management * public regulation
	Marginality	* poor exposure/knowledge	* pricing control * low-input crop
	Diversity	* production structure * gaps in growth rate * different requirement for input * difference in required time * different land types	
	Adaptation mechanism	* activity choice * indigenous method * local resource management system * lack of incentives by Government	
Limited technical support system	Inaccessibility	* cut off from mainstream * low mobility * low development	* readjustment of production * change land-use pattern
	Fragility	* low productivity * limited production options * no intensive land use * instability of production	* traditional method * low technique * invite technicians from outside * go outside to learn
	Marginality	* inferior quality of labour * no integration with other areas * high illiteracy * lack of Government support for R &D	* establishment of technical associations among farmers * specialized activities
	Diversity	* ecological variety * production structure * adaptability * ethnic groups	
	Adaptation mechanism	* land-use pattern * activity choice * mixed source of income * local resource management * risk adaptation	

Source: Study team

Environmental Degradation

Environmental degradation is a common and serious problem in mountain areas such as West Sichuan. According to the case study, there is a lot of evidence of environmental degradation and examples, in terms of the resource base, are given below.

In livestock-dominated farming systems (LDFS), the quality of grassland has deteriorated a great deal. During the period from 1960 to 1980 (on the basis of grassland surveys at county level), the average yield of grass decreased from 316 kg/*mu* to 250 kg/*mu*. In terms of the composition of grasses, the content of poisonous grass increased from 1.5 to 4 per cent. In horticulture-dominated farming systems (HDFS), the frequency of landslides increased. In the period from 1970 to 1980, landslides occurred from three to five times a year to seven to eight times a year (village case study). In crop-dominated farming systems (CDFS), forest land has been gradually replaced by crop land. The forest cover (at county level) decreased from 15 per cent of the total land area in 1970 to about five per cent by the end of the 1980s. In the mean time, floods are becoming less controllable. About 3,000 *mu* of cultivated land have been destroyed by floods during recent years.

Environmental degradation is caused by many factors. On the one hand, the mountain environment is ecologically fragile; natural hazards occur more often in the mountains than on the plains. On the other hand, overuse of resources or irrational use of resources accelerates the process of degradation.

Mountain farmers are well aware of what is happening on their land and the environmental changes in terms of area and time. They have responded both passively and positively to these changes.

Among their positive responses, farmers constructed terraces to stabilize their land, changed the land-use pattern to protect the resource base, and planted trees on marginal lands or slope lands to conserve the soil layers. Because the current production system in mountain areas is on an individual basis, the efforts made by farmers only partly solve or prevent the resource degradation on a very small scale. The local government now encourages farmers to construct terraces and irrigation schemes under an award policy for land construction (e.g., the local government gives 250–300 yuan/*mu* to each farmer for terracing). However, individual households have not enough labour surplus to take advantage of this. Among their negative responses, farmers have to abandon their degraded land, change their land-use patterns (low productivity crops replace high productivity crops), try their best to store food grain to ensure their subsistence needs when the harvest is bad, or migrate out.

Limited Available Land Per Capita

Another common constraint in mountain areas is the limited availability of land. Land is a basic resource, and, although the average land holding is larger than that in the plains, the quality of land is much worse in terms of accessibility, average size, thickness of soil layer, and irrigation facilities. What is more, land availability per capita in mountain areas has decreased sharply due to rapid population growth, land encroachment, irrational land use, and soil erosion.

The case study discovered that in LDFS (at county level), the grassland available per sheep unit decreased from approximately nine *mu*/sheep unit in 1976 to approximately six *mu*/sheep unit in 1986. The grassland available per capita (at village level) decreased from 458 *mu*/person in 1982 to 376 *mu*/person in 1988. In HDFS (at village level), the cultivated land per capita decreased from 1.84 *mu* in 1978 to 1.25 *mu* in 1988. In CDFS (at county level), crop land available per capita reduced from 1.7 *mu* in 1978 to 1.3 *mu* in 1986.

Because of the increasing shortage of land, traditional methods of mountain farming systems cannot produce enough to sustain the increased population and the increased needs of the people. To suit the changing conditions, mountain farmers responded in two ways. First, they increased the inputs per land unit (e.g., in CDFS, the average cash input for 500 kg of paddy increased from 12 yuan in 1980 to 129.5 yuan in 1988), adopted intensive management, reduced fallow land, increased the cropping index, and reduced green manure cultivation in an attempt to increase the total production. Second, they diversified into non-land-based activities or became involved in off-farm employment. So far, non-land-based activities, such as collecting medicinal herbs, or subsidiary occupations (e.g., transportation, baking bricks, small-scale manufacturing, handicrafts, and mining) are being developed. In CDFS, off-farm employment has become very common, and about 20 per cent of the labourers (case study village) are engaged in off-farm employment. Generally speaking, farmers believe that non-land-based activities and off-farm employment are the best outlets for the surplus rural labour force.

Inaccessibility of Resources

Mountain areas are rich in resources, but these resources are not easily accessible, and this results in severe shortages of energy and water. Fuel is a basic daily need and in mountain areas the lack of fuel has resulted in a reduction of inputs into the land. In LDFS, shrubs, grass, peat, and yak dung are used as fuel for cooking and heating. In the 1960s, farmers used grass peat as the major source of fuel, while yak dung was used as manure to spread on the grasslands. In 1980, the use rate of grass peat to total fuel was estimated as 50 per cent, and, by 1988, the use of grass peat had decreased to 10 per cent. Instead of grass peat, yak dung has become the major source of fuel. In CDFS, paddy straws were traditionally returned to the land as manure, but, because of deforestation, firewood collection is more and more difficult and farmers have to use paddy straw for fuel. In 1978, they used 30 per cent of all paddy straw on the land in the case study village; in 1988 the amount had fallen to 5 per cent. In cases of fuel shortage, farmers usually minimize its use by improving stoves, having simple food, or other means. The alternative is to devote their time to collecting firewood over long distances.

Water shortages, especially during the dry season, are a constraint on productivity. In mountain areas, the irrigation systems are inadequate. Farmers normally manage by taking both public and individual measures. In the public sector, the village community establishes certain regulations to allocate the amount of water each household can use and makes the price of water flexible according to the water available. In the private sector, readjustment of the cropping pattern (e.g., planting wheat instead of paddy) is normally the measure adopted.

The shortage of energy and water does not mean that there are no resources. The problem is how to make the resources accessible. In general, the improvement of transportation facilities or construction of irrigation schemes are long-term gain activities and need huge investments. However, individual farmers do not have the ability to afford such investments. So far, the credit system for mountain agriculture has not yet paid sufficient attention to this aspect.

Since 1978, the responsibility system has been adopted in the mountain areas of West Sichuan. This is an economic reform system characterized by:

- giving the right of decision-making to farmers in the management of their land,
- allowing farmers to run their land on a contract basis with tools partly belonging to themselves, while maintaining public ownership of land, and
- relaxing restrictions on privately conducted activities, including ownership of land.

The adoption of the responsibility system is particularly suited to the *diversified development of mountain agriculture*. Since the reform, substantial increases in agricultural production have taken place, although many negative side-effects have appeared gradually. *The major weakness appears to be the low capability of the production unit.*

In LDFS, *sheep raising is the traditional practice*. This tradition is slowly disappearing. The numbers of sheep in this system (village-level survey) declined by about 40 per cent from 1983 to 1989 and now most of the households keep only yaks and no sheep. This change is caused partly by the size of the production unit. Normally, sheep rearing needs labour because the sheep move very fast while grazing, whereas yak keeping is much easier as yaks can be sent out in the morning and collected in the evening. On the other hand, sheep are smaller than yaks and are poor at withstanding natural calamities. Because the current production unit is based on individual households (about four to six persons), activities that do not need much labour are preferred.

Before the introduction of the responsibility system, there were 150 mu of cultivated land for the production of livestock concentrates, by collective management, in the survey village. Grain production partly met the livestock needs. After the introduction of the responsibility system, due to decentralisation of the production system, the cultivated land was abandoned. Consequently, the locally produced concentrate was replaced by supplies from outside.

In HDFS, infrastructure plays a very important role in the farming system. During the 'People's Commune' period, terraces were constructed, irrigation systems dug, and a motorable road connecting the village to the county centre was built through collective efforts (case study village). After the introduction of the responsibility system, nothing has been done to improve the infrastructure. On the contrary, the former works were more or less destroyed. Many terrace walls have collapsed, many parts of irrigation channels have broken down, and the road is poorly maintained. At present, maintenance depends on a limited subsidy from the local government; the road is blocked. In 1978, the village owned 10 tractors, but, in 1989, there was only one left. Mechanical labour has been replaced by human labour now.

In CDFS, the lands are relatively flat, and it is possible to cultivate a large plot of land with machinery. After the introduction of the responsibility system, lands were divided among individuals, and cultivated land became too fragmented (5 to 10 times more than before). Decentralized management on small plots of land leads to difficulties in disease and pest control (for instance the effective area of diseases and pest control decreased from 95% in 1978 to 60% in 1988), and results in declining trends in the use of agricultural implements (e.g., 85% of the land was ploughed by tractor and 100% of the paddy was threshed by machine before the introduction of the responsibility system, whereas now, 40% of the land is ploughed by tractor and 60% of the paddy is threshed by machine). Farmers in the mountain areas realized the weakness of the small production unit and measures such as collectivization in certain activities, labour exchange, and formulating public regulations under village communities for common property were adopted.

Raw-Material-based Production

Mountain agriculture is generally raw-material oriented production. This kind of production needs large areas of land and more labour input. As a matter of fact, mountain farming systems are characterized by *sparsely distributed and diverse products*, and these can hardly be developed into a commercialized, market economy. There are many options in mountain production, such as food processing and medicinal plant processing, but, owing to factors such as poor mobility, poor market access, unskilled labour, and lack of technology, the choices for mountain farming systems are different from those of the plains.

Based on the case study, the input-output balances for several products in the mountain farming system were compared and it was concluded that mountain agro-production is not profitable (Table 18.4).

Table 18.4. Input-output balance of agro-products

Item	Gross input	Gross return	Net return
LDFS			
sheep (units)	30.4	37.3	6.9
yaks (head)	63.3	78.7	15.4
HDFS			
maize	231.5	331.0	99.5
wheat	233.0	288.5	55.5
potatoes	227.0	580.0	353.0
soybeans	101.0	125.0	24.0
pigs (head)	388.5	509.0	120.5
cattle (head)	394.0	268.0	−126.0
apples	407.5	546.0	138.5
CDFS			
paddy	237.5	456.8	219.3
wheat	220.5	287.5	67.0
barley	134.0	239.8	105.8
pigs (head)	248.8	285.0	36.2
buffaloes (head)	631.0	420.0	−211.0

Note: Calculated at current prices (1988–1989). In 1989 there were 4.72 yuan to one US dollar.

Source: Case study team.

According to the investigation, mountain farming is land-based and primary product-oriented, mainly for the purpose of local consumption. In fact, income generation depends largely on non-grain production (HDFS), subsidiary occupations (LDFS), and off-farm employment.

Frequent Policy Changes

Frequent changes in the basic policies resulted in the instability and discontinuity of mountain agro-development. Since the 1950s, mountain agriculture has undergone three phases of agrarian reforms. The first phase (1949–1957) started with the introduction of land reform and ended with the introduction of cooperatives. Land reform focussed mainly upon the redistribution of agricultural land, major agro-assets, and political power. In this period, farmers owned land as private property and managed production individually. The second phase of reform was called 'the People's Commune Period' (1958–1976). Under this reform, the ownership of land and of major assets shifted from individuals to the collectives. Farmers worked under the administration and were paid according to labour input; as a result, they were more workers than owners. The third phase of reform, introduced in 1977, is normally called 'The Responsibility System'. Under this reform, farmers have the right to manage land on a contract basis, and restrictions on the private ownership of assets have also been liberalized.

Farmers have experienced these changes in policies in the form of alternate gains and losses in the right to make decisions about and manage their own land. They experienced uncertainty about the future and looked upon the policy changes as 'policy risks'. Although the present policy stimulated agro-production, it did not stabilize it (e.g., an increasing number of items are taxed). To avoid losing benefits, farmers responded to the responsibility system in two ways. First, they chose activities that gave quick returns, and, as a result, green manure crop cultivation in CDFS (at village level) decreased from 56.3 per cent of the total land area in 1978 to 5.7 per cent in 1988. Green manure cropping has been replaced by wheat production; in the meantime, fallow land decreased sharply and the cropping index increased from 130 per cent to 170 per cent. The second way in which farmers responded was to reallocate the use of their income. In the case study villages, there are few farmers who use their money to purchase agro-machinery or to improve the basic land composition. On the contrary, most farmers are likely to use their money for house construction. Farmers believe that private house construction is the safest investment, because private houses are not considered to be public property.

Insufficient Supply of Input

At present, the input supply system for mountain agriculture is far from sufficient, and this hampers the potential for development. In LDFS, salt and animal feed concentrate are in extremely short supply. Through government channels, livestock farmers can get a few inputs on a quota basis (about one-fifth of the basic needs). In HDFS and CDFS, fertilizers and pesticides are not readily available from State markets. In response to the deficit in supply, farmers changed their land-use patterns, exchanged goods on unfavourable terms in the open market, or even reduced their inputs.

Limited Technical Support

In the mountain areas of West Sichuan, the technical support system was established during the period of the 'People's Commune' and was organized on a collective management basis. After 1978, the responsibility system was introduced. Decentralized management in production made the technical services more complex. Unfortunately, since the establishment of the Responsibility System, the technical support system in mountain areas has not kept up with the changing production system. The main problems are the lack of technicians and the lack of increasingly diversified support.

In LDFS, hybrid animals were introduced during the ''People's Commune'' period and proved to be productive even though keeping hybrid animals needs special facilities. During that period, collectives managed their livestock in herds according to the kinds of animal kept, and technical support was relatively easy to procure. After the responsibility system, because livestock were divided among individual households, the need for technical support increased and procurement became more complicated. Since farmers could not handle the new facilities and could not receive sufficient support, the composition of hybrid livestock in the farming system shifted from yak/sheep-dominated to yak-dominated. About 60 per cent of the households (in the survey village) were outside of the animal disease control system of the region.

In HDFS, farmers are enthusiastic about planting fruit trees and introducing new species, but, without appropriate technical guidance, the types of disease and pests increased from two, in 1978, to six, in 1988. In CDFS, farmers would like to plant hybrid paddy and to use small farm machines, but owing to the fact that most of the farmers are not provided with technical services, they are restricted to planting native species for which they can use and depend upon manual labour.

In dealing with the limited support system in mountain areas, farmers first choose traditional methods to avoid any possible risk. Low technical requirements and low-input activities are preferred. To improve the technical level is their second choice only. For example, in HDFS, several farmers' associations have been established recently. Farmers exchange knowledge, learn new techniques through demonstration activities, and, most important, acquire outside linkages.

Constraints Typical of Different Farming Systems

Besides the constraints common to each farming system, there are some typical constraints to different farming systems which are closely related to local conditions, production management, and policies adapted in the area. These typical constraints and their implications are shown in Table 18.5.

LDFS are usually located in the high mountain (plateau) and remote areas. Geographically, these areas are far away from the developed regions. For example, the distance from Hong Yuan County (area of the case study) to Chengdu City (the capital of Sichuan Province) is more than 900 km. Normal transportation needs three to four days by road. During the rainy season, landslides often block the roads for days or even months. Socioeconomically, LDFS areas are inhabited predominantly by minority groups. The historical development process, traditional culture, and the standards of living of the local people vary from those of the *Han*, and there is a big difference in methods of production and consumption. Due to remoteness and separation, LDFS are relatively isolated.

In terms of the local marketing system, LDFS areas are underdeveloped in the following ways:

- there are no regular markets in the area,
- terms of exchange between livestock products and industrial products are unfavourable,
- the State market, which is the major source for the local market, is quite far away,
- items available in the State markets are very simple and are not appropriate for production needs or for daily necessities, and
- the necessary inputs for livestock production are normally not available in the markets.

Lack of proper marketing systems in LDFS areas restricts the development of production. The rapid increase of livestock numbers and low rate of commodity production

Table 18.5. Constraints typical of different mountain farming systems

Constraints	Mt. Specificity	Implications	Response
Constraints typical of livestock-dominated farming systems			
Backward marketing system	Inaccessibility	* isolation from main stream * high transport cost * unfavourable exchange terms * low mobility	* self-sufficiency oriented production * collective trading exchange * dry processing * butter making
	Fragility	* slow circulation * limited production options	
	Adaptation mechanism	* land-use pattern activity choice * lack of incentive	
	Marginality	* marketing mechanism	
Constraints typical of horticulture-dominated farming systems			
Widespread pests and diseases	Inaccessibility	* poor market access	* traditional method * use of marginal land
	Fragility	* vulnerable to disease * instability of production	* mixed cropping * storage of food grain
	Marginality	* poor exposure * isolation from outside	* farmers' research association * collective foundation
	Adaptation mechanism	* activity choice * mixed source of income * risk adaptation	* invite technicians from outside * outside learning
Constraints typical of crop-dominated farming systems			
Unfavourable pricing system for grain production	Inaccessibility	* unfavourable terms of exchange	* off-farm employment
	Fragility	* limited production options * no incentives for land use	* reduce grain crop cultivation * selling grain to outside market
	Marginality	* marginal land * focus on subsidy	* reduce input on land * dryland farming to reduce risk
	Adaptation mechanism	* tendency to look for short-term gains * land use pattern * mixed source of income * risk adaptation * multi-cropping	

Source: Study team

(leading to grassland degradation) are partly caused by the lack of access to markets. In response to the limited marketing support system, mountain farmers adapted two kinds of strategies, namely the passive approach and the positive approach. In the passive approach, farmers manage their production not for commercial purposes but for local consumption. In the positive approach, farmers manage collectively and produce for the market, e.g., groups of farmers collect their livestock together and select representatives to sell the livestock. In addition, methods such as the rough processing of meat (as well as livestock hides and wool) and dairy products are also undertaken by most households. Low-weight, high-value products are preferred.

In HDFS, the lack of pest and disease control is the major constraint to production. In recent years, the number of pests and diseases has increased rapidly and this has affected the stability of production and the quality of fruit products. The reasons for the increase in pests and diseases are:

- the limited supply of pesticides and agricultural chemicals,
- the inadequate methods of control,
- the low quality of pesticides available in the local market, and
- lack of technicians to supervise pest/disease control (there is only one technician in the county for horticultural development and the approach to pest and disease control is decentralized).

Traditionally, the farmers rubbed limes on the branches of trees to protect them from pests and diseases. In the case of a serious disease, the farmers used to cut and burn down the trees. Recently, the farmers invited experts from outside the area and organized training courses in horticultural techniques through the farmers' association for horticultural development.

In CDFS, unfavourable pricing systems for grain products do not encourage the farmers to be more productive. Under the regional plan, CDFS areas (normally in the broad valley basins) are designated grain production bases. At the regional level, this programme aims to fulfill the regional needs for cereal crops. In doing so, the local government established a quota system for households for the planting of cereal crops. In comparison to the market-pricing system, the price of grain is lower than the prices of other kinds of crops, while the productivity of grain crops (in weight) is generally lower than that of other crops (e.g., fruit, cash crops). Although the local government provided some subsidies for grain production, farmers still feel that they lose out by cultivating grain. As a matter of fact, the climatic conditions and soil conditions in the area are suitable for vegetables, fruits, and many other cash crops. Overemphasis on grain cultivation without an appropriate adjustment in the pricing system will neither result in optimum use of the resource base nor encourage farmers to support the regional plan. When restrictions were placed on their farming choices, farmers in CDFS areas responded by decreasing their grain producing areas and by reducing input in grain crops, as long as they could harvest enough grain for their own consumption and for land revenue. Those farmers having surplus grain prefer to sell it outside the region where grain prices are higher.

Potentials/Possibilities of Mountain Farming Systems

The potentials/possibilities for mountain farming systems in West Sichuan are diversification, specialization, and productivity as shown in Table 18.6.

Table 18.6. Potentials/Possibilities of mountain farming systems in West Sichuan

Constraints	Mt. Specificity	Implication	Response
Diversification of production	Diversity	* resource base * production flow * management * zonal perspective * choice of production * sources of income techniques * high-value crop * suitable species * indigenous methods	* ensuring the basic needs * income-oriented approach * sectoral linkages * cooperatives * crop substitution * off-farm employment * kitchen garden cultivation
Specialization	'Niche'	* high-value crop or plant * resource availability * demand from outside market * favourable climate * zonal perspective	* use of marginal land * inter-cropping * collective management * incentives to adopt new techniques * specialized households
Productivity	Diversity	* climatic difference * land types or forms * multi-layer cropping	* market- oriented production * ensure the basic needs
	'Niche'	* high-value crop * local varieties * seasonal availability	* change land- use pattern * collective management

Source: Study team.

Conclusions

According to the case study, we can conclude that farmers' strategies in mountain areas are essentially based on a number of specific principles.

Ensuring Basic Needs (Self-reliance, Self-consumption)

Food, fuel, and shelter are the basic needs of the people. In mountain areas, these basic needs depend largely upon the natural resources and sources of income. To ensure the basic needs, mountain farmers manage their land resources by diversified activities and close linkages with different sectors in production (Figs. 18.1, 18.2). Generally speaking, the availability of basic needs affects the production process (inputs on land, improvement of techniques, allocation of farm labour, etc.).

Avoidance of Risk to Safeguard Basic Needs

Mountain environmental conditions are relatively fragile; natural hazards occur more frequently than in the plains. To reduce the impact of natural hazards, mountain people developed indigenous methods of managing land and resources. These involved both individual and collective efforts.

Frequent changes in policies placed the development of mountain agriculture at risk. This resulted in many short-term gain activities which pushed mountain farming systems towards unsustainability. To respond to policies and programmes, farmers adopted positive or passive approaches to protect their interests. Reliable policies and continuity in land ownership (including the right of making decisions about land use) will help in long-term development.

More Options to Generate Income

For a long time, mountain areas did not keep up with the development level of the plains. Because of this, mountain poverty hampered the sustainability of mountain farming systems. It is understandable that farmers should want to adopt an income-oriented strategy. There is no doubt that there are many options for the use of mountain resources; linkages between mountain agriculture and the mainstream of development are also important.

Suitable Techniques

Due to the high rate of illiteracy among mountain populations, the deficiency of the support systems, the fragmentation of lands, and poor infrastructure, mountain agriculture is not suitable for large-scale/high technology farming. Instead, low-input, small-scale, and easily handled techniques are more acceptable. Looking at it from this perspective, it is easy to see that many programmes failed in the mountain areas because planners did not analyse the feasibility of their plans in the context of specific conditions of mountain areas.

REFERENCES

Banskota, M. and N.S. Jodha. 'Mountain Agricultural Development Strategies: A Framework for Development Strategies.' Chapter 3, Vol. 1, this publication.

Dafu, Y., C. Goujie, W. Fei, Y. Side and F. Hong. *Farmers' Strategies and Sustainability of Mountain Agriculture in West Sichuan.* China: Institute of Mountain Disaster and Environment, Chinese Academy of Sciences, Chengdu (limited circulation), 1990.

N.S. Jodha. 'Mountain Perspective and Sustainability: A Framework for Development Strategies.' Chapter 2, Vol. 1, this publication.

19

FARMERS' STRATEGIES IN THE MIDDLE HILLS OF NEPAL

S. Shrestha and B. Katwal

Sugandha Shrestha is a member of the professional staff of the
Mountain Farming Systems Division of ICIMOD
and Bhimendra Katwal is an economist associated with APROSC

CONTENTS

INTRODUCTION

Background

Operational measures adopted by mountain farmers in Nepal have evolved successful resource management and production practices within relatively harsh environmental and socioeconomic conditions. Specific features of mountain areas, such as inaccessibility and diversity, both in topographical and climatic contexts, present formidable challenges. Over the centuries, mountain farmers have made a succession of adaptational responses to the mountain environment through the development and adoption of indigenous technologies and specific strategies. Terraced cultivation is an outstanding example of ethno-engineering. With the passage of time, increased pressure on the limited parcels of land available for agriculture, exerted by a rapidly increasing population, led to severe stress on the fragile environment. Environmental degradation has become a key concern in the search for sustainable development strategies in the mountain areas of Nepal. Of late, several negative changes relating to resource status, production flows, and resource management have been observed. They are often attributed to disregard of specific mountain conditions by public and private intervention agents in mountain areas. Farmers' strategies and responses to changes fall into the latter category.

The sustainability implications of farmers' strategies and responses need to be evaluated within the framework of mountain specificities such as inaccessibility, fragility, marginality, diversity, 'niche', and human adaptation mechanisms. Since many traditional resource management systems were compatible with their environment, it is important to examine these systems, as well as the changes in the responses of mountain farmers over time, in the light of development interventions such as improved accessibility and new technology. Such an assessment will provide a basis for evaluating the positive and negative effects of such changes on the sustainable development of mountain agriculture.

Focus of the Paper

This paper examines the quantitative dimensions of the mountain perspective. This was achieved by analysing the farming systems dominated by crops, livestock, and horticulture, in the light of their responses to the mountain specificities outlined above, in order to assess the relevance, effectiveness, and sustainability implications of farming strategies within these mountain conditions.

DESCRIPTION OF THE STUDY AREAS

Three out of four study areas were chosen from the Central Development Region (CDR); including one (Yelung) from Janakpur Zone and two (Bhadaure, Ekle Gaun) from Bagmati Zone. The fourth (Deupur Tapu) was in Gandaki Zone of the Western Development Region (WDR) (Annex 1).

Bhadaure village consists of about 100 households located east-west of Naubise bazaar which is two kilometres away. The bazaar is at the junction of two major highways (Tribhuvan, Prithvi) connecting Kathmandu with the *Terai* and the Pokhara Valley. The distance between Naubise and the capital is only about 30 km. The *Ekle Gaon* site is east of and about three hours' walk from Bahrabise, a commercial centre located 86 km northeast of Kathmandu Valley. The Arniko Highway, which links Kathmandu and Tibet, passes through Bahrabise. *Yelung* is six kilometres away, located to the east of Jiri bazaar which is connected by an all-weather road to Kathmandu. Development institutions are the government livestock centre, the Agricultural Development Bank, and the Nepal Food Corporation. *Deupur Tapu* is about 10 km from the Parbat District headquarters—Kushma; Lumle Agricultural Centre is nine kilometres from the village. The village is close to the main trekking route from Pokhara (about 40 km away) to Jomsom.

The total number of farm households for each cluster, representing each study site, are estimated at 100, 140, 130, and 70 for Bhadaure, Ekle Gaon, Yelung, and Deupur Tapu respectively.

MOUNTAIN PERSPECTIVE

The mountains are different from the plains due to the presence of certain typical characteristics or conditions of the former. These characteristics create certain adaptations and activities and permit long-term exploitation of resources; concurrently the same characteristics restrict the extent of adaptations and activities. The negative implications associated with the activities restricted by the mountain characteristics also damage the potential benefits ensuing from unrestricted activities. Therefore, a knowledge of mountain characteristics is important for handling the mountain areas in a sustainable manner. Keeping this in mind, the important mountain characteristics of the study areas have been identified and discussed in the subsequent sections.

Mountain Characteristics or Specificities

The important conditions characterizing mountain areas, which for operational purposes separate mountain habitats from other areas, are referred to here as "mountain specificities" (Jodha 1990). The prevelant features of the specificities or characteristics (inaccessibility, fragility, marginality, diversity, 'niche', and human adaptation mechanisms) in the study areas are briefly discussed below.

Most of the study areas are remote and are not easily accessible by road. The permanent market centres are far away from the villages. To compensate, there are temporary markets (locally know as 'hath bazaar') in the areas which take place, at the most, once a week. Poor communications and poor mobility are the general norms of the study areas. Therefore, the areas are *inaccessible* in terms of both physical and socioeconomic dimensions. High altitude, steep slope, and coarse and loose soil types, accompanied by high precipitation with skewed distribution patterns, have made the study areas more vulnerable to stress. The increased population pressure and high density of livestock have aggravated resource degradation, and this can be seen by the decreased size of forests and pasturelands as well as the denuded condition of the pasturelands due to overgrazing. The altitude of the study areas ranges from below 1,000 m to over 2,400 m. In addition, cultivation has extended on to lands having 30–45° slopes. These circumstances make the areas very *fragile*.

In this context, marginality appears to be the most important characteristic from the point of view of constraint, and this is followed by *diversity* and *fragility*. In the Naubise study area, inaccessibility has been rated as the weakest constraint, whereas the same characteristic is the greatest constraint in the Yelung and Deupur-Tapu study areas. The underlying characteristics of each mountain specificity not only are interrelated in many ways, but, within the mid-hills of Nepal, show substantial variability. The same applies to the potential of human adaptation mechanisms, which appear to be the most important specificity in Naubise and Dhuskun areas, as is diversity for Yelung and Deupur-Tapu. As far as diversity is concerned, its underlying characteristics result in constraints as well as potentials.

HANDLING OF MOUNTAIN SPECIFICITIES

With consummate or partial cognizance, or even without intention, both individual/private (farmers) and public (government) interventions seem to have responded to the above mountain characteristics. These responses either adapt to or modify the characteristics (specificities) through various operational measures (Table 19.1). The implications of the response can then be realized either in the form of an improved environment or a degraded one depending upon the compatibility or discord between the responses and mountain characteristics. This section focusses on the ways and means through which both the farmers, at micro-level, and the government, at macro-level, are handling mountain specificities in terms of both constraints and potentials.

Responses to Mountain Specificities as Constraints

Inaccessibility

Inaccessibility has created certain conditions that are generally characterized by poor mobility, high transport costs, poor communication and educational facilities, and so on. These conditions have then led to subsistence agriculture which is seen in all the study areas apart from Naubise. The latter is slowly being transformed into a commercial farming centre (on a limited scale) through the cultivation of vegetables, due to the presence of relatively better physical and other institutional infrastructure (Annex 1, Table 19.1). Inaccessibility appears to be the most prominent mountain characteristic, in the form of a constraint, at Yelung and Deupur Tapu. To offset high transport costs, farmers have emphasized the production of high-value low-weight agricultural products; ghee processing at Yelung and radish seed production at Deupur-Tapu are examples. Ghee sales alone account for 80 per cent and 30 per cent of the total farm and the total household income in Yelung. Farmers also cultivate mustard as a high-value cash crop, and this is a major source of cash income (22% of the total farm income) in Dhuskun. Due to poor communications, information (agricultural research and extension), and high transport costs, the people of Yelung have resorted to the use of local cultivars and traditional management practices. Farmers do not use chemical fertilizers, and dependency on external input supplies is almost non-existent. Because of the closed system, created by inaccessibility, there are strong linkages among the three important components (e.g., crops, livestock, and forestry) of the mountain farming system. Unlike in Naubise, where farmers obtain seeds, fertilizers, and agro-chemicals from outside, the people of Yelung and Dhuskun acquire all their inputs from their own farms although these inputs may not

Table 19.1. Farmers' responses to mountain specificities as constraints

Farmers' responses to mountain specificities	Unit	Study areas Naubise	Dhuskun	Yelung	Deupur-Tapu
Inaccessibility					
proportion of HH exchanging labour	%	90	95	50	85
proportion of HH sharing agri. tools and other implements	%	80	90	50	80
composting vs. fertilizer	ratio	1:2.22	1:0.35	1:0.0	1:0.49
local improved seeds	ratio	1:3.86	1:0.19	1:0.17	—
income (of total farm cash income) generated by oil seed, radish seed production	%	1	22	—	—
HH involved in ghee-making,	%	60	60	50	50
area under improved crops (of total operated land)	%	80	16	15	—
Fragility					
level terrace	%	97	95	95	94
sloping terrace	%	3	5	5	6
level terrace with bunds	%	40	30	—	93
level terrace without bunds	%	60	70	100	7
total no. of trees (e.g., fruit,	no./HH	118	19	96	65
fuelwood, fodder, timber)	no./ha	122	28	343	93
area (of total cropped area) under minor crops (e.g., millet, buckwheat, barley)	%	4	57	27	—
under millet with maize	%	4	57	19	—
area under monocropping	%	37	29	69	10
area under double-cropping	%	61	> 1	23	78
forest plantation	ha	50	130	—	5
adoption of improved stoves (of the total HH)	%	20	—	—	—
the present stall-feeding practice (% of total feed)	%	90	80	50	—
increment in stall-feeding over the last 30 yrs or so)	%	100	80	20	—
area under slash and burn cultivation (*bukma* system)	%	—	—	5	—
Marginality					
increased cropping intensity by farm size					
marginal	%	—	186	107	—
small	%	175	170	123	—
medium	%	182	182	116	—
large	%	157	158	—	—
plantation of fodder tree by farm size					
marginal	No./ha	—	10	250	—
small	cult. land	131	13	90	—
medium	cult. land	105	25	292	—
large	cult. land	82	23	—	—

income from high-value cash crops (% of total farm cash income)	%	1	22	—	10
off-farm income (% of total income)	%	45	97	65	40

Source: APROSC 1990 and Katwal and Shah 1990.

be sufficient. This situation is manifested by the high level of improved seed use and of fertilizer application in Naubise and the low level in Dhuskun and Yelung (Table 19.1).

The government has attempted to modify the inaccessibility by constructing a road from Kathmandu to Barahbise, which is four to five kilometres away from Dhuskum, and to Jiri, six to seven kilometres away from Yelung; Deupur-Tapu is about 40 km away from the road. However, apart from Naubise and Deupur-Tapu (to some extent) none of the study areas have easy access to good information services (agricultural extension and support services). The construction of mule tracks at Yelung and the installation of wooden bridges over small rivers are also examples of farmers' attempts to modify inaccessibility.

Fragility

The conditions created by fragility include the vulnerability to degradation of the natural resource base. In this context, the farmers, particularly in Dhuskun, have responded by constructing a long (over 1,000 m) series of uninterrupted and quite stable terraces having 35° slopes. This demonstrates an attempt to convert fragility into infragility (or less fragility). In general, terrace farming is an established norm of crop husbandry, and over 95 per cent of the total area is under level terraces in all the four different study sites. In addition, farmers have adapted to this mountain characteristic by means of several operational measures. Among these, the plantation of perennial crops on private land is increasing in all areas; at present about 100 to 300 trees per hectare of different types (fruits, fodder, fuelwood, and timber) have been grown in Naubise, Deupur-Tapu, and Yelung.

In an effort to increase national revenue through better management of the forests, the government nationalized the forests and pasturelands during the 1950s. Contrary to the objectives, forest encroachment and destruction reached its peak following this measure and more land became barren and degraded, i.e., without or with little vegetation. The Panchayat Government made a provision for Panchayat Forests (PF) and Panchayat Protected Forests (PPF). This approach may be considered to be a process for modifying fragility. It is no exaggeration to say that farmers have responded more positively to measures to prevent the degradation of forests and pastureland through different operational measures. Fodder and fuelwood have been planted in degraded forest areas in Dhuskun, bamboo trees within the protected forest areas in Deupur-Tapu, and improved stoves are used, although on a limited scale, in Naubise. About 20 per cent of the total sample households used improved stoves. In addition, there has been an increase in stall-feeding practices in all study areas.

Over the last 30 years, it is estimated (based on key informants) that the rate of stall-feeding has increased by 20 per cent (at Yelung) to 80 or even 100 per cent (at Deupur-Tapu and Naubise). Despite the population growth and its increasing demand for food, cultivation practices which help degradation are still intact in some parts of the middle mountains of Nepal, for example, in Yelung (Table 19.1).

Marginality

The poor resource base in terms of poor human resources (e.g., low literacy rate), as well as poor land, forests, pastures, and a weak institutional base are the causes of marginality. Farmers have converted marginal, sub-marginal land and even forests into cultivated land in order to meet the growing demand for food. Although there is no strong spatial evidence, the study findings indicate that marginal or small-scale farmers are most likely to have intensified crop production. The cropping intensities of marginal and small-scale farmers are higher by about 10 to 20 per cent than those of large-scale farmers (Table 19.1). This is an example of farmers' strategies in response to mountain characteristics (e.g., marginality) and the increased total of production permitted by marginality. However, this approach has increased the fragility of agricultural land and the degradation of the natural resource base owing to the low level of input into the maintenance of soil fertility. Overextraction from forests and pasturelands is also evident, although there is now an emerging trend towards planting fodder and other trees on private land. As the forest/pasture resources and land productivity declined, farmers began to resort to off-farm activities in order to supplement their incomes. This is evident from the fact that almost all cash income (97%) in Dhuskun, where the majority are marginal farmers, now comes from such off-farm activities (Table 19.1).

Diversity

Diversity is probably the only mountain characteristic that has appeared in the form of both constraints and potentials. This section will deal with constraints.

The higher cost of production of agricultural, livestock, and other products may be the single important constraint that this mountain characteristic has caused. One of the major disadvantages associated with diversity is the low replicability of a given technology. Farmers have evolved various strategies to deal with this circumstance also. In order to avoid a higher cost of production, people use those crops that require only low level and local inputs. An increased use of compost and local seeds are responses to both diversity and inaccessibility (Table 19.1). This has been reflected in the cultivation of millet over a large area (e.g., 60% and 20% of the total cultivated area in Dhuskun and Yelung).

Responses to Mountain Specificities as Potentials

Some of the dimensions of mountain specificities offer unique opportunities for harnessing potentials in a way that permits optimal use of the available natural as well as human resources. Diversity, as a specific characteristic, does not merely act as a constraint, since it also creates diverse opportunities in farm and off-farm sectors through crop diversification, crop-livestock integration, and various off-farm activities. The mountain areas are also endowed with what may be called micro-'niche' for the production of cereals, cash crops, horticultural crops, livestock raising, off-farm activities, and hydropower generation. The harnessing of such 'niche', however, is dependent upon a set of preconditions which include physical (climatic and edaphic), biological, sociocultural, and institutional aspects. These preconditions are affected by prominent mountain specificities acting as constraints. For example, inaccessibility, as a constraint, limits the scope for effective input and service delivery mechanisms which could be preconditions for harnessing a potential 'niche'. The human adaptation mechanism is another mountain specificity acting as a potential. The mountain farmer has not only devised various operational means through adaptation to mountain characteristics but has also been able to modify mountain characteristics through his ingenuity.

The development potentials created by diversity and the mountain farmers' responses to these opportunities, together with the interventions from the government, have been shown in Table 19.2.

Harnessing the Diversity of the Resource Base

Table 19.2 shows that farmers in all the four study areas have adopted diverse crops, which include cereals (6 types), vegetables (5—8 types), and fruits (2—10 types), in response to the varied micro-climatic variations that enable the cultivation of various types of major and minor crops. Similarly, the production of various plant materials, such as fodder (5—10 types), and fuelwood (5—10 types), on private and public land was observed to be an important indicator of farmers' responses. With limitations on the access to forest resources, following establishment of the protected forests, there has been a growing trend towards the planting of fodder and fuelwood trees on private land. A closely integrated crop-livestock farming system was observed at all the study sites, the integration being high in Dhuskun and Deupur-Tapu and low in Yelung and Naubise. The reason behind low integration in Yelung is the unfavourable climate for crop production, and thus livestock farming is very prominent in the area. By contrast, the farming system in Naubise has undergone a transformation from cereal to vegetable production due to the presence of very favourable market forces and due to the fact that the area is completely lacking in grazing land; the forest is also quite far away, degraded, and under the control of the government.

Off-farm activities were observed to be important contributors to both household cash income and employment. Various types of cottage industries (ghee-making, carpentry, bamboo work, metal work, and weaving), business, remittances from family members, wage labouring, portering, and working as guides for mountain tourists are the main off-farm activities.

Public interventions to harness diversity have emerged in the form of the emphasis on horticultural crops in the mid-hills and livestock keeping in the high mountains. However, there has been less emphasis on minor crops. The emerging concept of farming systems research, as implemented in Deupur-Tapu, is an important initiative in the mid-hills. The support to livestock development programmes, through the establishment of the cheese factory at Yelung, is an example of the right kind of public intervention to enable the optimal harnessing of comparative advantages in a relatively inaccessible area in the mountains.

Resource Modification and Upgradation

It is observed that mountain farmers have been able to modify several mountain characteristics (acting as constraints) through generations of endeavour by taming the difficult terrain through terracing and through the development of formal and informal organizations responsive to the needs of development. There have also been adaptations

Table 19.2. Responses to diversity as a potential

Responses	Farmers	Governments
1. *Farm activities*	1. *Use of multiple crops*	
— crop diversification	— major crops (maize, wheat, paddy, potatoes)	1. Emphasis on major cereal crops 2. Focus on horticultural crops
— production of various types of plant materials	— minor crops (millet, barley, buckwheat)	3. Emphasis on livestock in high hills (Y)
	— vegetables (5–8 types), fruits (2–10 types)	4. Emerging concept of farming systems research (DT)
— crop-livestock integration		
	2. Fodder (5–10 types), fuelwood (5–10 types), timber (4–8 types), tree plantations on private and public land	
	3. Adoption of integrated crop-livestock farming D, DT— high integration Y, N— low integration	
1. *Off-farm activities*		
— cottage industry	1. Ghee making (N, Y, DT)	1. Establishment of cheese factory (Y)
— quarrying (Y)	2. Carpentry, bamboo work (D, Y, DT), metal work (D)	
— remittance		
— business	3. Weaving (Y)	
— wage labour + portering		
— tourist or expedition guide		

Note: N = Naubise; D = Dhuskun; Y = Yelung; DT = Deupur-Tapu.

Source: APROSC 1990 and Katwal and Shah 1990.

through operational strategies and measures that have enabled mountain farmers to cope with constraints on the one hand and harness production potentials in a sustainable way on the other.

Fragility, which is a unique mountain specificity, has been largely modified through terracing. At all four study sites, cultivation is based on terraced farming and more than 90 per cent are level terraces. The development of community organizations to improve resource management is evident in the form of vegetable growers' associations and alpine pastureland users' groups. These offset some of the disadvantages of inaccessibility and marginality. In terms of human adaptations to mountain characteristics, an important operational measure is the widely adopted practice of sharing private resources, e.g., exchange of labour, and the sharing of draught power, implements, and petty trading activities.

Exploiting 'Niche'

The production pattern followed by some of the farmers demonstrated a gradual transformation towards the exploitation of 'niche'. Such transformations are widespread in the case of Naubise where a transition is taking place from a crop-based to a horticulture-based farming system. The close proximity of Naubise to Kathmandu and accessibility by road has facilitated the harnessing of a 'niche' for commercial vegetable farming. Intensive cultivation practices, indicated by the proportion of double cropped areas, are evident in Naubise (61%), Dhuskun (71%), and Deupur-Tapu (78%) as shown in Table 19.1.

Because the 'niche' of the Yelung area lies in livestock farming, farmers maintain a greater number of animals (livestock units—LSU) per household (7.4 LSU) than in Dhuskun (2.7 LSU), Deupur-Tapu (3.4 LSU), and Naubise (4.2 LSU). This is the reason why farmers in the area derive almost all (98%) farm cash income from livestock.

In Yelung, a large proportion (70%) of the cultivated area is monocropped, and this is basically due to the climatic conditions prevailing in the area.

An important response in the inaccessible area of Deupur-Tapu has been the adoption of low-weight and high-value cash crops (radish seed production) by the farmers in response to the technology, inputs, extension, and marketing services provided as an integrated package by the Lumle Agricultural Centre (LAC). Now about 10 per cent of the total farm cash income comes from selling radish seeds alone. Sale of oilseed (e.g., mustard) fetches about 20 per cent of the total farm cash income in Dhuskun. The successful adoption of radish seed cultivation in Deupur-Tapu shows the positive response of mountain farmers to development interventions that are based on recognition of mountain specificities.

Sectoral Interlinkages

Diversified and interlinked activities, as reflected through farming-forestry linkages, are an important feature of farming systems in all the study areas. The extent of interlinkages and their gradual weakening, due to increased pressure of population and the side effects of commercialization and modernization, were documented in the study villages. These issues have not been covered by this paper as they have been fully discussed in the paper by Yadav (Chapter 6), in this volume.

SUSTAINABILITY IMPLICATIONS

Farmers' Responses

This section is devoted to the implications of responses to the mountain characteristics (discussed in the previous section). Regarding the improvement or deterioration of mountain agriculture in particular and the environment in general, if improvement leads to sustainability[1] then deterioration leads to unsustainability. This will be examined, based on the positive and negative changes that have taken place over the last two or three decades. Hence, each of the four farming systems, namely, horticultural crop-based at Naubise, cereal crop-based at Dhuskun, livestock-based at Yelung, and crop and horticultural farming at Deupur-Tapu, were examined based on the positive or negative changes.

Negative Side Effects

Concerning the widespread negative changes, the crop-dominated farming system is apparently the most critical area in terms of unsustainability. Not only has the production flow (or extraction rate) been seriously affected (since the crop yields and biomass supplies are declining) but the resource base itself has been greatly damaged. Unlike Yelung, all marginal and sub-marginal lands have been brought under cultivation in Dhuskun. The cropping intensity is already considerably high (172%). Due to the lack of soil nutrients, stones and rocks have begun to surface on cultivated land. The supply of compost materials, from both private and public land, and manure from livestock have decreased significantly over the past 20 to 30 years. Due to inaccessibility, farmers cannot replace these with chemical fertilizers despite heavy subsidies.

Nor is the situation of Yelung encouraging either. The sustainability matrix (e.g., production flow, resource base) has been affected by degradation. Because of declining crop yields and population growth, per capita foodgrain production has declined. Therefore, more than one-third of the total foodgrain supply now comes from outside, i.e., through the Nepal Food Corporation branch office located at Jiri.

The time interval between one cultivation of the plot to another in connection with the *bukma* (or shifting cultivation) system has been shortened now to three to five years from six to seven years, and this method is still widely practised. The presence of this extensive cultivation practice may also be an indication that the farming conditions in Yelung have not deteriorated, although it has been realized that the resource base itself is under strain. Because the area under forest and the alpine pastureland is steadily decreasing, over time the genetic biodiversity has been reduced due to the fact that some of the nutritious grasses and fodder tree species are slowly disappearing. Goat flocks and chauri or yak herds are vanishing. Biodiversity, which is one of the important elements of sustainability, is dwindling, and this may lead to unsustainability in the long run. Some negative changes have also taken place in terms of water resources. Water mills (ghatta) that used to run almost throughout the year in the past now run for only eight to ten months. The present open and haphazard grazing practices on alpine pasturelands

---

1 Sustainability is the ability of a system to maintain a certain well-defined level of performance (output) over time, and, if required, to enhance the same, including linkages with other systems, without damaging the ecological in tegrity of the system. Sustainability is a dynamic rather than a static concept (see Jodha, Chapter 2).

and meadows will only lead to deterioration of the resource base and this has already been observed in some areas where there has been overgrazing. This seems to have happened because of the public intervention when the resource base was nationalized. However, farmers have now recently begun to manage the pasturelands and meadows through community management systems and this is a sign of sustainability.

The farming system depends on the local resource base in Yelung, an indication of sustainability, particularly in remote areas. Similarly, the natural resource base is not degraded as in the Dhuskun area. However, the presence of some negative changes will lead to unsustainability unless public policies and programmes compatible with mountain characteristics intervene to halt the process.

The farming system at Naubise has dramatically changed from being cereal crop-based to being horticultural crop-based over the past 10 years or so. Fruit and vegetable farming are becoming increasingly popular in the area, and this is bringing several positive changes to the system. Such changes include the increased level of nutrition of the people and the accumulation of assets (e.g., buying more land for cultivation and purchasing improved breeds of animals). This is all happening due to increased crop yields and fruit and vegetable production.

Weakening of Linkages

Despite the fact that linkages with one of the vital components of the farming system—forestry—have broken down to a great extent, the horticulture-dominated farming system is the most promising among the three systems studied. This is due to the presence of strong market forces which are supplementing or even replacing the forest biomass required for compost with chemical fertilizers, accompanied by improved farming practices and the planting of fodder and fuelwood trees on private land.

Dependency on forests is highest in Yelung and lowest in Naubise. Almost all the biomass required for crop production and livestock keeping comes from private land in the latter area. This has enabled farmers to practise stall-feeding. The situation has, in fact, brought some positive changes even in herd compositions. Farmers have begun to replace cattle with buffaloes which are more productive animals. Over the last two decades the buffalo-cattle ratio for a herd in Naubise has been improved by about 25 per cent in favour of buffalo. The ratio now is estimated to be 1:2.02 as opposed to the previous ratio of 1:2.66. Due to the increased level of fodder production, the intensity of stall-feeding is very high. Almost all of the total feed supply comes from stall-feeding. Even the firewood and fuelwood supplies from the farms are now quite substantial.

Some negative changes have also taken place, however. Nearby forests and grazing lands have been converted into cultivated land. Fragmentation of land is increasing, mainly due to population pressure, and the size of each parcel of land has been reduced (see Annex 2, Table 2). The volume of water in each kuwa (small well) has been decreased by almost one-third during the last three decades, although the same number of kuwa are still in existence. However, the changes in the resource base are generally positive and encouraging (Table 19.3).

Positive Changes

Similarly, some changes in production flow and utilization/management practices have been realized. The increased application rate of chemical fertilizer (from zero to

Table 19.3. Dominant characteristics of changes over time as indicators of sustainability and unsustainability in mountain agriculture

Dominant characteristics	
Sustainability	Unsustainability
1. Increased area under perennial crops (e.g., fruits, fodder, fuelwood trees)	1. Increased rate of converting marginal, sub-marginal land for cultivation
2. Decreased level of herd size	2. Extended land cultivation practices on steep slopes
3. Replacement of cattle by buffaloes	3. Reduced size of parcels and increased amount of land fragmentation
4. Increased rate of stall-feeding practices	4. Encroachment on forest, grazing, and pasturelands
5. Increased area under cash crops (e.g., oilseeds, vegetable seeds, and fresh vegetables)	5. Increased time for fetching water, fodder, fuelwood, and other plant materials
6. Increased level of biomass production on private land	6. Reduced level of compost and manure application
7. Improved management of natural resource base (e.g., forest, pastureland) through community participation	7. Increased use of chemical fertilizers
	8. Declining trend in crop yields
	9. Increasing process of marginalisation (differential access to credit, widening hunger gap, increased proportion of landless and marginal farmers)
	10. Reduced level of biodiversity

Source: APROSC 1990 and Kanwal and Shah 1990.

94 kg of nitrogen per ha of cultivated land in Naubise within two decades) and the use of agro-chemicals for storing grains are encouraging. Due to this external input supply, intensive cultivation practices (e.g., multiple cropping) are becoming more and more feasible (Table 19.1). Keeping in mind the population pressure, intensive cultivation practices are considered to be necessary. However, this option can be debated, particularly in mountain areas, because of the consequent high intensity of resource use. This practice could be feasible in the future as the supply of inputs (e.g., improved seeds, fertilizers, agro-chemicals, and support services) appears to be guaranteed and sustainable, because the Naubise area is not only accessible but also exposed to new agricultural technologies. Moreover, the largest market (Kathmandu Valley) in Nepal is just 30 km away from the study area.

Many of the changes taking place in Deupur-Tapu have positive implications in the context of sustainability. The increased production of biomass from private land can be expected to improve sustainability, especially when the extraction rates of fodder and compost materials from the forest have declined; and this is partially due to depletion as well as to the restrictions placed on protected forests. The tendency to plant fruit and fodder trees on private land has increased. The increased area of land under tree crops has the potential to reduce fragility. New cropping patterns, especially those in which leguminous crops are introduced, are a positive change which will help to improve the condition of the soil. Lentils broadcast on standing rice crops, just before harvest for instance, require no tillage and no additional labour at a time when farmers are busy harvesting rice. This type of practice (which is increasing) will certainly help to minimize fragility.

The other positive changes have been realized through the improved situation of food supplies due to increased crop yields. Cash income also has increased because of increased ghee production. On the other hand, this has caused more dependence on external inputs, especially chemical fertilizers, particularly for vegetable production and for the production of cereal crops. The sustainability of this practice is debatable because of the inaccessibility of the area. Timely availability of fertilizers and agro-chemicals is a problem commonly faced by the farmers in Deupur-Tapu, although there are institutional facilities (e.g., a cooperative) nearby. The present marketing and input supply problems may improve once the proposed road linking Pokhara and Baglung (which passes close to the village) is completed. In this context, vegetable seed (e.g., radish) production and sale provide an important source of cash income (10 per cent of the total farm cash income) and is the right kind of response to inaccessibility.

Unfortunately, the forestry-farming linkage, which is one of the vital elements of sustainability, is weak in this area.

Development Interventions

Negative Side Effects

The discussions in the previous section indicate that farmers' strategies or responses have brought both negative and positive effects and changes in the agricultural production patterns which directly influence the livelihood of mountain people. The strategies and responses that have evolved over the time period examined are due to two factors. The first involves their understanding of mountain characteristics and adaptations, and the second involves exogenous forces (e.g., population pressure, technology innovations, and other public interventions) that heavily affect the farming system. The second set of factors is beyond the farmers' control. Nationalization of the forests, for example, led to the disintegration of the community resource management system, and, consequently, the degradation of forest/pastureland because of haphazard use and lack of protection. Unlike in the past, villagers retained no feelings of ownership for the forest/pastureland once the government took over. This circumstance, in fact, encouraged people to convert common property resources into private property. Overgrazing, loss of biodiversity, and soil erosion as a result of some of the institutional interventions of the government surfaced conspicuously.

Another example of faulty public interventions can be seen in the case of irrigation management. When the government initiated the construction of irrigation channels without considering traditional management practices, farmers' traditional irrigation management practices just disappeared. This intervention invited a lot of conflicts among water users and such problems had not existed previously. Similarly, the government's emphasis on cereal crops (which are basically Terai crops) in the mountains was not successful. Public interventions must evolve from an examination of farmers' strategies and responses and from consideration of the inherent mountain characteristics.

Incorporation of the Mountain Perspective

Regarding the internalization of mountain characteristics, it is worthwhile to recall here that mountain specificities demonstrate variability from one area to another

(Table 19.2). The operational implications of variability are that we cannot generalize about mountain perspectives in the context of effective development interventions. Fragility will be the predominant characteristic in the Dhuskun area, whereas elsewhere this might not be the case. Hence, more effort should be given to minimizing fragility as a constraint in this area. In this context, perennial crops (e.g., fodder, fuelwood, and timber) and horticultural crop cultivation on private land will not only ameliorate the 'fragility' constraint but also generate cash income.

Similarly, development efforts should be directed towards solving the problem of marginality. In this context, 'marginality' can be alleviated by focussing on those activities that can improve the poor resource base (private or public). Emphasis on formal and informal education to increase the literacy rate, increasing fodder/fuelwood tree plantations on both private and public land to improve the land resource base, and cultivation of high-value cash crops in order to raise cash income from small pieces of land are some of the operational measures that are effective development interventions for tackling 'marginality'. Provision of cooperatives, formal credit, either through the establishment of an agricultural development bank or mobile banks, and an insurance system for crops and livestock are other important interventions. Assistance to farmers to reorganize in groups such as in the Small Farmers' Development Programme (SFDP) will increase both the resource base (human, land, livestock, and others) and accessibility in terms of required inputs (e.g., credits) and other information (e.g., agricultural extension). In the same way, diverse micro-'niche' or 'diversity' create conditions suitable for the localized and limited production of a number of agricultural and other products. The cost per unit of production tends to be high, and this damages the feasibility in terms of markets. Hence, development strategies should encourage farmers to cultivate crops requiring a low intensity of inputs. Maize, millet, buckwheat, and barley are possible options provided the climate is suitable. Research and development is also essential for the development of appropriate agricultural technologies.

'Inaccessibility' appears to be an important mountain characteristic in Yelung and Deupur-Tapu and not very important in Naubise and Dhuskun. Because of its closeness to the road and other institutional infrastructures, Naubise is a relatively accessible area in comparison to the other three study areas. Therefore, development interventions should improve the accessibility to remoter areas by improving or constructing trails and tracks or roads; establishing agricultural/livestock sub-centres, and stationing extension agents in these areas. Besides these improvements, interventions should strengthen the existing, appropriate production pattern of low-weight and high-value products (e.g., high-value cash crops, cardamons, vegetable seeds, purified butter). The installation of processing units for converting low-value, high-weight commodities into high-value, low-weight products would be the appropriate development intervention for harnessing benefits even in accessible areas. Similarly, the provision of mobile agricultural banks, trade schools, postoffices, and so on will improve the farmers' access to resources and information and reduce the intensity of this constraint, even without road construction, the conventional method of improving accessibility. Agricultural research and development should also be geared to the development of technologies (e.g., agroforestry) that provide backward and forward linkages among different components of the farming system, i.e., crop-livestock-forestry. For instance, farmers should be provided with high grain-straw ratio cultivars which will not only help to meet the increasing food demand but will also increase fodder supplies. This type of intervention will be effective, particularly in an area where the natural resource base is degraded. Naubise is one example of an area where such interventions have been successful.

Harnessing 'Niche'

The 'niche' or comparative advantages of any given area depend upon those activities or economic sectors that can perform to advantage in that area or region in comparison to other areas. Livestock, horticultural and other high-value cash crops, tourism, mining, and hydropower generation are well-known 'niche' of mountain areas.

'Niche' are based on biophysical characteristics and the potentials are there whether they are harnessed or not. The harnessing of 'niche' depends upon the creation of suitable conditions, and these may include improved access to new technologies, inputs, extension, credit, and linkages with the wider market economy through the development of a marketing system. Institutional and other support services play a crucial role in creating favourable conditions for various 'niche'. A brief description of comparative advantages existing in the four study areas and the prerequisites for harnessing them is given below.

Naubise

Naubise is an area suitable for cereals, vegetables, and livestock (cattle and buffalo). The area is now in a stage of transition towards commercial vegetable farming. Due to its proximity to Kathmandu, the major market centre, and access to institutional support services (Sajha, Agricultural Development Bank), favourable conditions exist for harnessing these 'niche' by adopting new production technologies such as improved seeds and fertilizers. Vegetable cultivation and raising improved breeds of livestock are also extremely viable in this area.

Dhuskun

This area has potential for the cultivation of cereal crops, both major (paddy, wheat, and maize) and minor (millet and buckwheat). Although most of the terraced land in Dhuskun has a high degree of slope, the land is stable and has good water retention qualities; hence it is suitable for irrigated rice farming. Dhuskun also has the potential for cattle raising and enhanced crop-livestock integration in the farming system.

Yelung

Yelung has a comparative advantage in livestock raising, especially sheep, goats, and yaks or chauri. The development of livestock in Yelung would require improved access to extension and support services (credit and inputs). The potentials of livestock development can be harnessed if suitable conditions in terms of breed improvement, provision of fodder saplings and forage crop seeds, and veterinary services are created. Yelung has the potential for growing high-value cash crops (e.g., cardamon) and temperate fruits.

Deupur-Tapu

In Deupur-Tapu, cereals, vegetables, fruits, and buffalo raising are the potentials. Paddy is the most important crop in terms of development, as most of the land in Deupur-

Tapu consists of bunded terraces for irrigated rice farming. Maize and wheat are other suitable crops, although limited access to irrigation in winter is a constraint for wheat. The conditions in Deupur-Tapu are suited to growing various types of vegetables and fruits. Vegetables grown in the off-season offer comparative advantages as they fetch higher prices. To fully exploit the potential for fruit production, a processing facility should be established, given the current inaccessibility of the area. Buffalo keeping is also a potential activity in Deupur-Tapu.

CONCLUSION AND RECOMMENDATIONS

With or without knowledge, farmers have responded effectively to mountain characteristics either by modifying them or adapting to them through various operational measures. The public interventions, which have also contributed to increasing crop yields and vegetable production in some areas, in general appear to have disregarded mountain specificities.

In this context, several negative changes have taken place in all the mountain farming study sites. Most of the forests and pasturelands either have disappeared (due to their conversion into cultivated land) or are degraded beyond regenerative capacity (at least in the short run) in some areas, resulting in a reduced level of compost and manure supplies for crop production. These circumstances have led to declining crop yields. Therefore, the increased out-migration rate and widening hunger gap are general outcomes of unsustainability in the study areas. This situation persists and it implies that mountain agriculture, in general, is becoming unsustainable.

Hence, it is recommended that several responses or strategies of farmers, associated with traditional resource management and farming systems and relevant for sustainable mountain agriculture, even in a changing environment brought about by increased demographic pressure, should be introduced along with appropriate technological innovations. Finally, any public intervention should evolve out of consideration of farmers' strategies and full awareness of mountain characteristics.

REFERENCES

Agricultural Products Services Centre (APROSC). *Farmers' Strategies and Sustainability of Mountain Agriculture* (Case Study). Kathmandu: APROSC, July 1990.

Jodha, N.S. 'Mountain Perspective and Sustainability: A Framework for Development Strategies.' Chapter 2, Vol. 1, this publication.

Katwal, B. and L. Shah. Case Study on Transformation of Mountain Agriculture. Kathmandu: APROSC, March 1990.

Yadav, Y. 'Farming-Forestry-Livestock Linkages: A Component of Mountain Farmers Strategies.' Chapter 6, Vol. 1, this publication.

ANNEX 1

Table 1. Details indicating inaccessibility characterizing the study areas

Characteristic	Unit	Study area: Naubise	Dhuskun	Yelung	Deupur-Tapu
Distance to nearest road	km	1–2	4	6	30
Distance to nearest bus terminal	km	1–2	4	6	30
Frequency of diurnal bus service	no.	>10	4–5	2–3	—
Distance to nearest market centre	km	2	4	6	3
Distance to Kathmandu	km	30	90	>100	250
Distance to nearest telephone or other communication centre (e.g., post office)	km	3	4	6	10
Distance to agriculture/livestock sub-centres	km	3	1–4	6	0
Distance to support services' centres (e.g., inputs, credit)	km	2	4	6	2–10
Literacy rate	%	64	49	48	54
Proportion of small and marginal farm households	%	33	43	93	28
Average size of landholding	ha	0.97	0.54	0.28	0.7
Distance to cultivated land					
—average	km	0.5	3	0.5	0.2
—range	km	(0–1.0)	(0–4)	(0–2)	(0–1.5)
Agriculture/livestock staff (JT/JTA)* stationed in the village	no.	1	1	—	3
Leader/progressive farmers	no.	—	2	—	3

Note: JT= Junior Technician; JTA = Junior Technical Assistant.

Source: APROSC 1990 and Katwal and Shah 1990.

Annex 1

Table 2. Details indicating fragility characterizing the study areas

Characteristic	Unit	Study area Naubise	Dhuskun	Yelung	Deupur-Tapu
Altitude —main settlement	m	1,000	1,650	1,800	970
—range	m	970–1100	1,050–2,200	1,600–2,600	800–1,000
Slope —dominant	0°	20	35	25	15
—range	0°	10–30	20–45	15–40	10–20
Soil type	texture	loam-silty loam	loam	silty loam	loam, silty loam
Soil depth	cm	—	50–100	—	—
Proportion of flat land	%	5	—	—1	2
Proportion of cultivated land				—	—
under terracing	%	100	100	100	100
Proportion of land under sloping				—	—
terraces	%	3	5	2	6
Proportion of land under level				—	—
terraces	%	97	95	98	94
—with bunds	%	40	30	—	93
—without bunds	%	60	70	100	7
Proportion of uncultivated[a] land to total private land	%	5	4	1	6
Fragmentation of land				—	—
—average	no.	5	4	12	n.a.
—range	no.	1–8	1–10	1–28	n.a.
Average size of parcel	ha	0.2	0.13	0.02	n.a.
Rainfall distribution pattern				—	
(proportion of rainfall occurring in June–Sept.)	%	85	85	85	80
Temporal temperature variation					
—minimum	0°C	7–21	6–19	(−1–17)	(6–22)
—maximum	0°C	19–35	18–29	13–25	21–33
Number of hailstorms					
—average	year	3–4	2–3	3	2
Drought frequency (interval period)	year	2–3	2–3	2	—
Flooding frequency/excess rainfall (interval period)	year	3–4	2–3	—	—
Vegetative cover of forest, pastureland, etc	%	70	50	>70	60
Landslides					
—area affected	ha	—	30	10	—
—frequency	year	—	2–3	4–5	—
Perennial crops (e.g., fruit, fodder, and other trees) per farm	no.	118	14	96	65
—trees/ha cultivated land	no.	122	26	343	93

[a] Indicates *kharbari* (land under thatch, trees and other plant materials).

Source: APROSC 1990 and Katwal and Shah 1990.

Annex 1

Table 3. Details indicating marginality characterizing the study areas

Characteristic	Unit	Study area: Naubise	Dhuskun	Yelung	Deupur-Tapu
Proportion of small and marginal farm households	%	33	40	90	28
Proportion of landless households	%	—	3	7	—
Distance to accessible natural resource base					
—forest	km	3–5	3–5	1–2	1
—pastureland/grazing land	km	5	3–5	1–2	—
—water (e.g., river)	hr	0.5	1–2	0.3	0.5
Distance to agriculture/livestock sub-centres	km	3	1–4	6	0
Distance to other support service centres (e.g., inputs, credits)	km	2	4	6	2–10
Literacy rate	%	64	49	48	54
Programmes under community management (e.g., forest mgt, vegetable association)	No.	4	20	1	1
Forest and pastureland area under community management	ha	2	25	300–500	n.a.
Farmer-managed irrigation schemes	no.	1	1	—	1
Proportion of off-farm employment (of the total)	%	32	18	13	n.a.
Proportion of off-farm income	%	45	97	65	40
amount of credit borrowed					
—institutional	Rs	2,960	787	3,367	2,492
—non-institutional	Rs	2,858	1,116	777	646
Three major cereal crops with percentage of total cultivated areas		maize 62%	maize 78%	wheat 38%	paddy 88%
		paddy 43%	millet 57%	maize 28%	maize 40%
		wheat 27%	paddy 32%	potato 28%	wheat 15%

Source: APROSC 1990 and Katwal and Shah 1990.

Annex 1

Table 4. Details indicating diversity characterizing the study areas

Characteristic	Unit	Study area: Naubise	Dhuskun	Yelung	Deupur-Tapu
Elevation range	m	970–1,100	1,050–2,200	1,600–4,000	800–1,000
Range of steepness (landscape)	0°	20	35	25	15
Types of soil	texture	loam to to silty loam	loam to sandy loam	loam	loam to silty loam
Variation in micro-climates,					
—range		tropical to sub-tropical	sub-tropical to cool temperate	warm tem-perate to alpine	sub-tropical
—no.		2	3	4	1
Variation in types of vegetation					
—different types of cereal crops	no.	5	4	6	4
—different types of cash crops	no.	2	1	—	2
—various horticultural crops					
—vegetables	no.	8	—	—	5
—fruit trees	no.	10	2	3	10
—different kinds of fodder trees	no.	10	5	8	6
—different kinds of fuelwood trees	no.	10	5	8	6
—different kinds of timber trees	no.	5	6	8	4
—berseem/cocksfoot forage grass	no.	—	—	2	2
—different economic farm groups (based on farm size)	no.	3	5	4	3
various ethnic groups	no.	6 Brahmin 75%	6 Chhetri 35% Thami 40%	3 Sherpa 90%	1 Brahmin

Source: APROSC 1990 and Katwal and Shah 1990.

Annex 1

Table 5. Details indicating 'niche' characterizing the study areas

Characteristic	Unit	Study area: Naubise	Dhuskun	Yelung	Deupur-Tapu
Crop diversification					
—different crops grown	No.	10	5	6	14
—proportion of monocropping	%	37	29	69	10
—proportion of double cropping	%	61	71	23	78
—proportion of triple cropping	%	2	—	—	12
—fallow land	%	—	—	8	—
Livestock farming					
—animals raised[a]					
—head	no	7.5	5.8	11.8	5.5
—LSU	no.	4.2	2.7	7.4	3.4
—types of animals kept	no.	3	3	5	4
Proportion of cash crops cultivated (sugarcane and mustard)	%	9	2	—	4
—water resource utilization					
—water mills (ghatta)	no.	1	11	5	—
—proportion of cultivated land under irrigation	%	52	31	—	88
—hydropower generation	kw	—	—	—	—
—cottage industry					
—proportion of households involved in weaving	%	—	5	10	—
—proportion of households involved in capentry, bamboo work	%	—	5	5	18
—rice mills	no.	3	—	—	—
—proportion of households involved in ghee processing for sale	%	10	—	20	50

[a]Poultry are excluded.

Source: APROSC 1990 and Katwal and Shah 1990.

Annex 1

Table 6. Details indicating human adaptions in the study areas

Characteristic	Unit	Study area			
		Naubise	Dhuskun	Yelung	Deupur-Tapu
Modification of mountain characteristics					
Terracing					
—level terrace	%	97	95	98	94
—sloping terrace (proportion of total land)	%	3	5	2	6
—level terrace with bunds	%	40	30	—	93
—level terrace without bunds	%	60	70	100	7
Organizing farmers' groups		3	19	1	1
—SFDP	no.	—	16	—	—
—farmers/vegetable growers' association	no.	1	—	—	1
—alpine pastureland users' group	no.	—	—	1	—
—forest management group	no.	2	3	—	—
Adaptation to mountain characteristics					
Crop diversification					
—crops grown	no.	10	5	6	14
—integrated crop-livestock farming (proportion of households adopting it)	%	100	100	100	100
—sharing of private resources proportion of household)					
—exchange of labour	%	90	95	50	85
—sharing of draft animals	%	90	90	30	25
—sharing of farm implements	%	80	90	50	80
—petty trading (proportion of households)	%	100	100	100	100
—commercial trading (proportion of households)					
—vegetables	%	80	—	—	70
—farmer-managed irrigation schemes	no.	1	1	—	1

Source: APROSC 1990 and Katwal and Shah 1990.

ANNEX 2

Table 1. Range of quantified indicators of inaccessibility by study area

Characteristic	Unit	Range of quantitative terms Low 1	 Medium 2	 High 3	 Very high 4
Distance to nearest road	km	< 2	2–5	5–10	> 10
Distance to nearest bus terminal	km	< 2	2–5	5–10	> 10
Frequency of diurnal bus services	no.	> 8	5–8	2–4	1
Distance to nearest telephone or other communication centre (e.g., post office)	km	< 2	2–5	5–10	> 10
Distance to agriculture/livestock sub-centres	km	"	"	"	"
Distance to support services centres (e.g., inputs, credit)	km	"	"	"	"
Literacy rate	%	> 60	50–60	40–50	< 40
Proportion of small and marginal farm households	%	< 30	30–40	50–70	> 70
Average size of land holding	ha	> 1.0	0.51–1.0	.20–0.5	< 0.2
Distance to cultivated land	km				
—average	km	< 0.5	0.5–1.0	1–3	> 3
—range	km				
Agriculture/livestock staff (JT/JTA) stationed in the village	no.	3	2	1	—
Leader/progressive farmers	no.	> 5	3–5	1–3	—

Source: APROSC 1990 and Katwal and Shah 1990.

Annex 2

Table 2. Range of quantified indicators of fragility by study area

Characteristic	Unit	Range of quantitative terms			
		Low 1	Medium 2	High 3	Very high 4
Altitude	m	<900	900–1500	1500–3000	>3000
Slope	0°	<10	10–20	20–30	>30
Soil type	texture	loam to silty loam	loam to sandy loam	sandy loam	silt sandy
Soil depth	cm	>50	30–50	10–30	<10
Proportion of flat land	%	>20	10–20	5–10	<5
Proportion of cultivated land under terracing	%	>95	90–95	80–90	<80
Proportion of land under level terrace	%	>95	90–95	80–90	<80
Proportion of land under sloping terrace	%	<5	5–10	10–20	>20
Fragmentation of land —average	no.	<4	4–6	6–8	>8
Average size of parcel	ha	<0.5	0.3–0.5	0.1–0.3	<0.1
Rainfall distribution pattern (proportion of rainfall occurring June-Sept.)	%	<50	50–70	70–80	>80
Temporal temperature variation (gap between min and max)	0°C	<10	10–15	15–20	>20
Number of hailstorms —average	year	>3	2–3	1–2	1
Drought frequency (interval period)	year	>3	2–3	1–2	1
Flooding frequency/excess rainfall (interval period)	year	>3	2–3	1–2	1
Vegetative cover of forest, pastureland, etc	%	>70	50–70	30–50	<30
Landslide —frequency	year	>5	3–5	2–3	1
perennial crops (e.g., fruits, fodder, and other trees) at farm —trees/ha cultivated land	no.	>200	100–200	50–100	<50

Source: APROSC 1990 and Katwal and Shah 1990.

Annex 2

Table 3. Range of quantified indicators of marginality by study area

Characteristic	Unit	Range of quantitative terms			
		Low 1	Medium 2	High 3	Very high 4
Proportion of small and marginal farm households	%	< 30	30–50	50–70	> 70
Proportion of landless households	%	< 3	3–5	5–10	> 10
Distance of accessible natural resource base					
—forest	km	< 1	1–3	3–5	> 5
—pastureland/grazing land	km	< 1	1–3	3–5	> 5
—water (e.g., river)	hr	< 0.5	0.5–1.0	1–1.5	> 1.5
Distance to agriculture/livestock sub-centres	km	< 2	2–5	5–10	> 10
Distance to other support service centres (e.g., inputs, credits)	km	< 2	2–5	5–10	> 10
Literacy rate	%	> 70	50–70	30–50	< 30
Programmes under community management (e.g., forest mgt, vegetable association)	no.	> 10	5–10	3–5	< 3
Forest and pastureland area under community management	ha	> 100	50–100	25–50	< 25
Farmer-managed irrigation schemes	no.	> 75	3–5	1–3	1
Proportion of off-farm employment (of the total)	%	> 40	30–40	20–30	< 20
Proportion of off-farm income	%	> 40	30–40	20–30	< 20
Ratio of institutional and non-institutional credit	ratio	> 1.0	0.75–1.0	0.5–0.75	< 0.5

Source: APROSC 1990 and Katwal and Shah 1990.

Annex 2

Table 4. Range of quantified indicators of diversity by study area

Characteristics	Unit	Range of quantitative terms			
		Low 1	Medium 2	High 3	Very high 4
Range of elevation	m	< 900	900–1500	1,500–3,000	> 3000
Range of steepness (landscape)	0°	> 10	10–20	20–30	> 30
Various types of soil	texture	silty sandy	loam	silty loam	silty loam
Variation in micro-climates,	no.	< 2	2–3	3–4	> 4
Variation in types of vegetations					
—different types of cereal crops	no.	< 3	3–5	5–8	> 8
—different types of cash crops	no.	< 2	2–3	3–5	> 5
—various horticultural crops					
—vegetables	no.	< 3	3–5	5–8	> 8
—fruit trees	no.	< 3	3–5	5–8	> 8
—different kinds of fodder trees	no.	< 5	5–8	8–10	> 10
—different kinds of timber trees	no.	< 5	5–8	8–10	> 10
—berseem/cocksfoot forage grass	no.	< 2	2–3	3–5	> 5
Different economic farm groups (based on farm size)	no.	< 2	2–3	3–4	> 4
Various ethnic groups	no.	< 3	3–4	4–5	> 5

Source: APROSC 1990 and Katwal and Shah 1990.

Annex 2

Table 5. Range of quantified indicators of 'niche' by study area

Characteristic	Unit	Range of quantitative terms			
		Low	Medium	High	Very high
Crop diversification					
—different crops grown	no.	< 3	3–5	5–10	> 10
—proportion double cropping	%	< 30	30–50	50–70	> 70
—livestock farming					
—animals raised[a]					
—head	no.	< 5	5–8	8–10	> 10
—LSU	no.	< 3	3–5	5–7	> 7
—types of animals kept	no.	< 2	2–3	3–5	> 5
Water resource utilization					
—water mills (ghatta)	no.	1	2–4	4–6	> 6
—proportion of areas under irrigation	%	< 20	20–40	40–60	> 60
—cottage industry					
—proportion of households involved in weaving	%	< 10	10–20	20–30	> 30
—proportion of households involved in bamboo work	%	—	5	5	18
—rice mills	no.	3	—	—	—
—proportion of households involved in ghee processing for sale	%	10	—	20	50

[a]Poultry are excluded.

Source: APROSC 1990 and Katwal and Shah 1990.

Annex 2

Table 6. Range of quantified indicators of human adaptation by study area

Characteristic	Unit	Range of quantitative terms			
		Low	Medium	High	Very high
		1	2	3	4
Modification of mountain characteristics					
Terracing					
—level terrace	%	< 80	80–90	90–95	> 95
—sloping terrace	%	> 20	10–20	5–10	< 5
—level terrace with bunds	%	< 20	20–30	30–40	> 40
—level terrace without bunds	%	> 80	70–80	60–70	< 60
Organising farmers' groups	no.	< 3	3–5	5–10	> 10
Adaptation to mountain characteristics					
Crop diversification					
—(crops grown)	no.	< 3	3–5	5–10	> 10
Integrated crop-livestock farming (proportion of households adopting)	%	> 70	70–80	80–90	> 90
sharing of private resources (proportion of households)					
—exchange of labour	%	< 50	50–70	70–90	> 90
—sharing of draft animals	%	< 50	50–70	70–90	> 90
—sharing of farm implements	%	< 50	50–70	70–90	> 90
Petty trading (proportion of households)	%	< 50	50–70	70–90	> 90
Farmer-managed irrigation schemes	no.	1	2–3	3–5	> 5

20

DIVERSITY OF FARMING SYSTEMS AND FARMERS' STRATEGIES IN THE MOUNTAIN VALLEY OF CHITRAL, PAKISTAN

Masood ul Mulk

Masood ul Mulk is a Regional Programme Monitoring Officer of the AKRSP, Chitral, Pakistan

CONTENTS

INTRODUCTION

This paper presents a descriptive account of Chitral District as being representative of a mountain area in the North West Frontier Province (NWFP) of Pakistan. The focus is on the presentation of features of the region in terms of mountain specificities, their implications, and farmers' and government's responses to them. The latter is done by describing specific farming practices which represent both traditional farming systems and adjustments to them resulting from pressures imposed by population increases, technology introductions, and changes in institutions.

The data which reflect features of these mountain farming systems were collected through intensive village studies conducted in a collaborative effort between the Aga Khan Rural Support Programme (AKRSP) and ICIMOD in 1990. The details of the procedure and methodology followed in these case studies are described elsewhere (Mulk 1990).

Of over 600 villages of Chitral, three were selected to represent varying types of farming systems characteristic of the agro-ecological zones of the NWFP. These three types are crop-dominated, crop-cum-livestock-dominated, and livestock-dominated farming systems. The features of these zones, as found in the villages of Chitral, are summarized in Table 20.1.

Table 20.1. Features of the three agro-ecological zones

Zonal characteristics	Zone 1	Zone 2	Zone 3
Altitude (m)	<1800	1800–2100	2101-2300
Rainfall (av/mm)	48	10	70
Soil type	Sandy loam, relatively fertile	Sandy loam with stones, poor in nutrients	Bare and shallow soils with very poor nutrients
Land holdings per household (ha)	1.0	.8	.5
Dominant activities	crops, horticulture	crops, livestock	livestock
Major crops	wheat, maize vegetables	wheat, alfalfa barley	alfalfa, wheat
Livestock/household			
number	6	25	29
type	cattle, goats	goats	goats, sheep
Representative village	Kesu	Marthing	Besti

Differences in village/farm characteristics, listed in items (4) to (7), represent farmers' adaptations to the varied conditions of the ecological zones, and can be further described as their strategies.

Source: Mulk 1990.

In Lower Chitral, below 1,800 m land holdings are larger, the climate allows cultivation of double crops, and rich forests of holly oak and deodar exist on higher mountain slopes. Better means of communication and the presence of a small urban market have produced significant changes in the crops grown. This farming system is a combination of crop and horticultural production, and is represented by Kesu Village.

Upper Chitral, with less rainfall, poor forest cover, single crop cultivation, poor means of communication and absence of an urban market, can be regarded as a crop-cum-livestock farming system. This system is represented by Marthing Village.

Some parts of Chitral, in the upper belt, have rich pastures, allowing larger numbers of livestock to be maintained per household. In such areas, land is extremely scarce and agriculture is only possible after bringing very difficult mountain slopes under cultivation. This is represented by Besti Village.

FEATURES OF THE STUDY AREA

Chitral District has an area of 14,850 square miles and a population of 266,000 (1989 estimates). The district is characterized by a rugged topography, with high mountain ranges penetrated by steep-sided, narrow valleys and rivers. Land access is via a number of high passes, the lowest of which is at an elevation of about 3,000 m. Vehicle access is via the Lowari Pass, and this is usually closed to vehicular traffic for six months in a year.

Settlements and agriculture are concentrated in the valleys and streams of the Chitral river system. The valleys attain a maximum width of 5 km. The land types of the area indicate that only 4 per cent is available for agriculture and 24 per cent is forest and grazing area. The rest is all barren mountain ranges.

The climate of the district is distinctly continental with summer temperatures ranging from very hot in the lower elevations of the main river valleys to warm in the uplands and cool at higher elevations.

The features of Chitral District can also be described in terms of mountain specificities (Jodha 1990).

Inaccessibility

Inaccessibility is the most dominant characteristic of Chitral. For centuries Chitral remained an isolated, independent state because of the harsh nature of its terrain. Accessibility, it may be pointed out, in the context of Chitral, has two aspects, first internal and secondly external.

In the past, inaccessibility, both within the district and outside it, has played a leading role in the evolution of farming systems. Communication between villages was an arduous task necessitating the evolution of subsistence-oriented farming systems. Every village strove to become self-sufficient in food. Irrigation channels were dug across difficult terrain to bring water and all possible land was placed under cultivation. These channels were difficult to build, giving rise to cooperative systems of ownership and management of channels. Each village had its own crop lands, forests, and pastures. The latter would be at considerable distances from the village and would provide grazing grounds for livestock in the summer months, thus saving invaluable fodder resources back home. This also explains why the villagers were growing crops as diverse as wheat, maize, millet, rice, and cotton. Fruits grown were largely those that could be stored for the winter.

In the last three decades, a network of roads has been laid down within the district. Today, except for Yarkhoon Valley, all other valleys are accessible by jeep. This has had a profound effect on the farming systems, which are now moving towards commercialization by exploiting each valley's comparative advantages.

Chitral, however, remains isolated for six months in a year when Lowari Pass closes in winter. This inaccessibility to the outside world remains a major constraint to development. Fruit production cannot be carried out on a large scale as the surpluses cannot reach markets outside the district.

Fragility

Chitral's environment is extremely fragile. It has been sparsely endowed with natural forests. Scanty rainfall does not permit the growth of luxurious vegetation to bind the soil. In the absence of dense population pressures and through careful preservation by communities, the environment was protected in the past against hazards such as soil erosion by rivers and floods. Similarly, over-grazing of forests by livestock and indiscriminate cutting of forests pose a serious threat to the environment. The farmers, whose livelihoods depend on crops and livestock systems, tend to keep many animals, causing over-grazing and denudation. Heavy downpours cause devastating floods. The in-migration of nomadic herdsmen, who seem insensitive to the environmental scenario, also threatens the sustainability of the farming systems.

Marginality

Chitral, in many respects, is a marginal area in comparison to other areas of Pakistan. The major resource of the area is land: almost 90 per cent of the population is dependent on agriculture for its sustenance, yet only 4 per cent of its total area is available for agriculture. Ninety per cent of the population has a landholding of less than 3 ha, and 70 per cent less than 1 ha. The district is poorly endowed with forests and pastures. The former face extinction at the hands of exploiters from both public and market forces. The quality of soil is poor, lacking in nutrients and nitrogen. The use of chemical fertilizer is leading to higher yields, but, in the long run, in the absence of crop rotation, the cost of these higher yields could be heavy.

The quality of both physical and economic infrastructure is poor. Investment levels in the social sectors are low. The people of Chitral have lived isolated lives for centuries. Lacking an innovative, competitive spirit, they generally find themselves outpaced by outsiders. The desire for off-farm employment inspires many to migrate seasonally but they end up with menial jobs in big cities.

Diversity

The area has tremendous diversity in terms of climate and topography. The diversity explains how pastures, forests, and croplands have been integrated into the farming systems to enable residents to make a living in the harsh environment and to make the best use of its ecological resources.

'Niche'

The area has an abundance of water flowing from the mountains which could be tapped for irrigation and for generating hydro-electricity; it also has spectacular scenery which attracts tourists. Fruit production benefits from the market demand for produce that is out of season in the plains. Many high areas offer ideal environments for wild animals such as markhors and snow leopards.

Adaptation Mechanisms

Over the centuries, the people of Chitral have evolved institutions, both formal and informal, which have enabled them to survive in this harsh environment. These institutions were for building and maintaining roads and irrigation channels, for providing collective labour at harvest time, and for common grazing arrangements in forests and pastures during the summer. The idea of a gram or 'collective will' of the community, expressed around the place of worship, overrode individual or selfish interests for the good of the community.

Over a thousand irrigation channels were constructed by illiterate people, using local technology, with no engineers to guide them. Similarly, fields on mountain slopes were terraced to prevent soil erosion. The crop rotation system ensured fertility of the soil.

PEOPLE'S ADJUSTMENTS TO MOUNTAIN AREAS

Differences in the features of farming systems among and within different regions are an indication of peoples' adaptations to environmental differences. These differences may be reflected in:

— demographic features
— occupation patterns
— position of assets, ie landholding, livestock holding
— farm operations
— land utilisation
— livestock composition
— land abandonment
— uneconomic and fragmented land
— responses to resource crisis
— indigenous practices
— cropping patterns
— income/expenditure patterns

These are described below.

Demographic Features of the Sample Households

The poor resource base of the area faces a rapidly increasing population. Forty-two per cent of the population is below 15 years in Kesu, 43 per cent in Marthing, and 50 per cent in Besti.

The literacy rates show significant trends. In Kesu the literacy rates are 33 per cent, in Besti 7 per cent, and in Marthing 21 per cent. The literacy rates are closely related to how early a village became accessible. These literacy rates, for a remote district, compare very favourably with general literacy rates in Pakistan. One reason which accounts for the high literacy rates is the desire for education among farm households with a view to finding off-farm employment for some of its members (Table 20.2).

Table 20.2. Demographic features of the sample households

Sites and farm size	Family size (no.)			Economically Active			Dependency Ratio	Literacy percentage		
	Male	Female	Total	Male	Female	Total		Male	Female	Total
Zone 1 (Kesu)	119	82	201	64	52	116	1.73	26	18	33%
Zone 3 (Besti)	140	114	254	71	56	127	2.00	11	2	7%
Zone 2 (Marthing)	122	117	239	75	60	135	1.78	29	14	21%

Source: Mulk 1990.

Occupation of Economically Active Members of Sample Households

An analysis of the data reveals that 55 per cent of the economically active population in Kesu find work at salaried jobs or business more important and profitable for family sustenance. This carries important implications for farm management. In Kesu Village, where 100 per cent of the farmers own less than 2 ha, 73 per cent of the farms are either cultivated on a share-cropping basis or given out to tenants. The consequences of the kinds of trends one sees in Kesu Village are far-reaching for the traditional collective management systems for irrigation channels and livestock. The absence of large numbers of people from the villages undermines these systems (Tables 20.3 and 20.4).

Table 20.3. Occupations of the economically active members of the sample households

Sites and farm sizes	Main occupation				Secondary occupation											
	Agri-culture		Salaried Job		Busi-ness		Labour		Contr-actor		Pens-ioner		Business		Labour	
	M	F	M	F	M	F	M	F	M	F	M	F	M	F	M	F
Zone 1 (Kesu)	30	52	22	0	7	0	3	0	2	0	7	0	1	0	0	0
Zone 3 (Besti)	58	55	13	1	0	0	0	0	0	0	0	0	1	0	16	0
Zone 2 (Marthing)	66	59	8	1	0	0	1	0	0	0	0	0	0	0	22	0

Source: Mulk 1990.

Table 20.4. Distribution of households by land ownership patterns (No. of households)

Site and farm sizes	Total no. of house-holds	Exclusive owner operators	Share cropping operators	Exclusive tenant
Zone 1 (Kesu)	30	8	17	5
Zone 3 (Besti)	30	30	0	0
Zone 2 (Marthing)	30	30	0	0

Source: Mulk 1990.

Land Size, Distribution, and Use at Household Level

Land holdings are very small. In Kesu Village, 100 per cent of land holdings are less than 18.50 *chak*, or 2 ha. In Besti and Marthing, 97 per cent of the households have less than 2 ha of land.

The average land holdings and use patterns show that the lowest land holding group in Kesu has 5.81 *chak* (0.63 ha) of land, while in Besti this figure is 4.05 *chak* (0.44 ha), and in Marthing 7.12 *chak* (0.77 ha). The cultivable land is even smaller, being 4.83 *chak*

(0.52 ha) in Kesu, 2.62 *chak* (0.28 ha) in Besti, and 4.55 *chak* (0.49 ha) in Marthing. Table 20.5 gives the average land-use pattern of all 30 households in the village.

Table 20.5. Average land use at household level (*chaks*)

Sites and farm sizes	Total land	Homestead and wool	(%)	Crop cultivation	(%)	Fodder and trees	(%)
Zone 1 (Kesu)	9.68	0.86	8	8.87	89	0.38	3
Zone 3 (Besti)	7.45	1.00	19	4.72	63	1.73	24
Zone 2 (Marthing)	9.33	0.67	7	5.92	64	2.68	29

Source: Mulk 1990.

Uneconomical and Fragmented Nature of Land Holdings

The number of irrigation channels handled by a household is an important indicator of the economical nature of land holdings. In Besti Village, the pressure on land is most intense; the average number of irrigation channels handled by a household varies from 8 to 20 among different farm sizes. In Kesu, the variation is between one and two irrigation channels per household while in Marthing Village an average of two irrigation channels are handled by each household.

Land fragmentation is another indicator of pressure on existing farming systems. In Besti Village, a journey of over one and a half hours on foot was required for 27 per cent of the households to travel between two fragments of land.

Land Abandonment

In Kesu Village it was found that over 30 per cent of the households had given up cultivating some of their lands because of higher yields on other parcels of land. In the recent past, to meet the growing needs of the population, marginal lands, which could only be cultivated if there was rainfall or moisture, had been given up after the introduction of chemical fertilizer and improved seeds.

In both Kesu and Marthing villages, a lack of water in the irrigation channels, despite water being available at the source, is a major reason for land abandonment.

Livestock Holdings

Livestock ownership is widespread, and constitutes an important part of farm production.

The average livestock holding in Kesu Village is about 6 animals, while it is 29 in Besti, and 25 in Marthing. The mean does not give a correct picture for the number of goats held by a household, as the data show considerable variation. A few households tend to have almost three times as many goats as others. There is not much variation in the data for cattle and sheep.

In both Besti and Marthing villages, the number of goats and sheep per household is still very high, showing the importance of these in the livestock system. At Kesu, the number of cattle has decreased considerably. As this village is easily accessible, tractors and threshers reached the village some years back. Most households reduced the number of cattle after the introduction of tractors and threshers. This has helped to improve animal health and nutrition as there are less animals to feed (Tables 20.6, 20.7, and 20.8).

Table 20.6. Livestock holdings of total households by farm size and study site

Sites and farm sizes	Cattle	Goats	Sheep	Donkey	Horse	Poultry
Zone 1 (Kesu)	93	62	4	7	0	372
Zone 3 (Besti)	336	687	681	0	0	885
Zone 2 (Marthing)	182	383	195	9	0	179

Each goat, sheep, donkey, horse and head of cattle has been considered a single livestock unit.

Source: Mulk 1990.

Table 20.7. Average livestock unit per household by farm size and study site

	Study Sites		
Farm sizes	Kesu	Besti	Marthing
0 to 9.25 *chak*	4.00	29.90	19.59
9.26 to 18.50 *chak*	7.73	24.38	40.14
> 18.50 *chak*	0.00	58.00	57.00
Average	5.53	29.36	25.63

9.25 *chak* = 1 hectare
Each animal is treated as a livestock unit. Poultry are excluded.

	Sample Size		
	Kesu	Besti	Marthing
0 to 9.25 *chak*	15	21	22
9.26 to 18.50 *chak*	15	8	7
> 18.50 *chak*	0	1	1

Source: Mulk 1990.

Table 20.8. Trends in livestock per household over a five-year period (no. of households)

	Trends		
Villages	Increase	Decrease	Same
Zone 1 (Kesu)	0	5	25
Zone 3 (Besti)	17	4	9
Zone 2 (Marthing)	0	0	30

Source: Mulk 1990.

Goats and sheep do not form an important part of the livestock holdings in Kesu. This is a trend in Lower Chitral where, because of the breakdown of indigenous management systems and because of opportunities for off-farm employment and education, households are keeping fewer goats and sheep (Tables 20.6, 20.7, and 20.8).

No new breeds of livestock have been introduced in any of the study villages. This, in itself, is not bad, New breeds find it difficult to adapt to this environment. The government has used the bull system for improving breeds in the district, but its benefits have been restricted to a few villages that lie in the neighbourhood of the veterinary dispensaries.

The use of medicine and vaccines for livestock is popular in both Marthing and

Besti, both of which are at considerable distances from veterinary hospitals. The system whereby livestock specialists are nominated by villagers, trained by the AKRSP, and paid and maintained by village organization members is very successful. This innovation is effective because these areas lie at a considerable distance from the road. The villagers have to develop self-sustaining systems for livestock protection against disease because the government can do very little to help them.

The stall-feeding of animals is increasing as the feed situation improves because of higher yields in crops after the introduction of chemical fertilizers.

Indigenous Livestock Management Systems

The indigenous management systems play an important role in enabling villagers to keep large livestock holdings. In both Besti and Marthing, we find that the indigenous management system is intact. Under this system, one individual from each household has to help in grazing the goats and sheep each day, irrespective of his herd size. The system, while apparently inequitable, benefits the villagers through what could be termed as an 'economy of scale', as only a few individuals are involved in looking after the collective animals.

In Lower Chitral (Kesu) 87 per cent of the households have an individual system of management, while the remaining 13 per cent use nomadic herdsmen. In contrast, in Besti, 100 per cent of the sample households use a collective management system. In Marthing, 87 per cent use a collective management system (Table 20.9).

Table 20.9. Goats/sheep management

	System of management		
	Individual herdsmen	Village	Nomadic
Zone 1 (Kesu)	26	0	4
Zone 3 (Besti)	0	30	0
Zone 2 (Marthing)	4	26	0

Source: Mulk 1990.

The breaking down of the collective management systems has played an important role in reducing the numbers of sheep and goats at the village level in Kesu, and, therefore, looks like a blessing as it reduces the pressure on natural resources. In actuality, what happens is that the villagers, unable to look after the livestock, hand over their management to nomadic herdsmen[1] These nomadic herdsmen living far away from the village gradually start misappropriating the animals. After a while the villagers start giving up their animals as they are unable to sustain the losses. The nomadic herdsmen become economically powerful, buy land, and start maintaining larger and larger herds.

[1] Nomadic herdsmen, or 'gujars' as they are locally known, do not speak the local Khowar language as their mother tongue. At some stage in Chitral's history they must have been invited to come and look after the livestock herds of important feudal officials. They did not possess landed property. In many places they took the livestock of different people to pastures and paid a tax called 'qalang' to the right holders. The nomadic herdsmen are considered to be very hardy people as their living conditions are extremely harsh.

Once they buy land, they possess the same rights to the natural resources as the rest of the villagers. Since they live outside the village, the traditional, informal village institutions which regulated the sustainable use of natural resources, with the help of social pressure, are unable to influence them. They show less reluctance to destroy pastures and forests.

It needs to be understood that as education levels rise, and off-farm opportunities increase, fewer individuals will be willing to contribute to these traditional systems, unless the villagers institutionalize their relationship on an equitable basis. If this is not done in the next decade, the nomadic herdsmen will take over all the pastures in other parts of Chitral too.

Livestock Feed

Goats and sheep are dependent on grazing in the pastures, the former in the more distant pastures. Cattle are sent to the pastures for about four months in the summer. These pastures, as indicated in Table 20.10, are at considerable distances from the villages. During winter, the period of stall-feeding has increased as the feed situation has improved over the years.

Table 20.10. Distance from village of summer pastures (no. of households)

Villages	Hours		
	3 Hours	5–7 Hours	8–12 Hours
Zone 1 (Kesu)	0	17	12
Zone 3 (Besti)	15	15	0
Zone 2 (Marthing)	0	30	0

Source: Mulk 1990.

The feed situation has improved in recent years in all the villages. Fewer households face feed problems now, and the number of months over which they occur has been reduced considerably. However, the need to improve the situation exists, as a considerable number of households report scarcity in winter months. Anecdotal evidence indicates that the shortage of livestock feed was a major cause of livestock deaths during winters in the past.

Cropping Patterns

Crops, besides providing grains for human consumption, also supply the feed for livestock.

In all the three villages wheat occupies over 31 per cent of the total cultivated area. It forms the staple diet and its straw is an important livestock feed. In Kesu, winter wheat is grown. In Marthing, an attempt has been made to introduce winter wheat, but its germination and yields have been poor. At present, out of the 36 per cent of cultivated area devoted to wheat, 31 per cent is for summer wheat. At Besti, 100 per cent of wheat grown is summer wheat. No new wheat seed has been introduced here. This emphasizes the need for research in to introducing new varieties (Table 20.11).

Maize, barley, and pulses occupy important positions among the crops grown.

In recent years, vegetables have become an important cash crop in Kesu Village.

Table 20.11. Crop area of total households and its distribution as a percentage of cultivated area (crop area in *Chak*)

Crops winter/ spring	Kesu		Besti		Marthing		Total	
	Crop area	% of cult. area	crop area	% of cult. area	Crop Area	% of cult. area	crop area	% of cult. area
Winter wheat	79	32	0	0	14	5	93	12
Spring wheat	0	0	81	31	80	31	161	21
Barley	18.50	8	15.50	5.70	32.50	12	66.50	9
Maize	52.50	22	10	4	20	8	82.50	11
Rice	2.50	10	0	0	0	2.50	1	
Pulses	14	5	1.00	0.30	0	0	15	1
Alfalfa	0	0	148	57	74	28	222	28
Clover	16.50	7	0	0	27	10	43.50	6
Summer vegetable	30.50	13	0	0	0	0	30.50	4
Winter vegetable	29.50	12	0	0	0	0	29.50	4
Millet	0	0	4.50	2	15.50	6	20	3
Total cultivated area (1).	243.00	100	260.00	100	263.00	100	766.00	100
Total cropland available (2)	251.00	—	141.50	—	177.50	—	569.50	
Cropping intensity (2/1)	98%	—	79%	—	107%	—	96%	
Total land of all households	290.25	—	223.50	—	280.00	—	793.75	
Cropland as a % of total land	86%	—	63%	—	63%	—	72%	
Fodder and trees as a % of total land	4%	—	23%	—	30%	—	19%	
Homestead as a % of total land	10%	—	14%	—	7%	—	9%	

*97% is grown on marginal lands where crops can grow, although very poorly. For this reason, the area under alfalfa has not been used for calculating cropping intensities.

Source: Mulk 1990.

Refugees camped in the neighbourhood of the village were instrumental in introducing commercial vegetable growing and innovations in cultivation techniques. A large proportion of the labour involved in cultivating vegetables is provided by the refugees who are working on leased land. The trend is spreading, but the market remains limited as no surpluses can be transported out of the area at economical prices (Table 20.11).

Fodder crops receive low priority. Clover is grown in Kesu and Marthing but does not exceed 7 per cent of the cultivated area. Alfalfa is grown on a considerable area in Marthing and Besti, but only on the poorer sloping land, fed by irrigation channels where other crops could not be grown. For this reason returns are low.

Cropping intensities are still low. In Kesu Village, cropping intensities could be as high as 140 per cent, as the climate is favourable, but this is not possible because there is a shortage of water for irrigation during the summer. While water is available at the source, the irrigation channel does bring this water to the fields. This explains why the total crop land available exceeds the cultivated area. In Besti Village, where the area under alfalfa has been ignored because it is grown on marginal lands, cropping intensities are even lower. The pressure on resources in Besti was explained earlier as being due to a shortage of land. The cultivable land is of poor quality and 100 per cent is not cultivated every year. Another factòr contributing to low cropping intensities at both Besti and Marthing is the climate, which does not allow a second crop (Table 20.11).

Seed rates are high in the higher belts because of poor germination, poor quality of seeds, and the need to obtain fodder through thinning excess vegetation. Most farmers retain their own seed for sowing but occasionally purchase seed from neighbours or get it from the Agricultural Development Authority of the NWFP (ADA) or the AKRSP, as they do in Marthing.

Considerable amounts of farmyard manure are used by farmers. During the summer months when livestock move to distant pastures, invaluable manure is lost. Animal wastes and crop residues form the important ingredients of farmyard manure.

RECENT CHANGES

Improved Varieties

Improved wheat seed (Pak-81) has been widely disseminated in Kesu Village. The dissemination has been gradual, and has been helped by the proximity and accessibility of the village to Chitral (ADA outlet) where these seeds are available. The sample households were unaware of frequently changing seeds or how to secure new seed. At Marthing, the AKRSP has been able to supply seeds of new varieties (Pak-81 and winter wheat), but they are not widely adopted because they require successful demonstration. Lack of adjustment to the taste, as well as germination problems, have hindered progress. In Kesu, maize seed is also of the new varieties, and clover and alfalfa seed from the area is considered to be of good quality. A breakdown of average yields of local and improved varieties of seeds is given in Table 20.12.

Chemical Fertilizers and Farmyard Manure

Insufficient farmyard manure was a major constraint in increasing crop yields. At Kesu, the introduction of chemical fertilizer about 10 years ago, and at Marthing and Besti in the last five years, has increased yields of both grain and straw significantly. The two main chemical fertilizers used in the area are urea and di-ammonium phosphate (Table 20.13).

At Kesu, there was little evidence of any liquidity problem in purchasing fertilizer as there are a considerable number of individuals in salaried jobs or off-farm employment.

The AKRSP has assisted the villagers of Besti and Marthing, which are comparatively poorer villages having low surpluses, by making fertilizer credit available against the savings of the village organizations. At Marthing Rs 70,000 and at Besti Rs 97,000 worth of credit has been taken by the village organizations in the last few years.

The government has also helped in the process by subsidizing and making fertilizer

Table 20.12. Average yields of local and improved varieties of seeds at different study sites (*maund/chak*)

Crops	KESU		BESTI		MARTHING	
	Local	Improved	Local	Improved**	Local	Improved**
Wheat	5	10*	4	6	4	8
Wheat Straw	7	10*	6	9	6	12
Barley	7	10	5	7	6	9
Barley Straw	7	10	5	7	6	9
Maize	8	12*	—	—	2	5
Maize Straw	24	36*	—	—	6	15
Alfalfa	17	—	14	—	17	—
Clover	20	—		—	15	—
Rice	12	15	—	—	—	—

*Improved varieties and use of chemical fertilizers account for increases in yields of wheat and maize only. In other crops, improved yields are due solely to the use of chemical fertilizers.

**No improved varieties. The improved yields are attributed to the use of chemical fertilizers.

One maund = 40 kg. There are 9.25 *chak* in a hectare

Source: Mulk 1990.

Table 20.13. Percentage of cropped area using chemical fertilizer (*chak*)

Crops winter/ spring	Kesu		Besti		Marthing		Total	
	Crop area	% of cult. area	crop area	% of cult. area	Crop Area	% of cult. area	crop area	% of cult. area
Winter wheat	79	100	0	0	14	100	93	100
Spring wheat	0	0	81	100	80	100	161	100
Barley	18.50	84	15.50	100	32.50	60	66.50	81
Maize	52.50	89	10	0	20.50	95	82.50	60
Pulses	14	0	1.00	0	0	0	16	0
Alfalfa	0	0	148	0	74	0	222	0
Clover	16.50	0	0	0	27	0	45	0
Summer vegetables	30.50	100	0	0	0	0	30.50	100
Winter vegetables	29.50	100	0	0	0	0	29.50	100
Millet	0	0	4.50	0	15.50	0	20	0
Rice	2.50	0	0	0	0	0	2.50	0

There are 9.25 *chak* in a hectare.

Source: Mulk 1990.

available at sales points near the village, paying transportation costs. The government spends Rs 2 million on the subsidy and transportation of fertilizer annually.

Farmyard manure is used by all farmers. The manure consists of livestock excreta, crop residues, and household wastes. A major expense in using farmyard manure is that of collecting it from livestock sheds and transporting it to the fields.

Pesticides

Pesticides are not widely used, although in Kesu all farmers used them while growing vegetables.

Crop Rotation

Crop rotation patterns have changed in the last two decades. In the past, prior to the introduction of chemical fertilizers, crop rotation was widely practised, as farmyard manure could be applied to a limited number of fields only. In Besti and Marthing, clover, millet, and barley were grown in alternation with wheat. In Kesu, barley was widely grown in the past. Farmyard manure was applied to the barley crop, making the soil rich and fertile in preparation for the wheat crop to be grown in the next season. Cultivating millet cleared many of the fields from weeds, but with the introduction of chemical fertilizer, more area is devoted to wheat and crop rotation is reduced. While crop yields have gone up, there are problems of excessive weeds in the fields.

In Kesu, some of the crops whose cultivated areas have been reduced over the years include rice, cotton, and barley. Cotton was the earliest to go out of fashion when cloth became available from outside. Less rice is grown as the village faces water shortages and the crop requires intensive labour. Barley was once an important ingredient of the diet. The area under barley had fallen, but it has risen again after the influx of refugees whose horses and cattle require feed. As it is not subsidized, it fetches a good price.

In Besti and Marthing, the area under barley has been reduced and that under wheat has increased. In the past, prior to the introduction of chemical fertilizer, barley was widely grown, as its straw yields are higher than those of wheat.

Natural Forests

Natural forests are found in Lower Chitral. They are sources of fuelwood and timber. Very small loads of grass are obtained from them for livestock. Holly oak trees provide fuelwood, and their leaves are used as fodder for goats. Timber from pine and deodar forests is extracted by the government on a commercial basis, and the local right holders are given 60 per cent of the revenues as royalty.

At Kesu, there are considerable oak forests that belong to the village as a whole. The villagers cut wood from these forests and use it as fuelwood. There are some households that have possessed exclusive rights to the forests across the river from feudal times and exclusively extract wood from it for commercial sale. All the villagers show an eager interest in watching the Forest Department cut the deodar 'State' forests to which they are right holders. Since it is uncertain how long the government will recognize these rights before withdrawing them, speculators play on the right holders' fears and buy their rights by paying them 20 per cent of their actual value immediately and reaping benefits later. Since even the largest land holders are not very sound financially, they easily fall prey to such temptations.

Some households do not have the labour force to get fuelwood from the forests. Others do it for them if they pay the price of the wood.

At Besti Village, there are no timber forests, but the villagers still get a great portion of their fuelwood from scattered, old trees of juniper and birch in the mountains. The government takes little interest in such forests, and villagers regulate their use. Considerable numbers of birch and willow trees grow along the streams in the village at Besti. At Marthing the forests, which were few, are now almost extinct.

The average number of trees per household is quite high in Besti and Marthing villages (Table 20.14), where the absence of any major natural forests has encouraged people to be active in planting on both private and public lands. At Kesu, the holly oak forests have not been counted as they are at considerable distance from the village and grow naturally.

Table 20.14. Change in the number of trees per household

	At present			5 years ago		10 years ago	
	Fodder/ fuel timber	fruit trees (no.)	Fruit bearing	Fodder/ fuel timber	Fruit trees	Fodder/ fuel timber	Fruit trees
Zone 1 (Kesu)	2,837	1,391	1,007	1,519	602	2,206	339
Zone 3 (Besti)	26,654	514	282	28,265	263	7,495	135
Zone 2 (Marthing)	30,398	906	547	24,587	662	31,015	438

Source: Mulk 1990.

The pattern over the years is indicative of the change in outlook (Table 20.14). At Kesu, the initial fall in the number of trees shows the cutting down of holly oak trees near the village, and the rise reflects plantings carried out recently by the government to compensate for fuelwood extracted by Afghan refugees. At both Besti and Marthing, well-organized village organizations had spear-headed plantation campaigns. The details of plantations at Besti and Marthing indicate that they planted 14,000 and 20,000 forest trees, respectively, between 1985 and 1990. The process has been aided by the action of village organizations at Besti to ban the free grazing of livestock on crop fields after the autumn harvest: the renewal of a past tradition. This has enabled new seedlings to survive and grow.

The valuable species of forest trees grown alongside irrigation channels indicate that timber trees, like poplars, are important in areas like Marthing where natural forests have died out.

In Chitral there are hardly any fodder trees on farms. In Kesu Village, holly oak trees are used for this purpose.

Fruit Trees

Fruits have been grown in Chitral Valley for centuries, but, in recent years, the availability of new varieties, whose fruits last longer, and a small but growing market have encouraged villagers to plant a variety of fruit trees. Apples, apricots, and pears are the major fruits produced. The fact that households have attempted to grow grapes in villages situated at high altitudes in Marthing and Besti is indicative of this enthusiasm.

Fruit cultivation is not labour intensive, and the returns are very high per *chak* compared to crops, but poor means of communication and high transport costs have hindered commercialization. Very few households use pesticides, and fruits are susceptible to many diseases.

New varieties of fruit trees have only been introduced in Kesu, mainly through an individual effort. New varieties of trees for higher altitudes have not been introduced. There are a few nurseries for fruit trees run by the government in the area. The AKRSP has distributed 32,552 fruit trees to Village Organizations between 1986 and 1990.

Labour Shortages

In an area where land holdings are small, and the population growth rate rapid, it is expected that labour shortages will be minor. But in Kesu, 33 per cent, in Besti, 100 per cent, and in Marthing, 40 per cent of all households reported labour shortages during some time of the year. Most households identified the wheat harvesting time as the period in which labour shortages occur. The introduction of tractors and threshers has helped to ease the problem.

Average Time to Fetch Forest Products

At Kesu Village, as the natural forests have been cut down, the time required for the collection of forest products has increased (Table 20.15). This pattern is also visible in Besti. But in Marthing, as plantations around the village have increased, dependence on the naturally growing trees in the mountains has decreased, reducing the time used in fetching such products.

Table 20.15. Average time taken per household to fetch forest products (wood and grass)

Study sites	Time Period (hours)		
	Present	10 years ago	20 years ago
Zone 1 (Kesu)	7	6	4
Zone 3 (Besti)	6	5	4
Zone 2 (Marthing)	3	5	5

Migration

Migration has been one way of reducing pressure on natural resources. Permanent migration outside the district hardly exists, but there is considerable movement within the district. At Besti, the pressure on land and river erosion have forced a considerable number of individuals to move into the village. In Marthing, too, a number of individuals have moved in. Seasonal out-migration outside the district is very high from Marthing, where individuals move outside the district in search of labour or education. At Kesu, where other opportunities for income earning exist, such movement is on a smaller scale. An analysis of trends in migration over a period of time in the study villages revealed that migration starts when a place becomes accessible.

Migration, as can be seen in Table 20.16, is mainly seasonal, and directed outside the village. The bulk of it takes place in winter from the villages of Besti and Marthing where hardly any crops grow at this time of the year. It is seen that most people working outside their village try to return on leave from their work at times when labour demand is at its peak on the farms. Out-migration, was seen to increase the work loads of elderly people, women, and children.

Seasonal migration for work means increased incomes in those households involved, enabling households to use new technologies such as threshing machines.

Table 20.16. Migration (no. of individuals)

	Out				In			
	Seasonal		Permanent		Seasonal		Permanent	
	Within	Outside	Within	Outside	Within	Outside	Within	Outside
Zone 1 (Kesu)	1	4	0	1	1	0	0	0
Zone 3 (Besti)	1	7	0	0	14	0	6	0
Zone 2 (Marthing)	0	23	0	0	6	0	0	0

Source: Mulk 1990.

Income

A study of the household incomes at the three case study sites indicates that at both Besti and Marthing the per capita income comes to about Rs 2,343 and Rs 2,525 respectively. At Kesu, the income is higher at Rs 5,047.

However, the farming sector still plays a significant role in the three villages and is a major contributor to farm income. It contributes 47 per cent of the income at Kesu, 66 per cent at Besti, and 62 per cent at Marthing. A major portion of this contribution is in non-cash form. Non-cash income at Kesu is 68 per cent, about 77 per cent at Besti, and almost 96 per cent at Marthing. The cash income generated on-farm at Kesu from the sale of vegetables is almost 96 per cent. Very small amounts of barley and maize are sold. The cash incomes in Besti from the farming sector are generated from the sale of livestock and *patti* making from sheep wool. At Marthing, the on-farm cash income comes from the sale of livestock products (Table 20.17). At Kesu, besides the off-farm opportunities for employment, a number of households sell wood from the forests.

Table 20.17. Income (Rs)

	Farm*			
	Non-cash	Cash	Non-cash	Total
Zone 1 (Kesu)	321,033	149,640	543,710	1,014,383
Zone 3 (Besti)	302,603	88,915	198,920	590,438
Zone 2 (Marthing)	357,670	16,242	230,000	603,912

*Wood from the forest is not fully accounted for because of insufficient data. Incomes would be higher if fully accounted for.

Source: Mulk 1990.

Expenditure

The expenditure patterns show that the per capita expenditures at Kesu, Besti, and Marthing come to Rs 1,388, Rs 1,011, and Rs 1,292, respectively. Average expenditure per household at Besti comes to Rs 9,300 per year. The farm sector generates a cash surplus of Rs 88,915 while the actual cash expenditure for 30 households exceeds Rs 279,000. Similarly, at Marthing, the expenditure per household comes to Rs 8,500 per year. The total cash generated on-farm comes to Rs 16,242, while cash requirements exceed Rs 255,000. At Kesu, the expenditure per household comes to about Rs 10,300 per year. The cash generated by the farming sector annually comes to Rs 149,640, while expenditures exceed Rs 309,000.

This shows that in all the farms in Chitral, there is a considerable need to work off-farm to generate the necessary cash required for purchasing commodities such as wheat, sugar, tea, and ghee and to pay education costs.

GOVERNMENT POLICIES AND THEIR IMPLICATIONS

An appreciation of mountain specificities means acknowledgement that mountains are a different habitat and require a different kind of attention. Mountain areas had preserved their environment and habitats in the past because of decentralized local governments which recognized and were sensitive to their needs. These have now been replaced by centralized structures which look upon the mountains as extensions of the plains.

In the past, except for constructing irrigation channels, the Government of Pakistan did not engage itself in agricultural development activities. Today, government departments in almost all fields of human activities exist in Chitral. Nevertheless, their whole approach is a centralized bureaucratic approach, indifferent to local realities.

The government policy in the plains is to extend the areas of agricultural land through highly visible, large-scale irrigation projects. This policy was initially applied to the mountain regions as well. Stress was laid on large schemes. In doing so, planners failed to realize that in an area like Chitral, there are over a thousand irrigation channels which supply water to small tracts of agricultural land which support a sizeable population. These irrigation channels have supported a large beneficiary base, and were previously entirely self-maintained. The new projects supplied water to land belonging to fewer individuals. As a result, although the irrigation projects were completed, it took years before land development could take place, and benefits be realized, as only a few took interest in it.

The policy of providing subsidised chemical fertilizer has helped increase the yields of crops in Chitral. But the tendency to overuse has also increased. The nutrient content of soil in Chitral is poor, and there is fear that it could be depleted. As a result of higher yields of wheat, through the use of chemical fertilizers, crop rotation has been given up in many areas. In the introduction of chemical fertilizers, the fragile characteristic of the mountain area was ignored.

As far as agricultural research goes, there has been no attempt to treat Chitral as a separate agro-ecological zone. The physical terrain, topography, climate, livelihood, and culture of the area are unique. Agricultural research has been dependent on an agricultural research station situated in a different agro-ecological zone, i.e., Swat. Chitral is administratively part of Swat, but agro-ecologically it is part of the Northern Areas. Both the government structure and policies ignore this. As a result, Chitrals particular needs go unattended.

The policy of reducing inaccessibility within the district, without doing it outside the district, has made the area dependent on outside inputs without its benefiting from its comparative advantage in fruit production.

The highly fragile nature of the environment has not won government sympathy. The destruction of forests by commercial logging, as is done in other regions of Pakistan, should not have been allowed here, as poor rainfall and over-grazing by goats do not permit the regeneration of forests. But, instead, the Forest Department thinks in terms of generating revenues.

Activities such as terracing which prevent soil erosion, but which require considerable labour and time, should be encouraged through incentives. The desire for off-farm employment does not allow people to get involved in these activities. But their contribution towards preservation of the environment is self-evident.

The human adaptation mechanisms expressed in local institutions and traditions have been ignored as a thing of the past. These institutions were based on the idea of cooperation and self help. They are extremely important for survival in these harsh environments. The government, by not investing in these institutions, fails to realize that they have a very important role in the preservation of the mountain habitat. The AKRSP has established village organisations with their capital base in 50 per cent of Chitral's villages. The government could use these village-level participatory institutions for its development work and help make them viable. Since, to make these institutions viable, it is essential for them to appreciate their obligations, the government would be contributing towards the long-term sustainability of agriculture if it avoids 'soft programmes', or programmes which do not place obligations on the communities.

Despite its physical and geographical isolation, Chitral enjoys no special rights in the Constitution. Many other backward regions of the country are given special quotas in federal services. The provision of such rights could help reduce the pressure for off-farm employment in the district.

The paucity of land has been a major constraint for agriculture in the region. An adaptive strategy adopted in the past by the farmers was to build their houses on mountain slopes and save the best land for agriculture. The big government complexes being built in the area try to buy the best land for their constructions. A major mountain specificity is thereby ignored.

REFERENCES

Jodha, N.S. A Framework for Integrated Mountain Development. ICIMOD-MFS Discussion Paper No. 1. Kathmandu: ICIMOD, 1990.

Mulk, M. ul. Farmers Strategies and Sustainability of Mountain Agriculture: A Case Study in Chitral, Pakistan. Report of the ICIMOD Commissioned Field Study, 1990.

21

DIVERSITY OF MOUNTAIN FARMING SYSTEMS IN HIMACHAL PRADESH, INDIA

J.P. Bhati, R. Singh, M.S. Rathore and L.R. Sharma

J.P. Bhati is a Professor in the Department of Economics, Himachal Pradesh University, Shimla; R. Singh is a Research Officer at the Agro-economic Research Centre of Himachal Pradesh University; M.S. Rathore is a Senior Research Fellow at the Institute of Development Studies, Jaipur; and L.R. Sharma is an Assistant Professor in the Department of Economics and Extension, Parmar University of Horticulture and Forestry, Solan, Himachal Pradesh, India

Oxford & IBH Publishing Co. Pvt. Ltd., New Delhi, India.

CONTENTS

INTRODUCTION

The two important features of conventional development approaches in mountain areas are: (1) the low priority accorded to mountains in comparison to the plains and (2) the specific situation of the mountains (Jodha 1990). The latter disregards the diversity or heterogeneity of mountain ecosystems and concentrates on a sectoral approach to development. Ineffectiveness or the uneven impact of public interventions has been the consequence of this approach. It would not be wrong to say that the farmer, who has to live with constraints and harness available opportunities, seems to understand mountain conditions better than policy makers and development experts. This is reflected by both the structural and the operational features of farming systems in different ecological zones within the mountains.

These features also represent different dimensions of what may be called strategies in mountain areas. The present paper, based on village-level and farm-level studies in different regions of Himachal Pradesh, illustrates the same.

The paper first presents a sketch of the diversity of the mountain areas in Himachal Pradesh by comparing broad characteristics of different ecological zones. This is followed by a discussion on structural and operational features of farming systems in these regions, as revealed by field studies in the selected villages. In particular, the paper emphasizes the inter-dependencies of different activities at the farm level that help sustain the farming systems.

The paper summarizes some of the public measures taken and describes farmers' responses to specific mountain conditions, such as inaccessibility, fragility, and marginality. It concludes with the enumeration of practices and measures adopted by the farmers to harness opportunity and to manage constraints in the study villages.

AGRO-CLIMATIC ZONES OF HIMACHAL PRADESH

On the basis of altitude, temperature, topography, rainfall, humidity, and crop diversity, the State Directorate of Agriculture has divided Himachal Pradesh into the following four zones.

(1) low hills: sub-tropical zone,
(2) mid hills: sub-humid zone,
(3) high hills: temperate wet zone, and
(4) high hills: temperate dry zone.

The salient features of these zones are shown in Table 21.1.

Table 21.1 clearly shows the differences in natural resource endowment of different zones along with human adaptations to them in terms of choice of land-use systems, dominant agricultural activities, and their constraints and opportunities. For analyses of the above issues and their quantification through enquiry into dominant farming systems in these zones, a detailed study on selected villages was conducted.

SAMPLED VILLAGES AND SAMPLE SIZE

By way of a brief on the methodology of the village studies it should be stated that in each of the zones described above, a village cluster was selected for detailed study. In

Table 21.1. General features of agro-climatic zones of Himachal Pradesh, India

Particulars	Low hill zone	Mid-hill zone	High hill wet zone	High hill dry zone
Altitude (m)	350–650	650–1800	1,800–2200	2,000–3500
Rainfall (cm)	80–150	150–300	100–150	<25 (No crops without irrigation)
Soils	Sandy loam, poor water retention capacity	Clay loam, acid and deficient in N and P	Shallow, acidic silt loam, deficient in N and P	Sandy loam, alkaline and low in fertility
Population density per km^2	234	143	63	5.6
Cultivated area (%)	37.4	25.1	20.2	3.0
Area under forests (%)	13.9	40.4	23.3	7.4
Area under pastures (%)	26.0	28.6	49.9	79.8
Livestock density/ km^2 (whole H.P.)	214	219	153	35
Livestock density per km^2 of forests and pasture land only	542	317	211	40
Major food crops	Wheat, maize, rice, oilseeds, sugarcane	Wheat, rice, maize, blackgram, ginger, oilseeds, pulses, off-season vegetables	Maize, wheat, pulses, millets, seed potato, temperate vegetable seeds, off-season vegetables	Barley, wheat, buckwheat, pea, seed potato, vegetable seeds
Fruit crops	Mango and citrus fruits	Stone fruits and citrus fruits	Apple and other temperate fruits	Dry fruits, hops, nuts
Key constraints	Severe pressure on agricultural land, land ratio declining below desired levels. Moisture stress in hot summer season.	Infertile, slopy lands and increasing human pressure on agricultural lands, to support land ratio falling below desired levels, limited water resources and irrigation.	Arable land scarcity. cold winters, limited water resources for irrigation on slopy lands.	Cold, dry climate, cultivation without irrigation, fragile lands, scarce natural vegetation.
Key potentials	Accessibility, snow-free, mild cold winters allowing multiple or double cropping. Suitable for mixed farming dominated by grain crops.	Accessibility, snow-free mild cold winters, allowing multiple or double cropping, rangelands available. Suitable for grain crops, livestock, and local resource-based industry. Mixed farming systems.	Cool, temperate climate. Balanced farm-forestry land ratio, fruits and vegetable cash focus on farming.	Dry summer cropping season allowing disease-free seed potatoes and cultivation. Low human pressure. Good for vegetable seed production.

N = Nitrogen, P = Phosphates.
Source: Compiled by author.

the mid-hill zone, two different farming systems were observed, hence two villages from this region were selected, one of them Narang Village, which practises a mixed farming system. This village is situated at an altitude of about 1,400 masl and lies in a *Tehsil* in the District of Solan. In the low hill zone, Gharyana Village, which is situated at an altitude of about 700 masl and lies in the Suni *Tehsil* of Shimla District, was selected for the study.

In the high hill temperate wet zone, Thanadhar Village in Shimla District, situated at an altitude of 2,100 masl, was selected. In the high hill, temperate dry zone, Chitkul Village, which lies at an altitude of 3,400 masl on the right bank of the River Baspa, at a distance of about 25 km from Sangla *Tehsil* in Kinnaur District, was selected.

From each village, 75 households, probably proportional to the sizes of the total landholdings of the village, were randomly selected for detailed survey. Data for the year 1987/88 were obtained on structured schedules on the basis of personal interviews. Detailed reports on all the five villages have been prepared and submitted to ICIMOD and are in the process of revision and finalization.

FARMING SYSTEMS IN DIFFERENT ZONES

In this section, we present an overview of the farming situation in each zone. Various types of farming system exist in Himachal Pradesh, but there are five main systems: mixed farming, remittance-supported mixed farming, vegetable-based farming, perennial plantation farming, and agro-pastoral farming. Each of these systems may vary in terms of input intensities and scale of operations. These, in turn, are influenced by climate, soil, vegetation, and other ecological factors, and by socioeconomic factors including access to market and relevant public interventions.

Level and Composition of Income

One of the ways to demonstrate differences in the farming systems is their net outcome, e.g., household income and its contributing activities. The data presented in Table 21.2 show that there are significant differences in the sources and magnitude of household incomes in different sub-zones of the Western Himalayan region. In the low hills, a mixed subsistence farming system exists, heavily supported by off-farm income (basically remittances). In the mid-hills, a mixed farming system, where livestock, farm forestry, vegetables, and off-farm activities are equal contributors to household income, prevails. The climate of the mid-hill zone is suitable for off-season vegetables and stone fruits. Fruits and vegetables are relatively more profitable and have higher employment potential than compared to cereal crops (Rathore et al. 1975, Singh and Bhati 1988, Tewari 1990). Wherever the weather is favourable and there is easy access to roads and markets, these activities have emerged as the main sources of household income and employment. Hence, in such locations, a farming system based on vegetables and stone fruits has become prominent.

Fruit cultivation (mainly apples) predominates in the high hill wet zone. In the high hill dry zone, livestock and off-farm activities are the main sources of sustenance. Due to the dominance of commercial crop cultivation (mainly apples) in the farming system, the per capita income in the high hill wet zone is much higher than in other zones where

Table 21.2. Per household source-wise income in different zones of Himachal Pradesh, India
(Amount in Rs)

Particulars	Low hill zone	Mid-hill zone		High hill zone	
	Mixed farming	Mixed farming	Veg-based farming	Wet zone	Dry zone
A. Income (Rs/family)					
1. Farm income	9,625 (45.7)	14,350 (78.0)	25,114 (26.2)	73,701 (7.4)	10,482 (43.7)
a) Cereal crops	(8.1)	(4.6)	(5.9)	(–)	(2.7)
b) Fruit crops	(3.2)	(3.1)	(21.1)	(78.3)	(0.4)
c) Vegetables	(1.8)	(12.4)	(21.5)	(–)	(–)
d) Livestock	(15.9)	(32.1)	(19.1)	(–)	(–)
2. Off-farm income	11,422 (54.3)	4,039 (22.0)	8,921 (26.2)	5.859 (7.4)	8,133 (43.7)
3. Total income	21,047 (100.0)	18,389 (100.0)	34,035 (100.0)	79,560 (100.0)	18,615 (100.0)
B. Total income (Rs/person)	2,748	2,842	4,862	13,194	2,482
C. Crop income (Rs/ha)	1,265	1,667	7,208	62,307	470
D. Farm size (ha)	2.19	2.22	2.29	1.00	1.24
E. Family size (persons)	7.66	6.47	7.00	6.03	7.50

Note: Figures in parentheses denote percentages to total family income.
Source: Field survey.

commercialized production is lacking. In 1987/88, per household agricultural income in the low hills was Rs 9,625 and this constituted 46 per cent of the total household income. In the mid-hill zone, where mixed farming is practised, the annual agricultural income was Rs 14,350 (78% of the total), and in the vegetable-based system, the annual agricultural income was Rs 25,114 (74% of the total). In the high hill wet zone, the annual agricultural income was Rs 73,701 (93% of the total) and in the high hill dry zone the agricultural income per farm household was Rs 10,482 or 56 per cent of the total annual household income. Hence, it is clear that in the low hill zone and in the high hill dry zone, 54 per cent and 44 per cent of the household incomes respectively are derived from non-agricultural sources.

Due to unfavourable weather conditions in the high hill dry zone, the productivity of crops is very low (Swarup and Singh 1988). In the low hill zone and the high hill dry zone, the rainfall is low and unreliable making crop cultivation risky, but sometimes it falls with such intensity that it causes soil erosion and imposes an expensive burden on soil conservation activities. Farming in the low hill zone and the high hill dry zone is, on the whole, subject to a greater degree of natural risks than in the mid-hill zone and the high hill wet zone, owing to the greater incidence of droughts and floods. Therefore, the low hill and the high hill dry farmers need (but rarely have) especially large capital reserves to enable them to surmount these disasters. The degree of risk is reduced in some areas by providing irrigation facilities.

Farm families rely on non-farm income and remittances to stabilize their annual incomes and consumption. Moreover, livestock rearing is the only feasible activity where there is uneven relief or a different climate (as in the case of the high hill dry zone) and where food production or the production of other agricultural products (e.g., wool) is difficult (Negi 1990, Swarup and Singh 1988). In the high hill temperate, dry zone the transhumance farming system is followed. Since the resource base is poor and the area remains snow-bound for about six months in a year (November to April), the farmers shift their livestock (sheep and goats) to the lower hills for grazing.

Family Size and Workforce

The average family size in the five farming systems under study ranges from 6 to 7.6 persons. The number of workers per family ranges from 3.3 (in the mixed farming system) to 4.4 (in the high hill dry zone). The proportion of household workers working on farms ranges from 83.3 to 92.5 per cent (Table 21.3).

Table 21.3. Per household family size, workforce and literacy rates in different zones of Himachal Pradesh, India

Particulars	Low hill zone	Mid-hill zone		High hill zone	
	Mixed Farming	Mixed Farming	Veg-based Farming	Wet zone	Dry zone
Family size (no.)	7.66	6.47	7.00	6.03	7.50
No. of workers per family	3.64	3.35	4.13	3.64	4.36
Literacy rate (%)	75	66	77	80	47
a) Males					
b) Females					
c) Total					
Occupational distribution of workers					
a) Farming	83.3	91.6	91.2	92.5	87.8
b) Services	16.3	7.6	7.3	6.8	10.6
c) Business	0.4	0.8	1.5	0.7	1.6

Source: Field survey.

The literacy rate was 66 to 80 per cent. It is clear that households earning substantial incomes from fruit farming, in the high hill wet zone, have small families, are more literate, have a greater degree of commercialization, and have a higher dependency on farm income, whereas households in the low hill zone and in the high hill dry zone have bigger families, lower incomes, a lower degree of literacy, and less dependency on farming.

Size of Land Holdings

The average land holding in the apple-based farming system (high hill wet zone) is one hectare, in the livestock-dominated farming system (high hill dry zone) it is 1.2 hectares, and in the remittance-based mixed farming system (low hills) it is 2.2 hectares. In the

Table 21.4. Comparative percentage-wise distribution of total land holdings

Percentage

Particulars	Low hill zone	Mid-hill zone		High hill zone	
	Mixed farming	Mixed farming village	Veg-based farming village	Wet zone (apple farming village)	Dry zone (livestock-based farming village)
Marginal (<1 ha)	24.3	36.0	41.1	66.1	56.4
Small (1–2 ha)	29.7	22.7	19.6	16.9	23.0
Medium (2–4 ha)	35.2	28.0	27.5	10.4	15.0
Large (>4 ha)	10.8	13.3	11.8	2.6	6.6
Total	100.0	100.0	100.0	100.0	100.0
Average size of holding (ha)	2.19	2.22	2.29	1.00	1.24

Source: Field survey.

mid-hill zone the land holding per household is about 2.2 hectares (Table 21.4).

The proportion of marginal-sized land holding to total holdings is higher (above 56%) in the high hill zone than in the low hill zone and the mid-hill zone (Table 21.4). The proportion of marginal-sized farms is 24 per cent in the low hill zone and 36 to 41 per cent in the mid-hill zone. The proportion of small farms accounts for 17 to 23 per cent of the total holdings in the high hill zone, for 20 to 23 per cent in the mid-hill zone, and for 30 per cent in the low hill zone. The proportion of large holdings (i.e., holdings above 4 ha) varies from 3 to 13 per cent of the total holdings in all the zones. In the low hill zone 35 per cent and in the mid-hill zone 29 per cent of the holdings are of medium size (i.e., of 2–4 ha). Nevertheless, the overwhelming majority of land holdings fall under small and marginal farm categories. Thus, viewed from the criterion of size of holding, the bulk of the households were characterized by economic marginality and it was obvious that the land alone could not sustain them.

Operational Dimensions

Crop Diversification

Cropping patterns under different farming systems are shown in Table 21.5. The important cereal crops are maize, wheat, and barley. Vegetable crops grown include potatoes, peas, capsicum, cauliflowers and tomatoes, and these are grown during the offseason period, in the plains. Millet is also an important crop in the high hill dry zone and only apples are grown in the sampled villages of the high hill wet zone. The 'niche' (comparative advantage) for fruits and vegetables was fully exploited by the farmers wherever natural conditions and the socioeconomic and institutional environment (prices of inputs and outputs, farm size, credit, extension services, and marketing facilities) were favourable. Furthermore, with a rise in elevation, crop diversity and the area under mixed crops decline. The area under sole (pure) cropping in the low hills was 32 per cent, in the mid-hills (mixed farming) 62 per cent, in the vegetable-dominated system of the mid-hills, it was 78 per cent, and in the high hills almost 100 per cent of the total cropped area.

Table 21.5. Distribution of gross cropped area to sole, mixed, and perennial crops in different zones of Himachal Pradesh, India

Crops and mixtures	Low hill zone Mixed farming	Mid-hill zone Mixed farming	Mid-hill zone Veg-based farming	High hill zone Wet	High hill zone Dry
Sole crops					
1. Maize	6.6	20.0	19.5	—	13.1
2. Tomato	5.9	5.5	13.2	—	—
3. Capsicum	—	0.1	3.8	—	—
4. Peas	—	—	16.1	—	2.7
5. Wheat	9.4	16.4	11.0	—	16.4
6. Barley	4.1	8.0	4.2	—	15.4
7. Millet	—	—	—	—	29.9
8. Potatoes	—	—	0.7	—	6.4
9. Others	5.9	12.2	10.3	0	14.4
10. Total	31.9	62.2	78.8	—	98.3
Mixed crops					
1. Maize + pulses + Sesame	32.5	10.4	—	—	—
2. Wheat + oilseed + peas	31.2	18.2	—	—	—
3. Peas + coriander		5.7	—	—	—
4. Potato + coriander	1.3	—	—	—	—
5. Others	1.56	—	—	—	—
6. Total	66.6	36.5	—	—	—
Fruit crops	1.5	1.3	21.2	100.0	1.7
Gross cropped area	100.0 (1.73)	100.0 (1.74)	100.0 (0.93)	100.0 (1.02)	
Cropping intensity	187	188	169	200	109

Note: Figures in parentheses denote per farm gross cropped area in hectares.
Source: Field survey.

Modern/Market-Purchased Inputs

As the proportion of cash crops increases, the per farm and per hectare application of fertilizers and pesticides increase. The proportion of the area under high yielding varieties (HYVs) of food crops also increases as farm incomes and cash crop production increases (Table 21.6). It should be pointed out that while the HYVs of wheat and rice have been adopted, the HYVs of maize evolved so far have not been adopted by the hill farmers. Local varieties of maize are more tasty and give higher yields than the HYVs of maize released in the region. The HYVs for maize, like those for wheat and rice, were developed in the plains, but, unlike the wheat and rice varieties developed, the maize varieties developed were not suitable for hill areas, and this suggested the need for local research on maize for the hills.

Table 21.6. Comparative annual application of modern inputs

	Low hill zone	Mid-hill zone		High hill zone	
Modern inputs	Mixed farming	Mixed farming	Veg-based farming	Wet (apple farming	Dry (livestock-based farming
Fertilizer (kg):					
Per farm	15	57	557	296	12
Per ha of of GCA	9	33	295	318	12
Plant protection material (Rs)					
Per farm	6	370	797	2,500	—
Per ha of GCA	3	213	422	2,689	—
HYV seeds (% area)					
Maize	33	48	100	N.A.	35
Paddy	84	84	N.A.	N.A.	70
Wheat	99	84	100	N.A.	95

N.A.= Not applicable. GCA = Gross Cropped Area
Source: Field survey.

In the commercial farming systems (i.e., in the vegetable-based and the apple-based systems) the use of pesticides and the application of plant protection materials are quite high. Due to reduction in cropping diversity and the replacement of traditional crops by new crops, the incidence of diseases, insects, and pests has increased. This has led to adoption of chemical plant protection measures on a large scale. Increased dependence on external sources and greater management complexity are other features of the scientific farming processes represented by the above two systems. The degree of risk and uncertainty (market risk and production risk) have also increased and this makes both these farming systems relatively unsustainable.

Observations of the farm households in commercial farming systems (i.e., in the apple-based system of the high hills and the vegetable-based system of the mid-hills) suggest that, as farmers adopt new methods of farming, their ideas also change. This is part of the transformation process in rapidly modernizing areas of the state.

Changes can be seen in terms of the attitudes of the farmers towards agriculture, towards natural environment, towards themselves, and towards their traditional partners; e.g., traders and government agencies. Such households no longer suffer from social, economic, and political marginality.

Livestock Numbers and Composition

Livestock numbers and composition in different zones are shown in Table 21.7. Differences represent adaptations to the diversity of different zones. In the high hill dry zone, 82 per cent of the livestock population consists of sheep and goats. On average, farmers keep 88 sheep or goats and the herds are of a migratory nature (transhumance). In winter, they move to the low hills and in summer they return to the high hill pastures, thereby making fuller use of resource diversity. In the high hill wet zone, (apple-based farming system) farmers of sample villages keep cross-bred cattle (each one having about two cows) and these are mostly stall-fed. Cross-bred cows are high-value animals and require high input and quality fodder; hence, only rich farmers can afford them (Negi 1990). The grazing of cross-bred cows is limited due to heavy body weight and risk of higher loss due to possible accidents (compared to low-value, local cows).

Table 21.7. Composition of livestock in different zones of Himachal Pradesh, India

	Low hill zone	Mid-hill zone		High hill zone	
Types of livestock	Mixed farming	Mixed farming	Veg-based farming	Wet	Dry
Cattle	53.2 (6.03)	69.4 (5.24)	68.3 (6.17)	98.4 (2.46)	6.0 (5.31)
Yaks	—	—	—	—	0.4 (0.30)
Buffaloes	10.4	16.2	18.0	—	—
Sheep and Goats	36.4 (4.13)	14.4 (1.09)	13.0 (1.18)	1.6 (0.04)	93.1 (81.94)
Mules and others	—	—	0.7 (0.06)	—	0.5 (0.45)
Total livestock	100.0 (11.34)	100.0 (7.55)	100.0 (9.04)	100.0 (2.5)	100.0 (88.00)

Note: Figures in parentheses are the number of livestock per farm.
Source: Field survey.

In the mid-hills, on average, a farmer keeps about eight animals out of which cattle account for about 70 per cent of all livestock. Buffaloes and sheep/goats respectively account for about 17 per cent and 13 per cent of all livestock. In this zone, animals are partly grazed and partly stall-fed. In the case of the low hills, the number of farm animals kept per farm is 11.3, out of which cattle account for 53 per cent, buffaloes 10 per cent, and sheep/goats 37 per cent.

Except for the high hill dry zone, the number of large animals in the livestock possessed by farmers is higher in all farming systems. Both manure and draft power are important needs necessitating the possession of large-animals. Under each of these systems a few small animals (sheep and goats) are also kept by farmers to meet their clothing and meat requirements and also as an asset that can easily be sold in times of emergency. Overstocking of animals and denudation of common grazing resources are two of the main problems of the livestock sector in the region (Bhati 1981, 1983).

INTERDEPENDENCY AMONG DIFFERENT FARMING SUB-SYSTEMS

Despite differences in the relative importance of the different components of the farming systems, the very presence of multiple components in all the zones illustrates the farmers' adaptations to diverse environmental and resource situations in the mountains. Furthermore, it is not only the diversity of land-based activities (cropping, animal husbandry, horticulture and farm forestry), but also their interlinkages which help to sustain these activities. The dependence of cropping on livestock, for manure and draft power as well as cash income, animal husbandry's dependence on crop by-products; and the role of the forests in reinforcing the above are well known in mountain areas. Use of forest land or common land as sources of fodder and other products is widespread despite the decline in or degradation of these resources.

The linkages indicated above finally lead to stability of farm income and consumption levels. Of course, with modernization of mountain agriculture some of these linkages are also disrupted or transformed. In the following discussion quantitative evidence on some dimensions of the above interdependencies are presented. The linkages of different production sectors are presented by indicating (1) the values of outputs generated by one sector and used as inputs by other sectors and (2) the ratio of such input values to per rupee of output generated by the input-using sectors. The situation is illustrated with reference to the dependence of farming systems on the livestock and forestry sectors only.

Dependence on Livestock Sector-generated Inputs

Table 21.8 provides details on the use of livestock inputs (such as manure and draft power) in different sub-sectors of farming in different zones of Himachal Pradesh. On the whole, the requirement of livestock input per rupee of output (i.e., input-output ratio) in the case of cereals is higher than in the case of vegetables. The input-output ratios also indicate that the requirement of livestock input in cereal and vegetable production is relatively higher in the mid-hill mixed-farming system and lower in the mid-hill vegetable-based, mixed farming system.

Table 21.8. Per farm annual imputed value of livestock sector generated inputs used in different sub-sectors

Sectors using input (manure draft power)	Low hill zone	Mid-hill zone		High hill zone	
	Mixed farming system	Mixed farming systems	Veg-based farming systems	Wet (apple farming systems)	Dry (livestock-dominated farming system)
Cereals	22.96 (0.34)	1,849 (0.40)	906	— (0.36)	1,297 (5.31)
Vegetables	209 (0.19)	1,012 (0.13)	2,199 (0.09)	—	—
Fruits	—	—	31 (0.001)	980 (0.01)	10 (0.03)
Total/farm	2,505 (0.29)	2,861 (0.22)	3,136 (0.11)	980 (0.01)	1,307 (0.34)
Total/ha	1,144	1,289	1,369	980	1,054

Note: Figures in parentheses denote input-output ratio, i.e., livestock sector-generated input per rupee of output in the other sector.

Source : Field survey.

In terms of absolute amount, mid-hill vegetable-based farms are observed to be using the highest amount of livestock input per farm (amounting to Rs 3,136), followed by the mid-hill mixed farms (worth Rs 2,861), the low hill mixed farms (worth Rs 2,505), the high hill dry zone farms (worth Rs 1,307), and the high hill wet zone farms (worth Rs 980). The same trend can also be observed in the case of per hectare use of livestock input by different farming systems. But as the input-output ratios show, the highest requirement for livestock sector-generated input use per rupee of output is found in the case of the high hill dry zone, livestock-dominated farming system. This is followed by the low hill mixed farms, the mid-hill mixed farms, the mid-hill, vegetable-based farms, and the high hill wet zone farms. On an average, the annual requirement of livestock sector-generated inputs per hectare of operated land is Rs 1,144 in the low hills, Rs 1,289 in the mid-hill mixed-farms, Rs 1,369 in the vegetable-based farming system, Rs 980 in the high hill wet zone and Rs 1,054 in the high hill dry zone. Seen in the context of the earlier discussion the sub-system's dependence on livestock-generated inputs declines with the degree of commercialization of the activity requiring greater use of marketed inputs.

Dependence on Forestry Sector-generated Inputs

Dependency on farm production activities on inputs obtained from farm and forestry, common property resources, and government forests is high (Table 21.9). Per farm annual value of inputs obtained from these resources (i.e., inputed value of grass and grazing, leaf fodder, and wood for implements, packing cases, and staking), were Rs 5,562 in the low hill zone, Rs 8,964 in the mid-hill mixed-farming zone, Rs 16,857 in the mid-hill vegetable-based farming zone, Rs 42,604 in the high hill wet zone (apple-based farming), and Rs 7,849 in the high hill dry zone (livestock-based farming).

It is quite clear that commercial farming (based on fruits or vegetables or both) requires more inputs from forests and common property resources. Since these resources are already showing signs of degradation, a rapid shift from traditional farming to commercial farming, due to the latter's 'servicing needs', will put more pressure on the natural (vegetation) resources unless some substitute for wood (used for packing and staking) is found. This may make the new and commercial farming systems unsustainable in the long run.

Examination of sector-wise dependence on natural resources shows that the livestock sector heavily depends upon common property resources (CPRs) in all the zones. Cereal crops are not so dependent on forests or CPRs, except for farm implements (e.g., ploughs, hand tools, etc).

The comparison of the use of forest and CPR-generated inputs per rupee of farm output in different farming systems reveals the following. The input-output ratio in the low hills is 0.25, in the mid-hills mixed farming system 0.32, in the mid-hills vegetable-based farming 0.27, in the high hills apple farming 0.42 and in the high hills dry zone 0.39. Viewed differently, 25 to 40 per cent of the input contribution, from the natural resource sector to the per rupee of output from different farming systems, indicates the undiminished role of the commons and public resources in the farming systems in Himachal Pradesh. Furthermore, persistent over-dependence on these resources and their rapid depletion, due to overexploitation, may make both traditional and commercial farming systems unsustainable in the long run.

Table 21.9. Per farm annual imputed value of inputs obtained from own trees and grassland, government forests, and community resources in different study areas in Himachal Pradesh, India

	Low hill zone	Mid-hill zone		High hill zone	
Sectors using inputs	Mixed farming	Mixed farming	Veg-based farming	Wet (apple farming)	Dry (livestock-based farming)
Livestock	4,972 (0.37)[a]	4,614 (0.32)	9,829 (0.58)	3,495 (0.26)	7,439 (0.46)
Cereals	540 (0.08)	333 (0.07)	182 (0.04)	—	260 (0.07)
Vegetables	35 (0.03)	3,881 (0.49)	3,582 (0.15)	—	—
Fruits	15 (0.02)	13 (0.02)	3,264 (0.19)	39,109 (0.45)	150 (0.53)
Total/farm	5,562 (0.25)	8,964 (0.32)	16,857 (0.27)	42,604 (0.42)	7,849 (0.39)
Total/ha	2,540	4,038	7,361	42,604	6,330
Dependence on public sources (%)[b]	23	68	82	97	100

[a]Figures in parentheses denote input-output rate, i.e., input per rupee of output.
[b]Indicates non-private resources, e.g., common property resources (pasture, community forest, and drainage), government forest.
Source: Field survey.

Pressure on natural common resources (forests and pastures) must be reduced. Emphasis should also be put on agroforestry so that the provision of wood and fodder becomes the responsibility of the farmers themselves. At present, the proportionate dependence on external resources for wood and fodder (government forests and CPRs) is 23 per cent in the low hill zone, 68 per cent in the mid-hill zone (mixed farming), 82 per cent in the mid-hill zone (vegetable-based farming), 97 per cent in the high hill zone (apple farming), and 100 per cent in the high hill dry zone (livestock-based farming). It is also clear that as commercialization increases, the dependence on external sources (which are beyond the control of the farmer) increases. This means the degree of uncertainty for these inputs also increases, which will make farm enterprises more vulnerable and unsustainable.

ADAPTATION AND ADJUSTMENT TO MOUNTAIN SPECIFICITIES

The types of farming systems and their structural and operational features represent farmers' adaptations to constraints and opportunities available in the mountain ecosystems. However, the above discussion on farming systems can be supplemented by specific measures constituting farmers' adjustment mechanisms. The field studies generated a number of qualitative observations in this respect. These have been summarized in Tables 21.10 to 21.12 while listing farmers' responses to selected mountain specificities (e.g., modification of a mountain specificity or adapting to it). We have also listed the public responses (i.e., development interventions) in relation to different mountain specificities. The three mountain specificities covered are inaccessibility, fragility, and marginality (Jodha, Chapter 2), which are major sources of constraints in mountain areas.

Table 21.10. Responses to inaccessibility as constraint through modification and adaptation

Inaccessibility and its implications and responding agency	Modification	Adaptation
Poor mobility and high transport costs		
a. Farmers' responses	1. Construction of mule tracks and wooden bridges	1. Self-help activities
		2. Use of local materials
		3. Head load/and/or use of pack animals
		4. Local trading
		5. Self-sufficiency in food items
		6. High-value, low-weight produce for sale (e.g., ghee, woollen cloth, oilseeds)
b. Govt. responses	1. Construction of roads and bridges	1. Transport subsidy (for input/output).
	2. Ropeways	2. Local processing facilities.
	3. Provision of growth centres	3. Storage, public distribution system.
	4. Public transport	
Weak agricultural extension and support services		
a. Farmers' responses	1. Provision of informal credit	1. Use of local breeds and seeds and indigenous knowledge for farm management
	2. Obtaining new information from local progressive farmers	2. Acquisition of radio for farm news and other information.
b. Govt. responses	1. Establishment of schools, banks, post offices, and agri/livestock centres	1. Farmers' training programmes
	2. Establishment of growth centres on area basis	2. Local research stations

Source: Field survey.

Tables 21.10 to 21.12 briefly sketch the government and farmer responses to constraints generated by three mountain characteristics. The government responses also take the forms of measures to harness positive features of mountain situation (e.g., diversity, 'niche'). Verma and Partap (Chapter 26) discuss most of these measures and their impacts. In the following discussion we elaborate on farmers' response vis-à-vis different opportunities and limitations created by mountain conditions. This listing of farmers practice is based on observations in the study villages (some of them are quantified also).

- Grasses are grown on the marginal and fragile portions of land holdings.
- Where moisture uncertainty is acute, mixed cropping is common.
- Where 'niche' or comparative advantages are clear-cut, farmers have adopted high-

Table 21.11. Responses to marginality (poor resource base) as a constraint through modification and adaptation

Inaccessibility (poor resource base) and its implications and responding agency	Modification	Adaptation
Human resource		
a. Farmers' responses	1. Access to new information and resources 2. Establishment of schools 3. Establishment of formal and informal schools 4. Anti-poverty programmes	1. Continue with traditional subsistence strategies
Land resource		
a. Farmers' responses	1. Conversion of marginal land into cultivated land, upgrading through irrigation, structural changes	1. Increased cropping intensity 2. Use of HYV seeds, fertilizers and other management practices to increase productivity 3. Cultivation of cash crops (e.g., fruits, vegetables)
b. Govt. responses	1. Public programme for land development,irrigation, etc.	1. Provision of improved agricultural technology
Livestock resources		
a. Farmers' responses	1. Adoption of improved animals. 2. Cattle replaced by buffaloes for milk production in low hills	1. Reduction in herd sizes 2. Stall-feeding 3. Plantation of fodder trees
b. Govt. responses	1. Support to livestock development (e.g., cross-bred cattle) 2. Support system for dairying, etc.	1. Provision of improved animal husbandry
Forest/pastureland		
a. Farmers' responses		1. Reduction in herd sizes 2. Stall-feeding
b. Govt. responses		1. Closure of forests from grazing land lopping
Institutions		
a. Farmers' responses		1. Informal credit system 2. Irrigation management committees
b. Govt. responses	1. Establishing banks and co-op. credit societies 2. Establishing agri/livestock extension centres 3. Establishing markets and marketing societies	
Income		
a. Farmers' responses	1. Shift towards high pay-off activities	1. Off-farm income (e.g., remittances, business, wage labour, cottage industry)
b. Govt. responses	1. Establishing small-scale industrial estates to generate off-farm employment 2. Anti-poverty programmes	

Source: Field survey.

Table 21.12. Responses to fragility (vulnerability to degradation of natural resource base) as a constraint through modification and adaptation

Fragility and its implications and responding agency	Modification	Adaptation
Cultivated land		
a. Farmers' responses	1. Terrace farming 2. Irrigation	1. Composting and manuring of land 2. Mixed cropping 3. Plantation of perennial crops
b. Govt. responses		1. Emphasis on soil conservation work 2. Emphasis on horticultural crops and farm forestry
Forest/pasture land		
a. Farmers' responses		1. Adoption of improved stoves 2. Using electricity for cooking 3. Stall-feeding 4. Reduction in herd size and improving breeds of livestock
b. Govt. responses	1. Forest regeneration programmes	1. Protecting forests by stopping entry of livestock and restriction on lopping

Source: Field survey.

value, remunerative activities, e.g., fruit and vegetable production, seed potatoes, cross-bred cows.

- When accessibility and rapid transport increase, farmers are encouraged to commercialize crop production.
- The choice of crops depends upon the suitability of agro-climatic conditions for particular crops. If the suitable crop has relatively less risk and less labour requirements then it results in total monoculture, as is the case with apples.
- Where the climate is such that crops alone cannot provide sufficient income, diversification of enterprises (crop, livestock, off-farm jobs) is more important. In the low hills, men move out to take up non-farm jobs, whereas in the high hill dry zone, animals are moved in search of new pastures.
- Where income is high, farmers avoid hard labour and less remunerative animals. They prefer few but better quality animals which are mostly stall-fed.
- Farmers in the low hills and mid-hills keep relatively more animals to provide sufficient manure for the crops. This is essential because, due to the washing away of top soils from slopy fields, most soils are poor in humus material.
- Farmers apply heavy doses of manure on vegetable crops, less on cereal crops, and least on fruit crops.
- Where water is scarce, efficient water management methods are used, e.g., applying water with plastic pipes and buckets and storing water in tanks near the fields.
- Farmers collect fodder and store it for the lean season. First, they collect fodder from CPRs and keep their own sources in reserve.
- Small and marginal farmers (having inadequate land holdings) increase their incomes by keeping more livestock that depend on CPRs for fodder.

- When growing a particular crop becomes difficult (due to diseases or shortages of staking material) farmers shift to the next best alternative crop for the same season.
- To avoid risk, due to fluctuations in prices and yields, farmers dispose of their prospective fruit crop in advance at blossoming time to pre-harvest contractors.
- When the surplus of milk, wool, cereals, or vegetables on a particular farm is small, the farmer sells the surplus produce to intermediaries, e.g., a village trader or other farmers, who transport it and sell in nearby or distant markets. However, as the total surplus increases, the farmer sells his own produce in the marketplace in order to get a better price. In some areas, cooperative marketing societies are becoming popular.
- When prices in the local markets are not remunerative, farmers individually or jointly send their produce to distant markets. In this context, market prices announced on the radio, during the agricultural information services, are very useful in helping farmers decide where to market a particular product.
- When green grasses are in short supply, farmers mix tree leaf fodder with dry grasses before feeding. The farmer gives good fodder to milch and draft animals while dry animals get the lower quality fodder (generally dry grasses) and are grazed throughout the year, although due to over-grazing, most pastures are heavily degraded. Milch animals receive supplemental stall-feeding.
- Commercialization of crops depends heavily upon external markets (both for obtaining inputs and for selling outputs), hence the degree of uncertainty increases and most farmers do not like this. They want higher incomes with least uncertainty.
- Larger family size increases the demand for food and this results in an extension of cropped areas by bringing grasslands (unsuitable lands) under the plough.
- Because of mutual dependence, farmers' decisions are greatly influenced by attitudes and relationships within the local communities in which they live. Members of rural communities commonly cooperate to do things that would be difficult or impossible for individuals working alone (e.g., making village paths or roads and small bridges, operating local irrigation facilities, and integrated pest management for fruit crops).
- Migratory shepherds face fodder shortages during the winter months when they migrate to the lower hills. Here, fodder on government pastures is scanty. Therefore, the shepherds hire panchayat lands on contract for grazing and also use private fallow lands. To supplement poor fodder supplies during this season, they feed salt and mustard oil to their flocks.
- The owners of small flocks send their flocks of sheep and goats to graze with other large flocks on a tenancy basis.
- Whenever there is a need for cash to buy food or other items, migratory shepherds sell their animals. Hence, animals are a form of capital savings for them.
- To remove the problem of inaccessibility, villagers themselves have built small roads and bridges to connect the village with nearby roads.
- Where transportation is a problem, farmers produce low-weight, high-value products (e.g., woollen clothes, dry fruits).
- At times of need they help each other by providing credit, labour exchange, and farm inputs (seed and manure).

The farmers' practices enumerated above provide a mix of measures designed to meet both the known and the unknown circumstances created by environmental features of the mountains as well as the demographic, institutional, and technological changes.

REFERENCES

Bhati, J.P. Population Resource Use and Ecological Stresses in the Western Himalayan. Ph.D. dissertation, University of Hawaii, Honolulu, 1981.

Bhati, J.P. 'Challenges for Rural Development in Hilly Areas—Case of Himachal Pradesh.' *Cooperative Perspective*, Pune, India, 18 (3), 101–117, 1983.

Jodha, N.S. A Framework for Integrated Mountain Development. Mountain Farming Systems, Discussion Paper No. 1. Kathmandu: ICIMOD, 1990.

Jodha, N.S. 'Mountain Perspective and Sustainability: A Framework for Development Strategies.' Chapter 2, Vol. 1, this publication.

Negi, G.C. Livestock Development in Himachal Pradesh: Retrospect and Prospect. Mountain Farming Systems, Discussion Paper No. 7, Kathmandu: ICIMOD, 1990.

Rathore, M.S., J.P. Bhati and R. Swarup. 'Resource Use Efficiency and Returns from Some Commercial Crops of Himachal Pradesh.' *Agricultural Situation in India*, 30 (7) 507–509, 1975.

Singh, R. and J.P. Bhati. 'Role of Vegetables in Increasing Employment and Income in Hill Agriculture: Case of Himachal Pradesh.' *Indian Journal of Agricultural Economics*, 43 (3): 516, 1988.

Swarup, R. and R. Singh. *Social Economy of a Tribal Village on Indo-Tibetan Border*. Delhi: Mittal Publications, 1988.

Tewari, S.C. Role of Off-season Vegetables in the Development of Hill Agriculture in Himachal Pradesh, India. Mountain Farming Systems, Discussion Paper No. 8: Kathmandu: ICIMOD, 1990.

Verma, L.R. and T. Partap. 'The Experiences of an Area-based Development Strategy in Himachal Pradesh, India.' Chapter 26, Vol. 2, this publication.

REFERENCES

Bhati, J.P. Population Resource Use and Ecological Stresses in the Western Himalayan. PhD dissertation. University of Hawaii, Honolulu, 1981.

Bhati, J.P. Challenges for Rural Development in Hilly Areas—Case of Himachal Pradesh. *Community Development Perspectives*. Pune, India. 18 (1): 104–112, 1983.

Jodha, N.S. A Framework for Integrated Mountain Development. Mountain Farming Systems, Discussion Paper No. 1. Kathmandu: ICIMOD, 1990.

Jodha, N.S. Mountain Perspective and Sustainability: A Framework for Development Strategies. Chapter 2, Vol. 1, this publication.

Jasrotia, O.C. Livestock Development in Himachal Pradesh: Retrospect and Prospect. Mountain Farming Systems. Discussion Paper No. 5. Kathmandu: ICIMOD, 1990.

Rathore, M.S., J.P. Bhati and R. Swarup. Resource Use Efficiency and Returns from Some Commercial Crops of Himachal Pradesh. *Agricultural Situation in India*. 30 (7): 507–509, 1975.

Singh, R. and J.P. Bhati. Role of Vegetables in Increasing Employment and Income in Hill Agriculture: Case of Himachal Pradesh. *Indian Journal of Agricultural Economics*. 13 (3): 316, 1988.

Swarup, R. and R. Singh. *Socio-Economics of a Tribal Village on Indo-Tibetan Border*. Delhi: Mittal Publications, 1989.

[illegible] Research Priorities in the Development of Hill Agriculture in Himachal Pradesh, India. Mountain Farming Systems. Discussion Paper No. [illegible]. Kathmandu: ICIMOD, 1991.

Verma, L.R. and T. Partap. The Experiences of an Area-based Development Strategy in Himachal Pradesh, India. Chapter 16, Vol. 2, this publication.

22

ANDEAN FARMING SYSTEMS: FARMERS' STRATEGIES AND RESPONSES

A. Camino

Alejandro Camino has been working on issues related to Andean Agrarian Ecology since the late 1970s, is associated with the Department of Anthropology of the Pontificia Universidad, Catolica del Peru, and is the Director of Development of the Fundacion Peruana para la Conservacion de la Naturaleza, Lima, Peru

CONTENTS

INTRODUCTION

Within the Andean Ranges, running from the extreme north to the southernmost tip of the South American continent, the Central Andes—stretching between the Equator and the Tropic of Capricorn—exhibit extreme conditions of heterogeneity and biodiversity, turning it into one of the most varied biomes on earth. Despite having numerous characteristics that set constraints and limitations for human utilization and which should have made it a marginal area for human subsistence, the region was an early cradle of civilization, supporting a peculiar and complex Amerindian imperial society with major achievements in all fields of culture. Most notably, farming and pastoralism attained an unsurpassed, all time height and substantial levels of productivity and production were achieved, as reflected by the massive accumulation of food and commodity surpluses.

By the time of the Spanish Conquest under the expanding rule of the Inca Empire, Andean society had evolved into an elaborate complex civilization based on a non-monetary redistributive economy (Murra 1975) which has been characterized as a stratified welfare state (Rowe 1946). Besides the magnificence of urban architecture and the wealth of the temples and the ruling classes, what most impressed the European conquerors, perhaps, was the vast amount of stored foodstuffs and manufactured goods, mostly kept in State warehouses (Garcilaso de la Vega 1959). Struck by the living conditions in the Andes, an early chronicler wrote that, as opposed to his former European experience, in these lands, hunger was unknown (Cobo 1964).

For years after the conquest, storehouses kept supplying feuding conquerors with food and cloth. Nearly 500 years have past, and, since then, the Andean countries have become net importers of food. In order to explain this dramatic change we need to understand two interrelated processes that characterized the colonial and republican experiences resulting from conquest and domination: socioeconomic change and the subsequent transformation of the traditional Andean subsistence system.

After nearly half a millennium of European colonization and contemporary modernization, the region has and is experiencing environmental degradation, encompassed by a significant drop in production and productivity with subsequent deterioration of human standards of living. Paradoxically, this process takes place side by side with the modernization and development of formerly marginal areas (Pacific coastal deserts and Amazonian tropical rainforest). In the context of this trend, the Andean mountainous region has been perceived as backward, traditional and unmodernised, and marginal to the economic mainstreams. Agronomic potential has been mostly disregarded in favour of the promotion of mining, in accordance with the perspectives of an export-oriented economy.

Only during the last two decades have academic researchers become interested in the Andean traditional farming and herding systems. The resulting increase of information triggered applied programmes aimed at both the recovery of paleo-technologies and the salvage and promotion of native food crops. Only most recently have government agencies started reconsidering their farming development policies for the highlands.

After reviewing some basic aspects of the Andean traditional subsistence strategies and their current status, this paper will examine the configuration and outcome of some pilot experiences for the development of mountain productive systems and review some of the changing national policies and their impact on mountain environments. Finally, we will attempt to define some broad guidelines to be considered for the promotion of alternatives addressed to assure sustainable development in the Andes.

THE CENTRAL ANDES: HETEROGENEITY AND BIODIVERSITY FROM A MOUNTAIN ENVIRONMENTAL PERSPECTIVE

Framed by the coastal desert and the world's largest tropical rainforest, along its north-to-south ranges, the Central Andes are subject to weather conditions set by the South Pacific and South Atlantic anticyclones. Above parallel 5°S, the coast receives increasing rainfall, gradually turning from desert to deciduous shrub forest and to tropical rainforest. South of 5°S, the narrow littoral becomes a dry, rocky sand dune desert interrupted every 20 to 60 km by seasonal rivers flowing from the highlands. Across the ranges, on its eastern piedmont facing Amazonia, high humidity and intensive precipitation predominates culminating in a gradual transition from high cloud forest to lowland rainforest.

In between this lie numerous high mountain ridges, upland plateaux, and broad and open, as well as deepend narrow valleys—all subject to a variety of meteorological conditions affected by longitude, latitude, and altitude. A gradient of life zones can be found at different altitudes along the east-west transect. Making this panorama still more complex, latitude accounts for further variation. Towards the Equator, rainfall becomes more abundant and, in general, mountains have a more gentle relief and become lower. In contrast, as one approaches the Tropic of Capricorn, the cordillera widens, rises to higher altitudes and becomes rougher and drier. Thus, while *paramo*[1] vegetation characterizes the northern biogeographical province, a dry area of desert-like *Puna*[2] dominates the landscape in the southwest. On the average, precipitation diminishes from north to south and from east to west. Mountain relief, range directions, and slope orientation dramatically intensify the degree of meteorological variability.

Originating in the Paleozoic Era, the Andes emerged during the Pliocene Pleistocene uplift. Active geological processes still keep remodelling and shaping the topography, forging a highly heterogeneous and fractioned distribution of soil types.

In terms of life zones, the Central Andes perhaps hold a world record. Bowman noted: 'Nowhere else in the Earth are greatest physical contrasts compressed within such small spaces' (1916). To this statement Carl Troll added (1968b): 'Nowhere else in the world have I seen a more striking example of climatoecological differentiation than in the Andean valleys'. Due to their location, the Central Andes do not experience significant seasonality, with less than 5°C of thermal variation on the average (Troll 1968b). However, diurnal nocturnal oscillations (as great as 20°C) are pronounced, particularly above 3,000 masl.

Enclosed by the South Pacific and South Atlantic anticyclones and subsequent wind movements, the region does experience a seasonal pattern of precipitation, particularly south of 5°S, with rainfalls intensifying from November through March, followed by a dry season which corresponds to the South American winter. Rainfall precipitation is heavier on the north eastern slopes (on the average, over 1,800 mm) and scarcer in the dry southwestern piedmont (less than 300 mm) as exemplified by the Upper Maraon

1 *Paramo*—moor or wasteland.
2 *Puna*—a high bleak plateau in the Peruvian Andes. The word is also used to refer to climatic zones.

(one of the wettest areas in the world) and the Atacama Highlands less than 3,000 km southeast (one of the driest spots on earth).

One of the most particular characteristics of Andean meteorological conditions is the rather short-term and long-range lack of regularity: rains may come in early or late, abundantly or in short supply, scattered or concentrated (Dollfus 1982, Guillete 1981, and Earls 1989). Floods, droughts, frost, hail, etc may come unexpectedly affecting both programming of productive activities and productivity. In addition to annual local variations, overall climatological conditions are cyclically affected by regional meteorological phenomena, such as *El Nino*, causing total disruption of biogeographical processes, provoking geological instability, and thus, dramatically affecting human populations.

Rugged topography, poor and unstable soils, limited biomass productivity, biogeographical heterogeneity, and meteorological unpredictability all constitute major obstacles for human adaptation to the Central Andean environments. As elsewhere in mountain contexts, these factors should have restricted this region to a marginal habitat for human subsistence. However, as mentioned previously, this was not the case. Furthermore, sociocultural evolution went through one of its main stages in the Andes, and this can be explained in terms of adequate environmental management and successful adaptation. Anthropological and environmental research in the last two decades have demonstrated that 'Andean technological evolution was not obstructed by environment, but actually accelerated in response to it' (Earls 1989:39, author's translation). Andean subsistence patterns took advantage of limiting factors that otherwise could have been severely restrictive. As noted elsewhere, 'the Andean environment may be understood as the local responses to such environmental constraints as low energy availability, cold, and hypoxia' (Baker and Little 1976). Actually, as we will review here, environmental restrictions in the Andes are complex and innumerable.

ANDEAN SUBSISTENCE STRATEGIES: DIVERSIFICATION IN TIME AND SPACE

Prehistoric Evolution of Subsistence Patterns

Early hunters and gatherers, who arrived in the Andean region some 15,000 to 20,000 years ago, found a highly diversified environment with abundant game resources discretely distributed in distinct habitats which, depending on each region, guaranteed their year-round sedentary exploitation (Rick 1983). Limited short-distance transhumance provided direct access to complementary resources, while interzonal trade and barter facilitated a broader resource base.

Population growth and the depletion of faunal resources caused by overhunting, among other interrelated factors, stimulated increased dependence on a more stable and reliable resource base, triggering the gradual domestication of two Andean camelids (the llama for transportation and the alpaca for wool), native tuber potatoes (*oca, olluco, izano*, etc.), and grains (chenopods, lupinus, and amaranthus). This gradual process possibly took place simultaneously in different zones along the neighbouring biomes, ultimately resulting in the full domestication of a substantial number of plants, all of which underwent further selection into a myriad of locally adapted cultivars. Furthermore, foreign crops (maize) were later adopted and equally diversified into many cultivars (Table 22.1). By

Table 22.1. Native Andean foods

Food	Botanical classification	Quechua term	Aymara term
Cereals			
Maize	*Zea mays*	*Sara*	*Tonko*
Quinoa	*Chenopodium quinoa*	*Kinua*	*Hupa*
Canihua	*Chenopodium pallidicaule*	*Qanawi*	*Qanawa*
Achita	*Amaranthus caudatus*	*Quihuicha*	*Koyo*
Legumes			
Common bean	*Phaseolus vulgaris*	*Purutu*	*Purutu*
Jack bean	*Canavalia ensiformis*	*Inchis*	*Pallar*
Lima bean	*Phaseolus lunatus*	*Pallar*	*Pallar*
Peanut	*Arachis hypogaea*	*Inchis*	*Chokopa*
Lupine	*Lupinus mutabilis*	*Tarwi*	*Tauri*
Tubers			
Achira	*Canna edulis*	*Achira*	*Achira*
Oca	*Oxalis tuberosa*	*Oqa*	*Oqa*
Mashua	*Tropaelum tuberosum*	*Anu*	*Isano*
Arracaba	*Arracacia xanthorrhiza*	*Raccacha*	*Rakacha*
Manioc	*Manihot utilissima*	*Ruma*	*Ruma*
Papa	*Solanum* sp.	*Papa, Akshu*	*Papa*
Ulluco	*Ullucus tuberosus*	*Ulluku*	*Colluku*
Sweet potato	*Ipomea batatas*	*Apichu*	*Apichu*
Yacon	*Polymnia sonchifolia*	*Yakon*	*Yakuma*
Fruits			
Chirimoya	*Annona cherimolia*	*Masa*	
Guanabana	*Annona muricata*	*Massasasamba*	
Avocado	*Persea americana*	*Palti*	
Pacae	*Inga feuillei*	*Paqay*	*Paqaya*
Passion fruit	*Passiflora linularis*	*Ccjoto*	*Apinkoya*
Tumbo	*Passsiflora mollissima* *P. quadrangularis*	*Tumpaka*	
Guava	*Psidium guajava*	*Sawintu*	*Sawintu*
Lucuma	*Lucuma obovata*	*Ruqma*	*Lukuma*
Pepino	*Solanum muricatum*	*Kachan*	*Kachuma*
Squashes and vegetables			
Zapallo	*Cucurbita maxima*	*Sapallu*	*Tumuna*
Achokcha	*Cyclanthera pedata*	*Achokcha*	
Condiments			
Paico	*Chenopodium ambrosiodis*	*Paiko*	*Paiko*
Chile	*Capsicum annum* *C. frutescens*	*Uchu*	*Waika*
Chile	*Capsicum pubescens*	*Rocoto*	*Chinchi*
Huacatai	*Tagetes minuta*	*Wacatay*	*Wakatay*
Narcotics			
Coca	*Erythroxylom coca*	*Coca*	

Maize was probably introduced into the Andes from its Mesoamerican centre of domestication. (Adapted from Gade 1975.)

2000 B.C., a fully developed agro-pastoral subsistence strategy characterized rural life in the Andes (MacNeish 1977).

Archaeological evidence demonstrates the success of the evolving subsistence strategies which repeatedly led to the emergence of developed civilizations. By 1000 B.C. a fully complex, stratified imperial society, with significant surplus production, existed throughout the Central Andes (Lanning 1967). Major social achievements were supported by a wide, ecologically diversified resource base, either directly exploited by highland states or controlled through exchange and trade. Whatever the case, the efficiency of the productive system was expressed in terms of substantial surplus production, transformation (through a freeze-drying technology), and storage of tuber crops. Equally important was the accumulation of alpaca fibre and cloth and low energy cost transportation on llama caravans of foodstuffs to feed full-time imperial armies (Troll 1935 and 1968a).

Andean Traditional Subsistence Strategies

Although highland subsistence patterns in the Andes exhibit some local and regional variation in specific aspects of their production systems, they all have overall characteristics common to peasant society throughout the region. The Andean subsistence strategy, based on simultaneous exploitation of differentiated production zones, primarily along the altitudinal axle, is usually referred to as 'vertical ecology' (Troll 1935 and Murra 1975).

According to the vertical ecology paradigm, direct control of as many ecological 'niche' across the vertical landscape is an ideal of Andean rural society (Murra 1975). Vertical zonation of camelid herding and agrarian activities is a response to highland constraints on production and to natural biogeographical conditions, and could have had its roots in the prehistoric origin of plant and animal domestication, and possibly even earlier (Murra 1985). Archaeological evidence suggests that State formation and expansion during the middle and late horizons (800–1100 B.C. and 1300–1500 B.C.) enhanced and consolidated the system, mostly through State-sponsored, direct control of distant resources.

Despite the great range of variability in the verticality patterns in the Andes, three basic types of system have been suggested (Brush 1974).

(1) Compact type: characteristic of steep valleys with acute environmental variations at short distances across the vertical gradient. This was mostly the case of peasant communities in the western and eastern piedmont valleys, where the latter persists, such as in Q'ero (Webster 1971), Uchucmarca (Brush 1977), and Cuyo-Cuyo (Camino et al. 1981) among others. Here, production zones, lie within a walking distance of one to three days.

(2) Archipelago type: structured around a human nucleus in highland valleys or plateaux with extensions in distant resource based 'islands' under its control and exploitation through satellite populations. This case has been well documented by ethno-historical research and is usually linked to centralized State administration (such as the Lupaqa Kingdom in the Titicaca Plateau or the Chupaycho Chiefdom in Huanuco).

(3) Extended type: typical of large inter-Andean valleys with a rather gentle gradient allowing continuous exploitation of contiguous altitudinal production zones. Here, rural communities distributed throughout the valleys tend to specialize in specific production zones and to participate in trade networks and weekly markets. The Vilcanota-

Urubamba (Gade 1975) and Mantaro valleys (Altamirano 1974) are good examples of this type.

Specific contemporary research on the 'vertical control' has emphasized functional relationships with other dimensions of social life such as kinship (Brush 1975), marriage (Platt 1976), political decisions (Bradby 1982), and rituals and ceremonies (Vallee 1972, Barrette 1972, and Isbell 1974).

Current research on Andean agro-ecosystems has conclusively clarified that 'verticality' is but the spatial expression of a specialized subsistence system adapted to the Andean highland tropics and aimed at diversification and risk minimization. Furthermore, the strategy is directed to increased diversification not just at the space axle but to temporal dimensions as well. I have stated elsewhere that 'time and space... in the Andean context [are] two intimately related variables, continuously interacting and intensively articulated' (Camino 1982:32).

The Spatial Dimensions

Andean subsistence systems organize space in clearly delineated production zones, a 'man-made thing on top of the natural variation of the environment' (Mayer 1985:47). Mayer states that rather than 'adaptations' to the environment, production zones 'are created, managed, and maintained, simplifying natural diversity'.

Distribution of productive activities in the rugged Andean landscape is partly determined by natural constraints and potentials, as well as by the opportunities offered by native domesticates and those of foreign origin but locally adapted, and also by the traditional appropriated technology to manage this. Zonification is the traditional, communal organization of space for productive purposes.

Barometric altitudinal criteria only provide a rough picture of zonation. It is the specifics of topographical relief, soil, and weather conditions, in each case, that determine the type of production zone.

In an attempt to schematize the common characteristics of Andean traditional zonification, six discrete production zones can be found throughout the Central Andes.

Pasture Lands

Pasture Lands distributed below the year-round permanent snow line (about 5,000 masl) and the uppermost limits of cultivation (4,000 to 4,300 masl). Traditionally devoted to camelid herding, nowadays they are open to adapted sheep and bovine cattle as well. Pastures are always communally owned and managed except for *haciendas* or large private land holdings.

Bitter Potato Production Zone (about 4,000 masl)

This rather narrow strip is usually intercalated with the lowermost parts of pastoral areas and the main tuber crop zone. This area is devoted to cultivation of bitter potatoes (*shiri*) which require transformation through traditional freeze-drying to make them edible and, simultaneously, storable. Sometimes *izano* or *mashua* (*Tropaeolum tuberosum*), barley, and, in some areas, the Andean chenopods (*quinoa, canihua*, and *maca* (*Lepidium meyenii*) are also cultivated. Sectoral fallowing determines one year of cultivation followed by five years of rest (*purun*). During fallowing, the area is open for grazing, thus enriching its soils with *taquia* (manure).

Tuber Crop Zone (3,800–4,000 to 3,000–3,200 masl)

This is the main production zone for the Andean staple potato (*Solanum tuberosum* sp.) and other secondary but not less important tuber crops (*oca: Oxalis tuberosus, olluco* or *illaco: Ollucus tuberosus*, and *izano* or *mashua: Tropaeolum tuberosum*). They are grown in many adapted cultivars following different patterns of crop association (Camino 1978 and Camino et al. 1981). European broad beans (*Vicia faba*), barley, and wheat are also grown in this zone in some regions.

This production zone is usually divided into six rotational cycles (*mandas, turnos*, and *o laymes*). Sectorial fallowing could take, depending on several factors, from one to three years, and, as in the previous case, offer the possibility to cyclically dedicate the area to grazing and enriching the soil.

Although potatoes predominate in the first rotational cycle, they are usually sown in association with *izano*. After the second year, crop associations are intensified. A third and, eventually, fourth year can be dedicated to broad beans, wheat, or barley, planted primarily as monocrops.

In the case of tuber crops, cultivar heterogeneity is emphasized and always desirable. A traditional Indian family may grow 20 to 50 potato cultivars, as well as a diversified stock of *ocas, ollucos*, and *izano*. A peasant community can keep 300 to 500 potato cultivars out of the total count for the Central Andes, which nowadays exceeds 13,000.

Depending upon the species, variety, and stage of rotation, tubers produced in this zone are mainly destined for direct consumption and secondarily freeze-dried for storage (*chuno, ccalla, lingli*, etc). Part of the crop is saved for seed and eventually a portion is traded or sold. The tuber zone is the most important agricultural area for Andean communities, as exemplified by the extensive terracing which can be found throughout this zone.

Transitional Zone/Home Orchards

Below the main tuber-growing zone, a narrow transitional zone precedes the lower maize fields. Human settlements are quite often located in this transitional area, and, therefore, family orchards for vegetables, medicinal plants, and experimental crops are enriched with human refuse and household animals' manure (conspicuously from the edible *cuy: Cavia porcellus L.*). These bonafide plots are intensively cultivated with a combination of elaborate forms of crop associations, bringing together highland and lowland cultivars. The area is not subject to fallowing, and, therefore, fields have to be carefully walled to prevent the unexpected intrusion of cattle.

Maize Production Zone (3,200 to 3,000 and 2,500 masl and below)

Whether grown under irrigation (western piedmont), rainfall (eastern slopes), or a combination (inter-Andean valleys), highly prized maize is intensively grown, normally without fallowing, thus demanding different strategies to assure soil fertility. As in the other zones, corn is always cultivated in association with other traditional crops: Andean squash (*Cucurbita* sp.), beans (*Phaseolus*), broad beans (*Vicia faba*), *yacon* (*Polymnia sorchifolia*), and *racacha* or *virraca* (*Arracacia xanthorriza*), among others of less relevance.

The Andean cultural importance of maize cultivation and ceremonial corn beer (*chicha*) has been noted (Murra 1975), and a fact which is reflected in the magnificent terrace systems which, were built for this crop, in the pre-Columbian period. As previously mentioned, native crops, germplasm variety, and plant heterogeneity were highly valued in indigenous society, a trait that helps explain the high degree of genetic diversification which maize attained in the Central Andes.

Coca Production Zone (below 2,000 masl)

Even if well outside our immediate concern, this area is of crucial importance. Coca (*Erythroxylon coca*, a mild stimulant, traditionally chewed) was, and is, a plant with a principal role in religion and ritual as well as in interzonal trade and barter (Instituto Indigenista Interamericano 1976). Highly esteemed, coca leaves acted as a quasi coin assuring ecological complementarity for those indigenous communities that, for some reason, had restricted access to one or more production zones.

Current international demand for cocaine, one of the 14 alkaloids of coca leaf, has seriously disrupted the Andean social patterns of coca use with drastic consequences for the traditional subsistence system (Camino 1989).

It should be noted that the vertical ideal of the Andean peasant, in addition to land division caused by inheritance, usually results in an extreme pattern of *chacras* (plot) dispersion. In order to assure the ideal self-sufficiency, the average peasant family tries to have direct access to land in each of the production zones as well as in each one of the rotation cycles of the zones with sectoral fallowing. To assure food crop self-sufficiency, the peasant family has to grow each crop, or a mixture of crops, through crop associations. Furthermore, due to constant risks, as exemplified by frequent crop losses due to frost, hail, drought, landslides, etc., the autarkic family attempts to hold as many discrete *chacras* as possible, in as many locations as possible, for each one of the rotational cycles of each production zone. Thus, ideally, one family will hold more than one tiny plot for every specific crop or crop association, at every stage of rotation, in each production zone. As we will examine later, this has an effect on the length of vegetative cycles and on the programming of the many agricultural tasks throughout the multivaried agro-ecosystem.

The vertical zonation of crops, as previously described in broad and general terms, is based on a supra-household system of production organization, clearly stipulated in tradition and strictly sanctioned in the regular communal administration of the system. *Ayllu*, the Andean traditional lineage, sets the criteria and rules of zonification, rotational cycles, use and management of pastures, and other communal lands to which all the peasant families have to adjust their herding and farming strategies and activities. It also plays a basic role in the design and execution of communal works for agriculture (irrigation canals, ditches, terraces, etc.) and other civil constructions (bridges, roads, etc.). It has been noted that 'there is a specific Andean collective form of organisation of production that, under varying and changing social and ecological conditions, will constantly generate technological solutions that bridge the gap between desired crops and the local environmental conditions that favor, limit, or impede production' (Mayer 1985:48).

The Time Axle

Altitude affects temperature, precipitation, and humidity. Temperature drops (on the average, 6°C for every 1,000 m of elevation), thus limiting productivity and slowing the growth rates of flora.

On the other hand, incoming solar radiation increases with altitude. Higher levels of cosmic radiation may explain the abundance of certain endemic plant life forms or the high index of variability within cultivated plant populations. Altitude increases unpredictable threats to agriculture in the form of frost, hail, drought, etc. Crop dispersion across vertical zonation, as explained earlier, becomes a way to reduce risk. Each crop, and in some cases each one of its cultivars, will be simultaneously grown in different sectors under varying geographical and meteorological conditions (slope and soil, sun exposure, humidity, precipitation) and, particularly at different altitudes, avoiding concentrations of the crop in just one homogeneous and contiguous sector, thus diversifying exposure to the recurrent risks of high altitude agriculture.

The ideal of spatial diversification and distribution of crops responds to a risk minimization strategy. Elsewhere, it has been pointed out that verticality is 'just a landscape architectural expression of a broader and more complex ideal which is directed to the diversification of the subsistence base in an unstable and unpredictable environment' (Camino 1982:29).

Along its temporal dimension, this same ideal is expressed in an equivalent manner. According to this temporal axle, effective and absolute limits for crop sowing and harvest affect the scheduling of the agricultural calendar. In the Bolivian and Peruvian Puna, over 4,000 masl, night frosts occur throughout the year (Troll 1968b:22). Agriculture, based on adapted cultivars, can only be conducted, taking this limit into consideration, where nightly frosts gradually diminish. Thus, effective and absolute limits along the time axle are determined by the natural growth cycle of cultivars.

In the process of plant domestication, Andean peasants selected short maturing varieties (*chauchas*) as well as others with frost-resistant traits, resulting in the extension of vegetative cycles. Some Andean agronomic practices are also directed to shortening or expanding vegetative cycles in order to adapt these to varying altitudes and to unpredictable delays or early presentation of rainfalls (or water availability in the case of irrigation agriculture).

Depending on these factors, the life-cycle of crops will vary accordingly, affecting production programming, and scheduling. Diversification along the temporal axle is also a desirable strategy as a response to risk due to unpredictable weather conditions.

For the purpose of sowing, harvesting, and other agricultural tasks in between, Andean tradition determines the scheduling of activities using the religious ritual calendar (a syncretic Andean-Christian blend). The ceremonial calendar operates as an agricultural calendar, fixing referential dates for the initiation and conclusion of each production zone. In this way, ritual festivals mark minimum and maximum limits for the onset and termination of each agricultural activity associated with a specific production zone and crop. However, in the context of unpredictability, a calendar that does not take into account and deal with predictable variations, and even chance, will not be effective or trustworthy. In both irrigated and rainfall agriculture, it is necessary to foretell, with some degree of certainty, the expected precipitation. This poses a major difficulty, due to Andean meteorological irregularity and the unpredictability of atmospheric phenomena and their impact on each zone and altitudinal level. This is the case when traditional Andean concepts, such as *tinkuy*, are invoked. *Tinkuy* refers to predestined encounters (rivers that meet, marriages, etc.). It also refers to the inevitable encounter of the sprouting plant and weather conditions (Fonseca 1966). In a pragmatic way, the concept is manipulated when applied to agricultural practices; spacing over time for sowing of tubers on a plot, even if it has a reduced surface. In this way, *tinkuy* will happen for each plant at a different stage of its growth cycle, assuring that, in the case of threats from hail or frost, at least a portion of the planted crop will be saved.

Each production zone has its peculiar characteristics and requirements, and, consequently, the scheduling of farming activities varies from zone to zone. To the diversity of vegetative cycles for each cultivar, usually grown in association with others, the modification that altitude will affect the duration of the maturation period is another factor to be added. In addition, we need to assess additional impacts: amount or opportunity of rainfall or water from irrigation, hours of sunlight, and eventual frost and hail. All of these will accelerate or slow down growth. Depending upon these factors and depending upon traditional criteria for weather prediction, peasants will move ahead or delay each agricultural activity.

Throughout the Central Andes, the peasants' lore identifies three discrete growing periods.

(1) *Naupa tarpuy*, or early sowing period (usually falls in August, depending upon the region).
(2) *Chaupi tarpuy*, or regular, middle sowing period (September).
(3) *Q'epa tarpuy*, or late planting period (October).

Accordingly, subsequent agricultural tasks follow each one of the periods. However, harvesting usually proceeds upwards from the lower areas, due to prolonged growth periods along the vertical gradient.

Concentration of sowing for potatoes or any of the other crops, in one or the other of the three planting periods, will depend upon weather forecasts which are normally formulated before the initiation of the agricultural year. Concentration of sowing activities will vary from year to year, depending upon these predictions.

If the forecast announces early rains, the bulk of sowing will take place in *Naupa*; if the rains are regular, most of the sowing will be done in *Chaupi*, and, subsequently, a prediction for late rains will transfer most of the sowing to the *Q'epa* period. In this way, through weather forecasts the agricultural calendar is constantly adjusted to changing meteorological conditions. However, some planting is always done, although with less intensity, in the other two periods. This operates as a mechanism to ensure at least some crop returns in the case of sudden changes or wrong forecasts.

Weather predictions are very important throughout the traditional peasant society. These are based on astronomical observations made on specified dates, careful monitoring of natural indicators (meteorological phenomena, fauna, and flora), or other cultural mechanisms (Camino 1982, Antunez de Mayolo 1976, and Cuba de Nordt 1971). Weather predictions not only allow for the programming of agricultural activities, but also facilitate the careful planning of extended family workforce distribution throughout production zones.

There are years in which all the production zones do not experience similar or homogeneous weather conditions, with the precedence of rain in one zone, delays in the other, and irregular hail distribution in yet another (Camino 1981), turning programming into a complex endeavour.

Andean subsistence strategies cannot operate on the basis of rigid calendars. Just as diversity characterizes the Andean environment, meteorological irregularity has demanded the development of flexible calendars, where a vast range of options facilitates the manipulation and management of temporal criteria in a way similar to the handling of diversity in the spatial transect.

ANDEAN TRADITIONAL TECHNOLOGY

Andean populations were capable of establishing major civilizations, achieving high and sustainable production, and maintaining vast and dense populations (Yamamoto 1985 and Dollfus 1982). This was possible through four basic and interrelated achievements:

(1) a comprehensive fine knowledge of the environment; its natural resources, their limitations, and potential;
(2) the breeding of a vast array of adapted domesticates;
(3) the development of innovative and appropriate technologies; and
(4) the consolidation of an adaptive social system which put a premium on austerity, severely sanctioning deviations from tradition.

As noted previously, diversification along time and space, a vital adaptation to a heterogeneous and unpredictable environment, is reflected in elaborate patterns for the spatial and temporal distribution of crops, adequate domesticates, appropriate management, and efficient programming. For successful adaptation, technologies had to be developed in several areas. Let us briefly examine the most prominent.

Soil Building and Management

The Incas and their predecessors were superb soil builders, as attested by their magnificent achievements in agricultural engineering. Terracing, in its many patterns and forms, attained full development on diverse types of relief, in diverse soils, and in diverse production zones (particularly the tuber, maize, and coca). Terraces, in addition to their contributions to erosion control and to the expansion of the agrarian frontier, created local micro-climates (Earls 1989), enhancing the potential of high altitude environments to sustain a variety of crops. Current estimates put the terraced slopes of Peru at one million ha, two-thirds of which are nowadays abandoned (Denevan 1986).

Other major developments were the vast ridged fields (*huaru-huaru*), still visible on the Lake Titicaca Plateau (3,803 masl), which, in the prehistoric past, covered 82,056 ha (Smith et al. 1981). These mounds were built on the annually flooded plains neighbouring the great lake. Archaeological estimates suggest that they supported dense populations (Denevan 1982 and Garaycochea, no date), by creating rich soils and special micro-climatic conditions which helped to drastically reduce the repeated effects of frosts on the plateau.

Cochas, or artificially ridged and drained natural depressions on the same plateau, were another device to expand agriculture beyond natural limiting conditions (Flores and Paz 1986 and Rozas 1986). An area of 384 km, north of the plateau, is still being farmed with *cocha* technology. Soil conditioning and management was not restricted to agricultural pursuits. Palacios (1977) has described the purposeful flooding of the high plateaux to create pastures for alpaca herding.

Use of manure and other natural fertilizers was common throughout the Andes. Different fertilization technologies for varied crops and production zones have been described for various regions (Camino 1982, Yamamoto 1985, Tapia and Flores 1984, and Winterhalder et al. 1974).

Irrigation

As indicated earlier, water availability and its appropriate timing is essential for high altitude agriculture, whether in the form of irrigation, rainfall, or both. Water availability becomes more critical on western slopes, where irrigation technology and management has attained greater development and where practices are strictly sanctioned by communal ceremonial rites. In the inter-Andean valleys, irrigation facilitates the temporal management of crops, making water available to plots before the onset of the rainy season in order to accelerate growth. On the eastern slopes, irrigation is mostly absent, the concern being rainfall prediction and the appropriate control of floods and droughts.

Water management, a complex Andean technology with substantial diverse regional patterns, is crucial and has received careful attention from researchers (Mayer 1985, Zegarra 1978, and Mitchell 1981). Canals, aqueducts, reservoirs, and other hydraulic constructions are still much in use wherever traditional agriculture persists.

Agricultural Tools

After four centuries of neglect of the traditional Andean agricultural tools in favour of European oxen-driven ploughs and motorized tractors, in recent years, investigative research on mountain agriculture and soil management has started reconsidering the role of the traditional Andean foot plough (*chakitaclla*), which has shown itself to be an adequate and versatile tool well suited to the Andean turf type of soils, assuring their conservation and preventing soil runoff (Donkin 1970). The type and direction of ridges and furrows opened with the *chakitaclla* (following the gradient contour or across it) vary from year to year according to water availability.

Different methods of turning the soil, planting seeds, banking soil on plants, etc, depend also on the type of soil and crop, the specific needs of each production zone, and weather conditions. Turning turfs upside down several weeks before sowing, on top of furrow, facilitates nitrogenation of the soil through the decomposition of grass-like cover vegetation.

As with many Andean technologies, *chakitaclla* ploughing is intimately linked to traditional social work arrangements (foot ploughing is done by pairs of kinsmen), and is deeply imbedded in the religious system and the spiritual significance of *Pachamama* (mother earth).

Native Crops

As indicated before, the Andean region has been one of the most fertile cradles of crop domestication. In addition to several tuber and root crops and their many cultivars, chenopods and a diversity of fruits and seed crops (*Phaseolus* sp., *Lupinus* sp., *Amaranths* sp., etc.) provide the peasant household with a rich and varied food base. Through directed selection for specific traits (agronomic, culinary, and symbolic) on either vegetatively or sexually reproduced crops, the Andean agriculturalists gradually developed an enormous variety of locally adapted cultivars that respond to specific requirements and conditions. Non-native crops of pre-Hispanic origin (i.e., maize) or of later introduction (broad beans) also became subject to this process.

With colonization, diversification of the cultivar base reversed with the traumatic substitution of mining for agriculture for mining as an economic priority. The process intensified with the development of urban monetary markets and their narrow range of demand for a limited variety of food crops.

Social Management of Production Systems

As stated throughout this paper, the Andean subsistence system is organized at the communal, supra-household level where nuclear and extended families, as members of an ethnic lineage (*ayllu*), have to adjust and participate in its social management (Mayer 1985 and Guillete 1981). Herding in communally owned pasture lands follows strict rules and severe sanctions are applied to transgressors. The same applies to the organization of the sectoral fallowing of tuber crops, the scheduling of agricultural activities in all production zones, irrigation, and the organization of communal work (*faena* or *minga*), among others.

Due to the complexity of the adaptive strategy, individual households cannot venture into an independent strategy, when aware of the risks involved. The whole productive system is intimately tied to all aspects of traditional culture and social life, and change can only take place successfully through consensus.

However, village organization is not static and, over time, it gradually adapts to changing conditions. One of the foremost authorities on Andean culture has conclusively stated: 'Without an understanding of local village organisation, the complexities of the verticality model cannot be completely understood... the household alone cannot by itself deal with all the technical and organisational problems of production in a given zone; it needs the concurrence of 'supra-household' organisation' (Mayer 1985:48).

DEVELOPMENT INTERVENTIONS: POLICIES, PROJECTS, AND CONCLUSIVE GUIDELINES

As has been stated before, colonization and the republican history of most Andean countries resulted in the neglect of highland habitats and their agro-pastoral traditional experiences. Viceroyalties and their administrative units and, later, emerging contemporary countries switched the introverted character of the Andean society into export-oriented, extrovert economies. Mining and other extractive interests became the dominant activities.

Simultaneously, new crops were introduced, some of which fitted adequately into the traditional adaptive strategies (broad beans) while others (forage grasses) brought disruption and environmental degradation. In some areas, substantial changes in inland tenure patterns and production strategies brought about the total collapse of the traditional subsistence system. In others, partial changes affected some aspects of the system and prompted adjustments, as in the case of the introduction of European livestock.

The last part of the previous century witnessed, for the first time, an increased world demand for an Andean commodity, in this case wool (sheep and alpaca), a phenomenon that had rather restricted duration and limited impact (Orlove 1977). After the 1930s, none of the Andean agro-pastoral products had any significance in the export economy, with the possible exception of alpaca wool.

Successive governments regarded the highland regions as marginal habitats for economic development, favouring plantation agriculture on the forested northern coast or, through irrigation, in the oasis type of valleys south of 5°S. The Amazon rainforest, erroneously perceived and promoted as the future bread basket of the region, suffered the consequences of State-promoted deforestation for colonization. It was only in the late 1960s that two simultaneous processes of academic reassessment of the Andean subsistence potentials initiated the reconsideration of what had formerly been an undiscussed paradigm. This resulted in two distinct research disciplines.

The Social Sciences, Particularly Anthropology

It was the anthropologists and the ethno-historians who, after intensive fieldwork and archival research, described the strategy that, as 'vertical ecology', explained the traditional Andean patterns of subsistence. Its rationale created an impact on agricultural researchers and developers. *Minifundismo* (the extreme division of properties), which had formerly been explained exclusively in terms of inheritance, suddenly acquired a new dimension. The autarkic orientation of Indian communities, previously perceived as a purely cultural trait and as a lack of market incentive, required an explanation that only ecological anthropology was able to provide.

Agronomy

A new generation of agronomists, some of whom had experienced the limitations and high energy costs of plantation agriculture in the lowlands, started turning their research interests to formerly neglected native crops which they saw as having the potential to solve the food problems of the region.

Both academic traditions had their *tinkuy* in the mid-1970s, giving birth to a new fertile paradigm: Andean traditional subsistence systems are worth consideration, since they may be able to solve the problem of food shortages in our contemporary nations.

However, due to the usual Latin American divorce between academicians and policy makers, this new perspective had, at first, very little impact on government policies, which remained resilient to change. Governments promoted a straightforward European style of modernization of agriculture in the highlands. The main concerns were to introduce new crops and technologies that had had successful histories elsewhere, without much consideration for the natural and social environment on which they were being imposed. An ideology of 'progress' prized and valued innovations, associated with foreign crops and technology, and rejected tradition which was associated with backwardness. This overall trend had very few exceptions, some of which existed as a result of isolated and restricted experiments.

During the late 1970s and early 1980s, Bolivia, Ecuador, and Peru started their first, significant applied experiments in development from a mountain environmental perspective. These began as pilot projects, promoted by scientists who had been researching on traditional crops and technologies. Some were privately funded by foreign development agencies and a few were able to receive a basic subsidy from the State. Let us review the main areas in which these experiments were conducted.

The Revigoration and Promotion of Native Crops

Aware of the nutrition potential of Andean chenopods (*quinoa* and *canihua*), a few projects for seed selection, improvement, and promotion were started with promising results on the Titicaca Plateau. The interest in these high-altitude crops expanded north, into Cuzco and Ayacucho, and south, into Bolivia, and gradually they attracted the attention of foreign researchers, mostly North Americans who started promoting their cultivation in the United States and Canada.

Tarhui or *chochos* (*Lupinus mutabilis*) soon followed because of its nitrogen-fixing properties and for the industrial production of oil. Breeders promptly developed 'sweet' varieties whose seeds are almost free of bitterness and need almost no washing. During the 1980s, *kiwicha* (*Amaranthus caudatus*), one of the most nutritious foods grown, was popularized by peasant farmers and gained the attention of decision makers.

Despite massive research on potatoes, apart from their use for germplasm purposes, nearly 14,000 Andean cultivars have been totally disregarded in the effort to salvage and promote native food crops. Genetic engineering and trait selection, conducted by the International Potato Centre in Lima, have assured the laboratory conservation of germplasm despite the alarming decline in the cultivation of native cultivars. The availability of engineered varieties, with a diversity of favourable traits, is drastically reducing the number of native varieties in the countryside while ensuring increased productivity. This is the factor which has probably led to the abandonment of any attempts to ensure the conservation of traditional cultivars in the field. However, the other Andean tuber crops, *oca, olluco*, and *izano* (see Table 22.1 for botanical identification), have not only received research attention but some benefit as well from the applied agrarian development programmes. Due to the fact that they have not been subject to any important genetic breeding, native cultivars remain vigorous in the rural Andes. Other Andean crops received less attention in the context of agricultural development. Their vast potential has recently been made public to a wide audience (Popenoe et al. 1989).

One of the main drawbacks of the numerous valuable attempts to salvage and promote native crops is the usual de-contextualization of the agronomers' or developers' efforts. In most cases, native crops are in little or no demand in urban markets, making it difficult to get the peasant or farmer interested in their cultivation for reasons beyond self-consumption. Urban dwellers are not familiar with them and do not even know their culinary uses or nutritional qualities.

Furthermore, promotion faces the constraint of the lack of seeds, mostly due to the restricted cultivation of these crops. In some cases, their traditional cultivation has already disappeared with subsequent ignorance on the part of the peasants concerning their cultivation.

Finally, their limited, but rather exclusive, demand (health stores, specialized exports) turns them into highly prized foods. However, the advances in the knowledge and promotion of Andean native crops certainly assure their of a promising future, with the possibility of their becoming once again important productive crops in high-altitude agriculture.

The Relaunching of Andean Paleo-technologies

Another significant development during this last decade has been the applied programmes to salvage and re-apply Andean traditional technologies which earlier research had proven to be valuable and promising. One of the outcomes of 'verticality' research was a renewed interest in terrace farming. Numerous field studies had proven the benefits of *andenes, cochas*, and *huaru-huaru* from the perspective of soil utilization and management, as well as from their potential in the expansion of the Andean agrarian frontier. It was previously stated that, out of the estimated one million ha of pre-Columbian terraces in Peru, 600,000 were abandoned. Furthermore, in contrast to the high costs of irrigation on the coast, or jungle colonization programmes in Amazonia, the cost for terrace rehabilitation was, on the average, quite low. While recovering one hectare of terrace has a cost of US $750 to US $1,000, the irrigation of a similar amount of land on the dry coast is US $15,000 to US $20,000 (CEPAL 1988).

Several experiments on terrace rehabilitation have been conducted in recent years (de la Torre and Burga 1986), all of which, despite difficulties and drawbacks, have proven fruitful. Among the many advantages and benefits, terrace agriculture:

- expands the agrarian frontier at a comparatively low cost,
- increases slope productivity,
- protects and regulates watersheds, and
- increases work opportunities for the unemployed or underemployed rural labour force and creates better production conditions promoting highland settlement and preventing migration to urban slums or to Amazonia for coca cultivation.

In general terms, the same benefits can be extended to the rehabilitation of ridge fields (*huaru-huaru*) and *cochas* on the Titicaca Plateau.

However, soil rehabilitation, through terrace or ridge field reconstruction, has several problems that need to be solved. Among them, the most recurrent have been:

- the lack of global macro-economic policies in support of agricultural activities,
- the involvement of rural communities through active participation in the planning and execution stages,
- the need to ensure that complementary engineering works are also integrated into the rehabilitation strategy (i.e., irrigation canals, water reservoirs),
- the proper cultivation of adequate crops, and
- the provision of technical advice in the architectural aspects of terrace reconstruction, particularly in those communities where traditional knowledge on the subject has been lost.

A few projects have focussed on the evaluation and reintroduction of traditional tools as well as on adequate technologies, some of which have succeeded and others of which have failed depending on several factors related to agronomic or social conditions. Agricultural tools, as with all technological devices, do not operate in environmental or social vacuums and need to be carefully evaluated from an integral perspective. Efficiency needs to be measured, not just in terms of environmental adequacy but in terms of productivity and social acceptance as well.

Recent years have seen an increased concern on the part of governments and international organizations in adapting development policies and actions to existing natural and social conditions in mountain environments, and this, it seems, will be the trend for the coming decade. Success or failure in promoting development, from a mountain perspective, will ultimately depend upon both the formulation of adequate policies to suit characteristic mountain environmental conditions and upon their consistency with the social and cultural context in which they are meant to operate.

REFERENCES

Altamirano, T. 'Estructura Regional y Actividad Empresarial en un Valle Peruano.' In *Revista de Ciencias Sociales*. Huancayo: Universidad Nacional del Cento, 1974.

Baker, P. and M.A. Little (eds.). *Man in the Andes: A Multidisciplinary Study of High Altitude Quechua*. Stroudsberg, Pennsylvania: Dowden, Hutchison, and Ross, Inc., 1976.

Barrette, C. 'Aspects de l'Ethno-ecologie d'un Village a Andin.' *Revue Canadiene de Sociologie et Anthropologie* 9(3):255–267, 1972.

Bowman, I. *The Andes of Southern Peru*. New York: 1916.

Bradby, B. 'Resistance to Capitalism' in the Peruvian Andes. In D. Lehman (ed.), *Ecology and Exchange in the Andes*, Cambridge Studies in Social Anthropology, 1982. Cambridge: Cambridge University Press, 1982.

Brush, S.B. 'El Lugar del Hombre en el Ecosistema Andino.' *Revista del Museo Nacional*, 40:279–303, 1974.

Brush, S.B. 'Parentesco y Agricultura en un Pueblo Andino Peruano.' *America Indigena* 35(2):367–390, 1975.

Brush, S.B. *Mountain, Field and Family: The Economy and Human Ecology of an Andean Valley*. Pennsylvania: University of Pennsylvania Press, 1977.

Camino, A. 'Monocultivo y Policultivos en las Montanas Tropicales.' In *ler Congreso Internacional sobre Cultivos Andinos*, pp. 44–51, Lima, 1978.

Camino, A., et al. 'Flexibilidad Calendrica en la Agricultura Tradicional de las Vertientes Orientales de los Andes.' In La Tecnologa en el Mundo Andino H. Letchman, and A.M. Soldi, (eds.), *La Tecnologa en el Mundo Andino*. Mexico: UNAM, 1981.

Camino, A. 'Tiempo y Espacio en la Estrategia de Subsistencia Andina: Un caso en las Vertientes Orientales Sud-Peruanas.' In *El Hombre y su Ambiente en los Andes Centrales, Senri Ethnological Studies* No. 10, Osaka, 1982.

Camino, A. 'Coca: del Uso Tradicional al Narcotrafico.' In D. Garcia Sayan (ed.), *Coca, Cocaina y Narcotrfico*, Lima: Comision Andina de Juristas, 1989.

CEPAL (Comision Economica para America Latina y el Caribe). *Gestin para el Desarrollo de Cuencas de Alta Montana en la Zona Andina*. Santiago de Chile: Naciones Unidas, 1988.

Cobo, B. *Historia del Nuero Mundo*. 2 vols. Madrid: Biblioteca de Autores Espanoles, 1964.

de Malcyo, A.R. *Meteorologa Inka, Epoca de Siembra y Seleccin de Cultivar*. Lima: 1976.

de Nordt, C. 'Las Cabanuelas.' *Allpanchis*, Cusco 3:45–46, 1971.

de la Torre, C., and M. Burga (eds.). *Andenes y Camellones en el Peru Andino*. Lima: CONCYTEC, 1986.

de la Vega, I.G. *Comentarios Reales de los Incas*. Lima: UNMS, 1959.

Denevan, W.M. 'Hydraulic Agriculture in the American Tropics: Forms, Measures, and Recent Research.' In K. Flannery (ed.), *Maya Subsistence*. New York: Academic Press, 1982.

Denevan, W.M. 'Abandono de Terrazas en el Peru Andino: Extension, Causas y Propuestas de Restauracion.' In C. De la Torre and M. Burga (eds.), Andenes y Camellones en el Per Andino.Lima: CONCYTEC, 1986.

Dollfus, O. 'Development of Land-Use Patterns in the Central Andes.' *Mountain Research and Development*, 2(1):39–48, 1982.

Donkin, P.A. 'Pre-Columbian Field Implements and Their Distribution in the Highlands of Middle and South America.' *Anthropos* (USA) Vol. 65:505–529, 1970.

Earls, *Planificacion Agricola Inca. Bases para un Manejo Ciberneticio de un Sistema de Andenes*. Lima: U. del Pacifico, COFIDE, 1989.

Fonseca. 'La Comunidad de Cauri y la Quebrada de Chaupiwaranga.' *Cuadernos de Investigacion*. Antropologia 1, Huanuco pp. 22–23, 1966.

Flores O.J. and F.P. Paz. 'Agricultura en Lagunas (gocha).' In C. de la Torre and M. Burga (eds.), Andenes y Camellones en el Peru Andino, Lima: CONCYTEC, 1986.

Gade, *Plants, Man and Land in the Vilcanota Valley of Peru*. The Hague: Dr. W. Junk B., 1975.

Garaycochea, I. 'Potencial Agricola de los Camellones en el Altiplano Puneno.' In C. de la Torre and M. Burga (eds.), Andenes y Camellones en el Peru Andino . Lima: CONCYTEC, no date.

Guillete, D. 'Agrarian Ecology and Peasant Production in the Central Andes.' *Mountain Research and Development*, (1):19–28, 1981.

Instituto Indigenista Interamericano. *La Coca Andina. Vision Indigena de Una Planta Satanizada* Mexico: J. Boldo i Clement, 1976.

Isbell, B.J. 'Parentesco Andino y Reciprocidad: Kuyaq: Los que nos Aman.' In *Reciprocidad e Intercambio en los Andes Peruanos*. Lima: EP, 1974.

Lanning, E.P. *Peru Before the Incas*. Englewood Cliffs, N.J.: Prentice-Hall Inc., 1967.

MacNeish R.S. 'The Beginning of Agriculture in Central Peru.' In C. Reed (ed.), *Origins of Agriculture*. The Hague: Mouton, 1977.

Mayer, E. 'Production Zones.' In S. Masuda (ed.), *Andean Ecology and Civilisation* Tokyo: University of Tokyo Press, 1985.

Mitchell, W. 'La Agricultura de Riego en la Sierra Central de los Andes: Implicaciones para el desarrollo Delestado.' In H. Letchman and A.M. Soldi (eds.). *La Tecnologia en el Mundo Andino*. Mexico: UNAM, 1981.

Murra, J.V. *Formaciones Economicas y Politicas del Mundo Andino*. Lima: IEP, 1975.

Murra, J.V. 'El Archipelago Vertical Revisited.' In *Andean Ecology and Civilization*, Tokyo: University of Tokyo Press, 1985.

Orlove, B. *Alpacas, Sheep and Men: The Wool Export Economy and Regional Society in Southern Peru*. New York: Academic Press, 1977.

Palacios, F. 'Pastizales de Regadio para Alpacas.' In F.O. Jorge (ed.). *Pastores de Puna*. Lima: IEP, 1977.

Platt, T. 'Espejos y Maiz: Temas de la Estructura Simbolica Andina.' In *Cuadernos de Investigacion*, 10. La Paz: CIPCA, 1976.

Popenoe, H., et al. *Lost Crop of the Incas*. Washington, D.C.: Bostid, National Academy Press, 1989.

Rick, J.W. *Cronologia, Clima y Subsistencia en el Preceramico Peruano*. Lima: INDEA, 1983.

Rowe, J.H. 'Inca Culture at the Time of the Spanish Conquest.' In *Handbook of South American Indians* (Vol. 2), Washington, D.C.: Smithsonian Institution, pp. 183–230, 1946.

Rozas, J.W. 'El Sistema de Cultivo en Qocha.' In C de la Torre and M. Burga, (eds.), Andenes y Camellones en el Peru Andino, Lima: CONCYTEC, 1986.

Smith, C.T., et al. 'Antiguos Campos de Camellones en la Region del Lago Titicaca.' *Geographical Review*, 134:353–367, 1981.

Tapia, N.M. and O.J. Flores. *Pastoreo y Pastizales de los Andes del Sur del Peru*. Lima, 1984.

Troll, C. 'The Cordillera of the Tropical Americas. Aspects of Climatic, Phytogeographical and Agrarian Ecology.'

Troll, C. 'Los Fundamentos Geograficos de los Civilizaciones Andinas y del Imperio Incaico.' *Revista Universitaria*, Arequipa, 9:129–182, 1968a.

Troll, C. (ed). *Geoecology of the Mountainous Region of the Tropical Americas*. Bonn: Dummler Verlag, 1968b.

Vallee, L. 'Cycle Ecologique et Cycle Ritual: le Cas d'un Village Andin.' *Revue Cannadiene de Sociologie et Anthropologie* 9(3):238–254, 1972.

Webster, S. 'An Indigenous Quechua Community in Exploitation of Multiple Ecological Zones.' In *Actas y Memorias* In del XXXIX Congreso Internacional de Americanistas, 3:174–183, Lima, 1971.

Winterhalder, B., et al. 'Dung as an Essential Resource in a Highland Peruvian Community.' *Human Ecology* 2(2): 89–104, 1974.

Yamamoto, N. 'The Ecological Complementarity of Agro-Pastoralism: Some Comments.' In S. Masuda (ed.), *Andean Ecology and Civilization*. Tokyo: University of Tokyo Press, 1985.

Zegarra, J. 'Irrigation y Tecnicas de Riego en el Peru preColombino.' In R. Ravines (ed.), *Tecnologoa Andina*, Lima: IEP, 1978.

23

INDIGENOUS FARMING TECHNOLOGIES AND ENVIRONMENT: EXPERIENCES IN BHUTAN

A.K. Gupta and K. Ura

Anil Gupta is a Professor at the Centre for Management in Agriculture, the Indian Institute of Management, Ahmedabad and Karma Ura is a Senior Planning Officer, Planning Commission, the Royal Government of Bhutan, Thimphu

CONTENTS

INTRODUCTION

How did an unmodernized, isolated mountain country succeed in keeping 64 per cent of its area under forests? How did it succeed in avoiding widespread deprivation and maintain a fairly sustainable ecological balance? What problems does it face in its anxiety to keep its cultural core intact and at the same time improve the living standards of its people? Buddhism plays an important role in blending culture with technology.

This chapter deals with the experiences of Bhutan, a relatively small, land-locked Himalayan country which has established common property and other collective institutions for resource management.

The text is divided into seven parts. Part 1 deals with the culture of conservation and Buddhist ethics. Part 2 includes a discussion on the framework of institutional emergence in mountain societies. Rules and principles have to be established while dealing with the boundaries of moral and ethical responsibilities towards the environment. Examples of specific institutions for managing water, grazing land, forests, and labour contributions for public and common works are provided in Part 3. The institutional innovations are related to technological innovations. The culture of this innovative ethic is reviewed in Part 4. Specific examples based on water-driven prayer wheels, architecture, education, alignment of irrigation channels, movement of livestock, prevention of the diffusion of animal diseases, generation of cropping systems, and management of shifting cultivation are covered in this part. Part 5 includes issues that are emerging in the process of technological transfers within the above context.

The major risks and how they are perceived by the people are discussed in Part 6. The role played by collective institutions, moral responsibilities, and non-monetized reciprocities are highlighted. The policy implications for sustainable mountain development are listed in Part 7.

We believe that bureaucratic or market institutions have often failed to conserve natural resources everywhere in the world. The Western concept of resource conservation through complete closure is neither scientifically nor environmentally sustainable. The principles of maintaining social and environmental diversity and complexity through innovative institutions are available in the framework of Buddhist ethics. The rules that can guide the behaviour of individuals and groups have to be developed in the changing technological and political and economic environment. Bhutan's experience can be helpful even for developed countries.

Why do formal development models tend to 'destroy the only cultures that have proved able to thrive in these (isolated and harsh) environments?' (McNeely 1989). The question, raised in the recent report of the World Commission on Environment and Development, has been asked in Bhutan ever since the programmed efforts to reduce isolation started a few decades ago. Blending indigenous technology, which has evolved in the crucible of culture and local environment, with alien concepts, techniques, and tools requires adapting institutions as well. The relationship between natural sources and the people has been forged within moral, cultural, politico-economic, and ecological boundaries. Respect for these boundaries, by different communities and social groups, was ensured historically through a set of formal and informal rules and norms. It is obvious that the same rules will not help regulate human 'needs' and 'wants' in characteristically different economic and institutional environments. The experience of Bhutan offers lessons for late 'starters' and early 'stoppers' in the game of sustainable development. Bhutan seeks to resolve the creative conflicts in policy choices in the light of Buddhist ethics and values.

New institutional arrangements sometimes mirror these values but occasionally fail to do so. The choice of technology and the definition of sustainability are pursued in the manner illustrated in the Zen story—a simple and subtle beginning and an enigmatic ending, leaving enough scope and opportunity for multiple interpretations. The basic belief is that there is no need to rush through the so-called transformative (and modernizing) modes of development. The regenerative, interdependent, and eco-adaptive alternatives must have precedence. But these alternatives do not always have precedence and problems are resolved experimentally, intuitively, and, at times, through debates in the national assembly and at local level.

PART ONE: CULTURE OF CONSERVATION—BUDDHIST ETHICS

There is, in general, a great distaste in Bhutan concerning the killing of animals to gratify one's tastes. Buddhist culture requires its followers to recognize that animals have the capacity to suffer and that one should not differentiate between the human and the non-human species as far as suffering goes. It is a philosophically refined culture, which, in principle, rejects differentiation among species, i.e., putting the status of human beings above that of other animals. This is not tolerated even for propitiatory purposes. After the advent of Buddhism, animal sacrifices were replaced by symbolic rituals that used, if required, animal forms made of flour or other edibles. Any erosion of this principle leads directly to the exploitation of the animals in our interest. Given that taking life for food is considered mostly repugnant, the society has a tendency to hold those who raise beef and pork in low esteem. In modern Bhutan, the meat industry is, by and large, inhibited by this principle of not taking life, although there are signs that such beliefs are yielding to commercial considerations.

Eating meat, vital for survival amidst a harsh climate, poor soil conditions, and a very limited scope for agriculture produced guilt (Ekvall 1989). Giving back to nature in some form what has been taken away was thus institutionalized as a compensatory or purifying ethic. The concept of accumulating virtues (be it by retreat into a monastery for a few years or by contributing land, labour, or wealth for the common good) provided an ethic that counteracted or counterbalanced the compulsion to kill animals for their meat.

There are various other strands in the national culture that allow some degree of culling. The semi-nomadic yak herders (the *Brockpas*) do not frown upon culling a few male yaks during the year. Yak meat and dairy goods are the only basis of exchange for cultivators, and piggeries are accepted as a part of subsistence animal husbandry. There is often also a gulf between the ideals of the religion and the pragmatism of daily life. In what is perhaps a device to have one's cake and eat it too, non-vegetarianism is not a taboo but the slaughtering of animals is. It can be argued that the interaction of the demand and the supply of meat unites the butcher and the consumer into one single moral persona. Their economic and moral existences appear interwoven.

The paddy farmers in Paro Valley used to enter into contracts with the pastoralists of Ha, a hilly district, to exchange paddy for cheese, butter, and meat. The ecological context provided a basis for economic exchanges across space and time. The interdependence, therefore, between not only butcher and consumer but also cultivator and pastoralist, monk and ordinary people, constituted several layers of consciousness or a number of multiple interfaces underlying the conservation ethic. The retreat into monasteries could regenerate the individual consciousness. Leaving lands fallow under swidden cultivation systems permitted the regeneration of the natural system. The system required both social sanctions and faith in the natural and spiritual order.

The principle of respecting life in all forms extends to wildlife. Protecting the natural environment in which wildlife occurs is the reverse side of the same coin. The Bhutanese have seldom been keen on trapping, hunting, and shooting animals and game-birds. Wildlife has never been regarded as game. Very rarely do trophies of prize catches adorn the walls of private or public mansions. These conservationist attitudes were, in an Asokan move, legally formalized and reinforced by restrictions on the hunting and killing of wild animals. One may kill a wild animal if it poses a threat to life and property and if the fact that it is a threat to life and property can be proved. Even this narrow latitude in the rules is not often applied given the prevailing aversion to killing. Farmers have sleepless nights when they make noises to keep off wild boars, deer, and bears and prevent them from destroying their crops, but they would rarely resort to killing them.

However, when farmers do attack wild animals causing repeated damage to crops, they are in a moral dilemma if the animals do not die in the field but run away into the forest. International advisors have suggested that people inform the forest departments if shooting an animal becomes necessary (Laushe 1987). The faith that administrative policing can provide more safety for wildlife than a voluntary conservation ethic is perhaps misplaced. The issue of protecting the habitats of wildlife to the extent of totally excluding human beings has been debated. IUCN shares the belief that '*even the ecosystems which appear most natural*' have been significantly altered by humans at some point in the past. Exclusion of 'human influence from the ''natural'' ecosystem, as in the strictly protected national parks, can lead to a situation that has not occurred for thousand of years' and, it is feared, might 'have unknown ecological implications; the devastating fires that hit Yellowstone National Park in 1988 are a dramatic example of what can happen when nature is allowed to take her own course without human intervention.' (McNeely 1989). The Buddhist ethics of cohabitation and peaceful coexistence might thus need reinterpretation and reassertion in the light of international advice for wildlife protection parks exclusive of any human interface. Retribution for greed and reward for restraint are inherent in a genuine conservation ethic. IUCN studies show that this ethic may even be superior on scientific grounds.

The Bhutanese have been conscious of their green heritage and its distinctiveness, compared to Tibet for example, for a very long time. Epithets such as 'the southern land of medicinal herbs' and 'the land of cypresses' characterized this distinction. Both Buddhist and pre-Buddhist (Bon and Animism) beliefs promoted a cautious attitude to the environment. They suggested that there were spirits inhabiting 'three parts of the world—the sky, atmosphere, and the earth.' (Vigoda 1988). The mountains, rivers, lakes, streams, forests, rocks, and soil were believed to be the domains of different spirits. Pollution and disturbance to these sites were believed to be the cause of deaths and diseases.

It might be surmised that nature conservation was part of the traditional Bhutanese culture. The religious basis of traditional life bred in the people a respect for life in all its forms and for their ecological 'niche'. The constant interaction of the people with the natural environment would have naturally sharpened their intuitive insight into the use of resources from the environment, and this tended to make them practitioners of sustainable resource use.

PART TWO: RULES, PRINCIPLES, AND CULTURE—A FRAMEWORK FOR INSTITUTIONAL EMERGENCE IN MOUNTAIN REGIONS

Jodha (1990), while outlining the mountain perspective, has highlighted two important interlinkages among mountain specificities: (1) commonality of causative factor and (2) shared consequence of disturbance to each other. In the first case, it is suggested that besides the climatic factors, 'degrees of diversity, fragility, marginality, human adaptation, and inaccessibility are, in different measures, directly linked to factors such as elevation, slope angle, slope orientation, and exposure.' The second interlinkage is the 'externality' imposed by disturbance in one sub-system over the rate and extent of regeneration of other sub-systems.

The relationship between objective, ecological conditions and subjective, human perception and response is mediated by the cultural and institutional memory of society. Which consequences of disturbances are considered to be externalities is in fact the outcome of cultural consciousness.The boundary of responsibility towards the shared consequences, we could argue, is one of the important outcomes of historical institutional development. Once the State and its authority supersedes the authority of religion, village, or community institutions, conflicts between the historically desired perception of this boundary and the legally, administratively, or politically legislated limits of these boundaries are bound to arise.

How these conflicts are resolved can be studied only after the relationships interlinking culture, institutions, technology, and ecology are properly appreciated (Gupta 1986). The socio-ecological perspective makes two assumptions: (1) ecological conditions define the range of economic enterprises managed by different classes of households and (2) the scale and mix (or portfolio) are determined by the access to factor and product markets, non-monetized kinship, and other exchange networks; public, private, or communal risk-adjustment strategies; and the perception of and response to various environmental and social risks. The stakes that different classes have in environmental preservation are modified by the surplus, subsistence, or deficit condition of the household budget on one hand and by the institutional context on the other.

Once beliefs about what is considered 'natural' or even 'spiritual' are shared widely, cultural codes are institutionalized. Thus, certain conflicts about choice of technology or scale or mix of resource use do not arise because duality between cause and effect is resolved through oneness with the phenomenon. Thus, if the share of birds in the crops is considered as justified as men's own, the historical selection of varieties with loose-set grains, easy for birds to pick, can be understood.

In other cases, technological and institutional innovations help generate rules and principles (Khalil 1989), cemented through social sanctions, individual repentance, and sometimes accidental (but believed to be mythical or godly) dispensation of justice or retribution. Rules do not have to be devised always. They could arise, at least in certain cases, through the logic of circumstances. Khalil cites the case of traffic rules. If there were no rules, the result might not be chaos. Anyone who has seen a huge congregation around religious centres on festive occasions will have noticed how crowds arrive at an order through a period of adjustment. It is suggested, 'pedestrians or cars interact for no purpose (when they move on a rule-less road); rules arise because they are in each other's way.' (Khalil 1989, 11).

The principles are supposed to specify purpose. The destination of pedestrians is determined by the purpose. The possibility of reaching a given destination may of course depend upon the evolution and observance of rules. The grammar, syntax, vocabulary, and style facilitate communication. It is the idea, will, and need for a relationship between sender and receiver that constitutes communication. Buddhist ethics are 'an ideology with a set of principles dressed up as rules'. The ideology could spell out the limits or boundaries of communal or individual rights irrespective, as Khalil says, of the substantive principle at hand. A scheme of rationalization is considered necessary for achieving this. Scientific theories may serve such a purpose in modern times. It is argued that religious mythology served these needs in pre-modern times. We, however, believe that, even in the current context, proper bridges between so-called scientific theories (generalizable, consistent, and verifiable) and a set of historical beliefs or cultural codes (be they of religious or mythological origin) are necessary.

The restraint on individual wants and the acceptance of social and ecological limits in deference to the claims of the next generation may become difficult to institutionalize through simply using only the so-called rational rules.

The long time frame within which returns should be appraised and the multiplicity of vectors on which utility is assessed become acceptable, perhaps through what Jodha calls the limits or potential of shared consequence or externality.

The public choice theorists often equate principles with rules and emphasize a narrow definition of justice, i.e., procedural fairness. Their contention is that 'as long as the rules of the game are fair, one should not manipulate the outcomes. Distribution of income, for instance is one of those outcomes which the State should not tamper with' (Khalil 1989). The substantive principles such as primacy of equality over an unequal system or of collective rights over the need for an individual autonomy can not be resolved mathematically. Under what circumstances the State should circumscribe individual rights and autonomy would thus be resolved politically and not 'scientifically'.

The legitimacy of such restrictions on individual needs or wants, our contention is, may spring from respect for ecological principles and the cultural basis of generating a socio-ecological ethic. It justifies the primacy of these principles over certain other rules dealing with individual preference in life-styles.

Thus, an affluent person cannot justify the violation of certain rules regarding the clearance of areas near roads to establish apple orchards or the killing of animals or the killing of trees as long as he or she can pay the necessary taxes or penalties. Such a procedurally just system, we submit, may violate the substantative principles that sustainable development, in a high-risk mountain area, requires that priority be given to (1) the rights of the community over the individual, (2) the non-consumptive over the consumptive use, (3) a consistent consumption below maximum sustainable yield (also called internal bioethics) over a consumption up to the maximum, and (4) a slow, steady growth rate with a homeostatic advantage over a high, volatile, and unstable growth rate in resource use.

Mountain societies have evolved unique ways of maintaining social amity amidst conflicting resource demands and experiences. In the transhumant community of Meraak, a festival is celebrated every year in honour of their patron deity *Ama Jomo*. It is a peculiar feature of this occasion that normal inhibitions are temporarily disregarded so that bawdy jokes can be shared, even between a father and his daugher. Catharsis at the community level is a traditional institution for settling *emotive* accounts in the short run so that *moral accounts* can be settled over the long term.

Cultural norms can help to counteract some of the 'rational' (in the short term), but non-sustainable, resource-use strategies. A pastoral group can evolve the norm of spending the most time on patches with the highest rate of return; or it can evolve norms of mobility even if resource supplies do not warrant it. Norms can be guided by the need to avoid intermixing yak herds with cattle herds (as we shall see later) in order to prevent disease transfer. In some cases, hunting tribes have used a randomization rule to overcome the tendency to hunt where the maximum game is likely to be found (Stocks 1987). For instance, there may be a water source to which animals come at a particular time in the day. 'Rational' strategy might imply hunting the animals when they come there. Some tribes use different ways of deciding in which direction to hunt by circulating a stone tied to a sling of rope and then throwing it. The direction for that day's expedition would be the direction in which the stone went. Such norms dictate that hunting groups consider the periods of scarcity and the periods of abundance with equanimity. It is also expected that norms of sharing will emerge to take care of the hardship periods.

Centralized exchange networks sometimes compensate for geographic diversity. In a study of Torbel, a small mountain village in Switzerland, the adaptations of the community to the diversity of resources and the uncertainties of the environment have been explained through expansive, intensive processes of regulation (Sahlins 1988). Building irrigation channels though inter-village coordination was the expansionary strategy. The fertilization of meadows, the repair or modification of irrigation channels, and the recovery of washed away soils were described as the intensification processes. Regulation referred to preventing outsiders from settling in the village, limiting the numbers of grazing stock, limiting wood cutting, and establishing democratic methods of centralizing power.

Mixing fluctuating scarcities with geographic diversities and resource concentration can lead to complex social structures. This is seen among three groups of people in the Zagros Mountains of Pakistan: Pathan farmers, Kohistani farmers, and Gujar herders. The farmers using rich soil and water resources exchanged goods with the herders who exploited dispersed grasslands. Two of the groups shared a resource. The Gujar herders used Kohistani pastures in the winter when the latter fed their cattle from other resources (Barth 1956). Mutual dependence among cultivators and herders, in some cases, can be mediated by State control as in the Middle East (Bates 1974). A segmented, polycentric, integrated network on the other hand concedes the autonomy of different sub-groups, the diversity of their goals, and the multiplicity of leadership or potential for leadership. Gerlach and Palmer (1981:350) suggest several possible ways in which strategies of environmental adaptation can be classified. e.g., diversification, mutual sharing, self-regulation, selective specialisation etc.

The Bhutanese experience not only illustrates the use of several of these strategies but also expands the framework of adaptation through interdependence. As with Japanese villagers, the Bhutanese may not rely entirely on socialization as a means of ensuring behaviour that avoids the tragedy of the commons (McKean 1984). They do not rely only on material incentives or disincentives.

The maintenance of law and order, in the past, was achieved through reflection on a total of 26 deeds—10 good, fruitful deeds and 16 livelihood deeds. Every family considered it a privilege to have a *lama* in one's family. Bonds between the religious (the other-worldly life) and day-to-day life were forged through unresolved contradictions between ideals and the programmatic imperatives of survival. The doctrine of omission or passive indifference thus permitted acquiescence to the killing of animals by others and the eating of meat. However, other features of Buddhist logic required compensatory measures to appease the nature gods so that one set of compromises were off-set (now or in the next life) by another set of good deeds. The calculus of survival imposed the need for conservation as intensely as it provided a practical way out of the contradictions.

The conservation ethic has been often defined inadequately, leading to the inappropriate design of reinforcement mechanisms. The re-emergence of the conservation ethic after various religious symbols, signs, and restraints fail to check resource depletion. It is observed that the game shortage was dealt with not by controlling invaders, nor by 'developing' the conservation ethic, but by moving out. The error here seems to be that 'moving out' is not considered to be a part of the conservation ethic.

The cyclical nature of natural forces and their interaction is a fundamental concern of Eastern philosophy. The search for existence, absence, dormancy, or activation of different rules will suffer if this dimension of resource management institutions is missed. Not all rules can be invoked at all times for all types of resources.

There are 'episodic' institutions that differ from 'concurrent' institutions. The episodic institution may surface only under certain critical circumstances. Certain principles may be transformed into rules, if the purpose has been served. In any case, these rules may also mutate as the changes in the supply of resources or availability of technology take place. The transmutation of rules, and of associated social organizations, in Bhutan, has occurred in different ways in various resource situations. The spatial, seasonal, sectoral, and social variabilities in different resource regimes illustrate how the pressures of contemporary social values fuse with or contradict the traditional norms and institutions.

PART THREE: INSTITUTIONAL CONTEXT FOR RESOURCE MANAGEMENT—WATER, LAND, LABOUR, LIVESTOCK, AND FORESTS

Notwithstanding the efforts of some scientists to divide mountain regions into different agro-ecological zones, we feel that ecological variability within a valley or slope is so high that any broad centralization on this basis will be meaningless. Micro-ecological 'niche' have to be considered an inalienable feature of any mountain region. The correspondence between these 'niche' characteristics and the norms of institutional governance of resource use is complex. It would be difficult to predict or suggest, with the knowledge available to those institutions, how rules would vary among different 'niche'.

However, available evidence does indicate that study of this relationship could be quite rewarding in terms of designing experiments for sustainable mountain development.

Water Management

Availability of water seems to have been one of the main bases for settlement distribution in the past. There are instances where once the stream dried or diverted, due to landslides or other factors, the people abandoned the settlements. The abandoned ruins can still be seen. The need to manage water distribution, whether in a community of dispersed households or in a densely clustered settlement, can be considered to be one of the prime movers behind societal evolution.

The emergence of leaders to organize social groups for the cooperative use of a resource was one consequence of water-based technological change. Synchronization of farming needs, due to a given terrain and climate imperatives, was another.

Depending upon the scarcity or sufficiency of water, cooperation in following certain rules for harnessing and maintaining water management structures, such as channels, became critical. There are instances where villagers have even brought water from streams located 25–30 km away from a particular settlement. The rules for maintaining water channels and distributing water may vary sometimes within valleys and among valleys a short distance away, due to micro-environmental variation in the nature of the resource, the availability of labour, or just the difference in power structure of respective social groups. Rules have emerged for the following operations:

- desilting canals before farming season,
- repair of minor breaches or major dislocations due to landslides or disturbances by animals,
- distribution of water over space, seasons, crops, and for varying operations,
- waste disposal in streams or rivers, and
- transportation of wooden logs in rivers or streams.

Institutional arrangements for different functions have varied in different places. Each household may supply equal labour (one or two persons per household) for annual maintenance functions, or the contribution of labour may be in proportion to the land holdings within the command area and the distance of the fields from the canal outlet.

The bases for water-sharing, identified after a study of 21 farmer-managed irrigation systems in Nepal (Pradhan 1989), were as described below:

- Allocation on the basis of original investment or shares purchased; water is sold or purchased independently of land ownership. The transaction has to be registered with the irrigation committee. The obligation to contribute resources towards the maintenance of a channel is determined on the basis of the size of the share.
- The number of hours of water received is proportional to the investment made.
- Allocation of water on the basis of the size of the command area; the size of the notch in the proportioning weir servicing each field or set of them is fixed according to the size of command area.
- Water-sharing on the basis of labour contributed for maintenance by a village or household.
- Distribution on the basis of type of land.
- Distribution on demand in systems where wat r is not scarce.

In Bhutan, several more innovative, institutional arrangements have been noticed. In a village with very limited water, the need for providing standing water in different paddy fields, almost at the same time, was met by an imaginative use of the rule of randomization. Conflicts were anticipated due to synchronicity of demand. The farmers collected as many sticks as the number of claimants. After writing or etching a serial number on each stick, they were shuffled and each person was asked to pick one. The lottery system determined the sequence of users.

Each received water for a fixed duration, in some cases, one day, and in other cases less or more. The cycle is repeated in the same sequence. There are also situations where the sequences are written into contracts so that people know with certainty when their turn is due. Pooling or exchange of labour, bullocks, and implements is also facilitated through known sequence.

In another village, the elders met and decided the sequence, duration, and labour supply for maintenance and protection. One person was elected to monitor the distribution. The size of the hole in the weir varied depending upon the allocation. New settlers were generally given the last turn.

Conflicts did arise once in a while. The defaulters could be asked to (1) give a gift to the persons who suffered and (2) miss their turn in favour of the aggrieved parties. If the offense was repeated, the village people would destroy the contour bunds in the paddy field of the offender.

In another case, supervisory duty was assigned to different people all along the canal length. Each person had to complete his turn and place a flag signifying the completion of his task. The next person began where the previous one left off.

Conflicts also arose over the terracing of new fields in the command areas of old channels. The punishments for laxity in monitoring could be as severe as for the offense of stealing water. In general, people reported that the collective institution of water-sharing and canal maintenance was respected. Whenever a conflict went to court or to the higher level officials, an attempt was first made to persuade people to arrive at a solution among themselves, even though they may have already tried once or twice to do so. Arbitration was avoided until it was absolutely necessary. There were of course areas where these traditional institutions were coming under strain.

There were instances where a local person was attached as an apprentice to the fitter while an irrigation system was being laid out. The idea was that this person would be able to show the people how to repair the structures whenever the need arose.

Pasture Management

The role of livestock in hill farming systems has been recognized as very crucial for sustainable resource management (Hardwood 1989, Richards 1985). While cultivation dominates at altitudes between 7,000 and 10,000 feet, livestock is the major means of livelihood in settlements above 10,000 feet.

Several factors influence the evolution of property rights and have a bearing on choice of technology and environmental stability. Because of the widespread prevalence of livestock in rural Bhutan, a large number of rules and institutions have evolved around animal management. Trespassing on cultivated lands by animals and the vulnerability of crops to livestock are negative sides in an otherwise complementary interaction between crops and livestock. Animals have to be confined, or watched closely, if the farmers do not fence their land. Fencing land is capital intensive and full-time animal tending is labour intensive. Neither of these options is economically attractive for the small, rural farmers in Bhutan.

The problems created by stray animals on cultivated fields are very serious and a social institution has developed to deal with them. Two or three households are elected every year as 'crop police', and it is their function to monitor the movements of livestock straying into crops. The crop police round up the animals caught in a field and maintain a register of estimated damages to be compensated for by the owner of the animal. A part of the fines levied on the owner of the animal accrues to the 'crop police' as service charges and the remainder is paid to the injured party. The revenue from service charges increases according to the number of 'catches' and the amount of fine imposed. Households that round up the stray animals get a share of the fine money. Revenue for the service increases as the frequency of catches increases. This feature gives an incentive to the crop police to maximize their catches. The accounts are often settled at the end of the year. At higher altitudes there are hardly any problems caused by animals and thus there is not much need for protection. At lower altitudes, however, there are serious problems.

The gathering of leaves from a common grove for composting of animal manure or for bedding for animals is a regular activity. Here again, individuals cannot strip the forage unilaterally. Forage is collected from a common grove only on a publicly declared day so that all community members get fair shares.

Institutions evolve when the conflicts in resource use are inherent in the nature of the demand and supply situation, and there are considerable externalities involved. In general, we have found some evidence for the evolution of such institutions more often than not in higher-risk environments. The case of *auran* lands, left for gods and goddesses in Rajasthan, is a good example. While *aurans* did exist in other places, they were still being maintained in more arid parts of the region when they had begun to disappear elsewhere.

The boundaries of the grazing territories of different villages and individuals within them were demarcated by cliffs, rivers, or streams. A village could lie within the domain of the grazing rights of another (even non-resident) individual. However, the villagers could object if the grazing rights were leased out to outsiders without consulting the villagers.

The herders moving through the territories of different villages lying on a migration route needed permission for passage as well as for stays of varying durations. If the fields were sown with crops, the herders had to send advance intimation to the villagers so that the latter could organize protection. Damages had to be paid for any loss of crops.

Yak herds always graze in alpine regions (strictly speaking the term 'yak' is used for the male of the species). Cattle migrate over vast distances to sub-tropical regions in the winter and back again to the alpine region around the fourth Bhutanese month. As the cattle arrive from the sub-tropical regions, yak herds must vacate the pastures by going still higher up. The yak herds must leave about a month before the cattle arrive, to avoid contact between yaks and cattle. This serves two purposes: (1) the grasses can regenerate and (2) the possibility of certain diseases being transferred from the cattle to the yaks is minimized. The movement of yaks and cattle is based on tacit information and annual routine. Cattle herders need not inform yak herders of their arrival.

Breeding was a common activity for which institutions had sprung up. In some places, e.g., the *Sambhaka* and *Yabobhog gewogs* (blocks), people are not allowed to bring in bulls of any breed except Siri. A pure strain of the Siri has been maintained here for a long time.

Disease control is another area in which village institutions have evolved. A person can be fined for bringing a diseased animal near the village before it is sufficently clear of infection. Foot and mouth disease has been a very serious source of mortality in cattle. Arrangements have been worked out, e.g., in Komatanga Village of Wangdi District, to prevent its diffusion. If the village cattle were infected with this disease, two outposts were set up outside the village at locations that outsiders had to pass to enter the village. People from the village took turns to man these posts. No animals from the outside were allowed to enter the village lest they also become infected. This was a case of an institution emerging not to optimize returns to the individual or the village but to generate positive externality. Obviously, if everybody used such methods the diffusion of disease would decrease.

Other traditional methods of disease control are also used. In some villages, rancid pork (two to five years old) was boiled and fed to cattle and also used as an ointment. Perhaps this acted as a vaccination measure. A few of the large herds of cattle kept male goats which were supposed to prevent a particular kind of epidemic. The smell of the goat was assumed to keep away the vectors of this disease.

Trade and potential and intensity of resource use along migratory routes during different seasons have not been adequately studied. The scope for using market instruments for encouraging sustainable resource use can be identified only if barter markets and the money exchange system are studied.

There are private grazing rights in the names of individuals. One household in Central Bhutan could have such rights in different places in the east and south of Bhutan. These rights do not imply any right over trees, rocks, or soil. People living in villages where northerners have pasture rights in effect can request the owners for permission to use these resources in exchange for gifts or a small fee. There are several reciprocal arrangements arrived at by visiting pastoralists and settled land and livestock owners. The herders offer to use the grazing land and in return look after the livestock of pasture owners. Conflicts around grazing rights have frequently arisen in different places. Some of the most celebrated court cases deal with conflicts about pasture and grazing rights.

Fencing by the Government Forest Department was also not accepted easily by the people. In some cases, the level of degradation was intense and animals were considered to be the major culprits. The farmers living in the valleys or on the lower slopes felt that fencing would obstruct water sources and the repair of stream channels. Social fencing, which worked for so long, was becoming weaker in several places due to the pressures of the commercial economy and imbalances in public policies.

Forest Institutions

There are several positive and negative interactions among forests, crops, and livestock. For instance, rights and responsibilities towards the forest among cultivators and pastoralists include cleaning the *Sokshing* (community forest); collecting firewood on a particular day *(ashi)*; prevention and control of forest fires; the sustainable management of *tsheri* lands (shifting cultivation practices); prevention of the intrusion of cattle herds into reserve forests; the protection of wildlife; the lopping of trees (with prior permission) for domestic use; collection of shingles (flat thin pieces of wood cut for roofing) every few years to replace and repair the roof (the old shingles are piled in the field to decompose as manure or burned to prepare fields for the cultivation of buckwheat); and the collection of leaves and pine needles to prepare bedding for animals.

It was noted in the National Assembly a few years ago that large-scale outbreaks of fire had become frequent. The forest guards were found inadequate. It was suggested that a person from each village be appointed as a fire-watcher to work under the village headman. The fire-watcher would not be obliged to contribute compulsory labour. People found responsible for starting fires were to be imprisoned for three years. If the villagers could not locate the culprits or did not help to extinguish the fire, they could be penalized.

There is no doubt that the fires caused a lot of damage to the forests, no matter whether they were started to clear land for apple orchards or for crops. However, doubts have been expressed about the damage caused since the 1860s and it has been suggested that it may well have been the practice to burn parts of the forests that were suitable for grazing.

H. Cleghorn, in a 'Report upon the Forests of the Punjab and the Western Himalaya', in 1862, observed:

With regard to the conflagrations which are universally described as being so destructive, according to my observation they are almost in every instance, wilfully caused. The practice is very common in all parts of India, where there are extensive tracts of waste or prairie land used for grazing. At the end of the rains the ripe grass dries up, forming an innutritious fodder upon which the cattle soon fall off, and the most ready remedy is to apply fire, and burn the withered straw in order that the young grass shoots, which spring up immediately after, may be accessible for browsing. Firing the grass jungle is universally practised in the prairie 'khadur' lands along the Terai, where bullocks and buffaloes are grazed and wherever binjarahs take their cattle in the cold weather. The same object leads to it in the hill districts. The paharees will bide their time patiently for wind and weather suited to a favourable spread of the conflagration. In very many instances, the dry withered grass is an evil, for which burning is the only cure. Under these circumstances, it appears to me questionable whether any amount of injunction, or penal enactment will be effective against a practice which is ingrained with the wants and the immemorial usages of the people. The best plan would be to have the plantations in situations not liable to the risk of fire and the sites best adapted in other respects for planting would be of that character.

Another British officer, writing on the forests of Kullu, observed, in 1851, that the dry grass was fired to produce new tender shoots of grass and in the process killed myriads of young trees. He suggested that the old sites of certain important forests could be marked and saved from this hazard. He favoured conservation over plantation and added, 'we must trust to the law of nature, which constantly provides for the maintenance of any products by efforts, increased in proportion to the danger of extension....' The suggestion was that villagers should have the charge of the forests and be allowed to collect an *ad valorem* tax on wood cut provided they did not burn the grass and prevented the trees from being cut before maturity (Thornton 1968).

Pastoralists in Bumthang observed that less fire burning of the soil means less productivity. Land was burned to fertilize the soil for growing buckwheat. After that, the fields were converted into pastures, followed by shrubs, and then wheat was cultivated. The practice of growing buckwheat on such soils was actually quite regenerative. Even in cultivated fields, the wood was collected and burned with mounds of sand over twigs in the winter before sowing buckwheat.

It is true that public policy, at present, assumes that forest fires are an unmitigated disaster. Therefore, a series of measures are provided to prevent them. At the same time, historically, fire-setting was a part of resource management strategy. There may be a case for scientific analysis of the conditions under which the burning of forests used for pastures can be treated as a part of management practices and, at the same time, other forests might be protected from this practice.

The relationships (tested by regression equations) between such variables as terrace cultivation, population, pine forest, and degraded forests show that yaks are found to be predominant in the regions dominated by blue pine forest. Cattle are found in regions having a high population density and terrace cultivation (Ura 1988).

The rights of people, in not merely conserving forests but also enhancing the productivity of their farms and herds, go hand in hand with their responsibilities in the matter. A country with 64 per cent of its area under forests, and with a clear policy of maintaining it at that level, has to blend culture, technology, and environment in a socially and administratively acceptable manner.

Labour Contributions to the Creation of Public and Common Goods

One of the most important features of mountain societies throughout the Himalaya is the traditional practice of voluntary contributions of labour for the management of common properties such as water tanks, irrigation channels, pastures, and mule tracks. The Bhutanese Government has made it absolutely clear that the entire development process has to take place with a clear understanding that potential consumers of facilities must pay, at least partially, for them in the form of labour or cash. While cash crops fees are levied in urban settlements, labour taxes are still in vogue in rural areas. Several types of labour taxes were instituted in Bhutan; a few are mentioned here.

A Tax for Local Public Utilities (Shabto Lemi)

Potential users of public utilities such as the domestic water supply, basic health unit, and primary schools are expected to contribute labour for construction. Non-attendance is fined.

Renovation of Administrative Buildings (Dzongsel Woola)

The seats of the administrative and religious authorities are located in the same building in each regional unit. For maintenance of this building, every household is expected to contribute at least one day of manual labour per year for which a compensatory wage rate is also paid.

Renovation of By-ways (Chusel Lamsel)

The community takes care of the irrigation channels, mule tracks, bridges, etc. A few days of labour, depending upon the magnitude of the work, must be contributed. This labour is used under the guidance of the village chief.

Housing Tax

Housing tax is charged at the rate of 15 man-days per year per household for creating or maintaining a public utility but not necessarily a local public utility. Since it is payable in cash at the rate of Nu 25 per day, most urban house-owners do not contribute in labour.

Roughly speaking, each household contributes about 41 days per year, amounting to approximately 43 million *Nugtrum* worth of contributions from the people (Ura 1988). This would amount to an excess of 10%—a significant proportion over and above cash taxes—of domestic revenue in 1986/87. This practice not only ensures the people's participation and reduction in the wage component, but it also ensures minimum reliance on immigrant labour. In addition to this, labour contributions also reinforce cultural values that can only be cemented through collective work for the common good. Most developing countries, by delinking the creation of public or common utilities from their maintenance and regulation by the local communities, have faced tremendous problems in maintaining these investments. On one hand, maintenance requires large bureaucracies, and, on the other hand, people become more and more dependent upon the State for the provision of various services. For a country that does not have much to export or that cannot generate much revenue from internal taxes such an approach inevitably leads to a high budget deficit often accompanied by high foreign debt. This, in turn, not only compromises the autonomy of a society but also limits the extent to which the State can subsidise essential public services. More and more revenue is spent by such countries on maintaining their own public bureaucracies, and this inevitably alienates the people from the State.

Bhutan has very consciously decided to avoid such an eventuality. It has not merely decentralized the majority of functions but has also reduced exemptions from paying this labour tax to different categories of people. Given the labour scarcity in several parts of the country, the choice of a slow, eco-adaptive path of economic development seems the only sustainable alternative.

However, notwithstanding the merits of labour tax, labour tax is a system that is biased against rural people. The bias would be increased over time as labour costs rise and the opportunity cost of labour rises. Indeed, urban residents are not subject to any of the labour taxes, except the housing tax (*Gungda Woola*), and it is a tax haven, in addition to having utilities and amenities constructed at the government's expense (Ura 1988).

PART FOUR: CULTURE OF INNOVATION

Sustainability in high-risk environments requires both technical and institutional innovations. We have looked at only a few of the institutional innovations so far. The technical innovations provide a clue as to how a society generated an ethic that was progressive but not necessarily guided by the pursuit of accumulation.

Water-driven Prayer Wheels

Use of wind and water energy to keep prayer wheels moving led to very innovative mechanical improvements. The energy of streams and small rivers is harnessed through the alignment of horizontal and vertical gears. The chants are block-printed on Bhutanese Daphne bark paper and Hrolled in cylindrical shapes. Reams of chants are encased in a bronze or leather cover and placed on water wheels to be turned by the running streams. It is difficult to suggest whether this led to the development of water-driven flour mills or vice versa.

Architecture and Education

The construction of *Dzongs* (fort-monasteries) seems no less than a miracle given the limited availability of building materials and absence of scaffolding, pulleys, pins and nails, formal survey techniques, and blueprints.

Some of the more innovative school teachers have used these buildings to impart basic skills to children about reading, writing, arithmetic, and resource literacy. The children are taken to these buildings and asked to measure them. In the process of measuring, they are shown the use of different types of woods in different parts of the building. The relationship of this wood to the tree it came from and the soil and slope on which the tree grew is made obvious to the students. It is not surprising, therefore, that such students develop a great pride in their culture, one which provides rules for respecting religion, environment, and common institutions.

Tangible evidence of the enculturation process is the almost cent per cent return rate of the students who go abroad for higher education. Apart from a few, all the students have come back to Bhutan—an achievement which perhaps no other country can claim. Lest the fusion between modern ideas and traditional values not take place, every student who returns is expected to spend about six months in a village assignment to unlearn and realign his or her coordinates of the cultural maps of the country. The new education policy is re-emphasizing the relationship between resources and their place in one's life.

Bridges

Isolated settlements on the far sides of the rivers have always had difficulties of accessibility and transportation. Since long-distance trading for material exchange plays a lesser role in the mountain environment, 'niche'-specific adaptive strategies had to be developed to fulfill all the cultural needs. The degree of interaction is highly compromised by intercommunal isolation. These problems lead to various technological solutions. One of the most remarkable engineering feats, promoted by the mountain condition, was the construction of iron-chain and wooden bridges across the country as early as the 15th century. George Bogle, who visited Bhutan in 1774, noted that 'the bridges were either entirely of wood or entirely of iron. The wooden bridges are very common and are from 30 to 70 feet long'' (Clements 1986). The iron-chain bridges were attributed to Saint Thangthong Gyalpo (1395–1464). Building them became as much a religious task as a technical one. Bridge building is considered a virtuous act because it removes people's obstacles.

Handloom Textiles

Almost all the raw materials for the handloom sector, which was once thriving in Bhutan, were derived locally. This sector is now subjected to great competition from manufactured clothes. Rural women generate substantial income from this activity. The traditional skills of dyeing and designing were based on local wool, tree cotton, and sericulture. Slightly rough but lasting materials were also woven from a species of nettle. For fine clothes endi-silk is used, and a wide variety of natural dyes such as rubia and *zhim* are used.

Alignment of Irrigation Channels

The science underlying the techniques of aligning channels across different terrains, soil types, and slopes remains to be properly understood. While there are channels constructed by the government which have failed after a few years, traditional channels have been maintained, in some cases, for several hundred years. The institutional context of this technology is an important basis for the sustainability of the channels. The same technology without the support of common property institutions might fail as easily as the modern irrigation channels. The variety of materials which are used, ranging from tree trunks to various kinds of lining, also show extremely innovative efforts.

Mobility of Livestock and Feed and Fodder Practices

As mentioned earlier, movement across altitudes and of different species is essentially organized by two different occupational groups: the semi-nomadic pastoralists in Laya, Linghsi, Mera, and Sakteng and the sedentary agriculturalists in other parts of Bhutan. In addition to the institutional arrangements guiding the movements of cattle and yaks, there are several other innovative indigenous practices of veterinary medicine, food and fodder mixtures, and livestock management. Given the fact that providing medical facilities to large numbers of people in the interior mountain regions has been so difficult, the possibility of building up an animal husbandry health infrastructure is quite remote in the near future.

Therefore, reliance on indigenous knowledge systems is necessary. Some studies have been undertaken to find out the scientific basis of traditional fodders. For instance, it was found that willow leaves (*Salix babylonica*), when eaten by sheep, support the energy and protein requirements for body maintenance as well as growth. Willow leaves are recommended as the most suitable fodder for ruminant livestock in and around Thimphu and the temperate regions of Bhutan. The leaves are so rich and palatable that they can sustain the animals without any additional supplement. There is no significant difference between the metabolizable energy concentration of willow leaves and that of lucerne, vetch, fodder maize, and fodder oats. It is unfortunate that development programmes for the mountains often ignore the accumulated wisdom of centuries of experimentation. In this case, out of a large number of possible leaves and grasses, only a few have been selected by people living in different regions. The emphasis on cultivated fodders should, in fact, be less than on tree fodders in fragile regions, because tree ecology is more sustainable than the cultivated crop ecology for higher altitudes as well as steep slopes. In the sub-tropical parts of Bhutan, leaves from fodder trees complement grazing. In sub-alpine and alpine regions, wheat is grown as green fodder for cattle and horses. The animals go through nutritional stress during the long winter in the alpine regions. Buckwheat straws and even some crops are used as supplementary feed.

Cropping Systems

Several innovative uses of weeds and other materials have been attempted for fertilizing the crops, plant protection, or seed storage. The farming systems in Bumthang District in North Central Bhutan are discussed first followed by descriptions of some other innovative practices. In Bumthang, agriculture and animal husbandry are closely intertwined. The grazing of livestock appears secondary to cultivation in the lower settlement valleys, while at higher elevation the reverse is the case. Potato cultivation is expanding rapidly and replacing buckwheat, which, in addition to wheat and barley, is the major crop at higher elevations in Bumthang. Wheat and barley are grown in the flatter parts of the valley and buckwheat on its steep slopes. Potatoes can be grown on any gradient but there is a preference for the roadside land and land closer to the houses. Turnips, radishes, onions, and cabbages are also grown.

Irrigation plays a very limited role in these farming systems. There is a limited use of fertilizer for winter wheat, barley, and buckwheat. The soil needs to have a sufficiency of phosphorus and, of course, nitrogen. Soil fertility is enhanced by collecting mounds of soil over twigs, branches, and dung, and these are then burned to produce ash.

While the normal land holding is about 20 acres per household, only a fraction of this is cultivated in any given year. A plot is cultivated only once between 6 and 10 years and there is a regeneration of weeds on such plots in the intervening period. It is in this context that some people have argued for differential norms for land holdings, in different parts of the country, by taking into account the productivity and the frequency of cultivation.

Diversity of crops is one aspect of risk adjustment, and diversity of livestock species is another. The agriculture is predominantly labour intensive and seeds are replaced every third year or so. Wheat and common buckwheat are alternated in the same field every year. Common buckwheat is a short duration crop and is known to deplete soil nutrients. Buckwheat is rotated with wheat, which requires less nutrients. Compost is applied for buckwheat plantation but the residual effect is obtained by the wheat planted in the same field next year without adding compost.

The crops in the highlands attract a large numbers of pests, including bears and wild boars. Buckwheat is exceptionally vulnerable to bears. To protect everyone's crops, all the farmers collectively cultivate buckwheat in one single area. During the maturation period, it is not uncommon to hear shrieks and howls from the farmers to scare the wild animals away. There are reports of several experiments on farmers' fields, conducted in different locations, that have been destroyed by wild animals.

Winter wheat is sown in November, weeded in April or May, and harvested in September. Bhutan has three species of buckwheat. The common buckwheat, or bitter variety, is sown in April, weeded in July, and harvested in August. Common buckwheat needs a very short maturation period. It is obvious from the overlapping growing period, between wheat and buckwheat, that none of the areas in Bumthang have double cropping. This particular crop combination, besides being consistent with the agro-climatic conditions, allows the labour, draft animals, and management to be spread fairly evenly over time. Planting and weeding of both wheat and buckwheat are staggered so that labour resources are used more efficiently.

Since the early 1970s, potatoes have been cultivated as a cash crop. They are planted in March/April and harvested in August/September. Weeding is done twice, the first time in May and the second time in June. In terms of labour, management, and draft power requirements, potato cultivation competes with buckwheat cultivation and thus the expansion of potato cultivation is taking place by partially displacing buckwheat.

Preparation of Manure

The dung found on the communal grazing land is collected by the communities and divided among themselves. Dung on private pastures is collected exclusively by the owner. The ash of the burnt dung is applied back on to the soil. The cattle or the flock of sheep are penned on the fields to enrich the soil by their droppings. There are customary exchanges among visiting herds and the owners of cultivated land in some areas.

Animal sheds are built on the ground floor of two- to three-storeyed buildings for keeping horses, swine, cattle, and their livestock. The straw leaves collected from the forest or pasture lands are spread on the floor of the cattle shed and are mixed with cattle urine and dung. Farmers have recognized the difference in the quality of different manures. Oak leaves are considered to make the most potent compost, followed by pine needles. Waste from horses is the preferred manure. While the waste from swine, is considered quite fertile, it creates a problem. The swine often excrete undigested seeds and thus their manure encourages the growth of weeds when applied to the field. A combination of horse waste and oak leaves forms a superior compost.

The leaves of wormweed (Artemesia) are chopped and used for mulching and composting with chillie plantation. It is also assumed that they have some plant protection properties. The boiled extract of green leaves of this weed is given to the animals to control flatulence.

Shifting Cultivation

A study on this practice carried out in Pemagatshel District in Eastern Bhutan has revealed the socio-ecological compulsions that lead to the dominance of shifting cultivation. The district is the most backward district.

The majority of farmers in this district consume 90 per cent of their grain produce. About 30 to 50 per cent of the foodgrain is converted into liquor for domestic consumption. Farmers have permanent houses and shifting cultivation is practised either individually or by groups of households. The land may also be owned or leased. It was estimated that about 32 per cent of the total cultivated land was under some form of shifting cultivation in Bhutan, although in the case of Pemagatshel District it goes up to 79 per cent of the cultivated land. The importance of shifting cultivation can be assessed by looking at the following figures: only one per cent of the cultivated land is under paddy (irrigated) cultivation and 20 per cent of cultivated land is under permanent dryland (rain-fed) cultivation, maize and buckwheat being the two principal crops grown. Farmers also grow wheat, barley, pulses, mustard, and potatoes as winter crops.

The burning of slashed vegetation, after the land has lain fallow for three to eight years, is organized by one or several fire specialists (*mesungpal*). Farmers recognize the correlation between soil fertility and the duration of the fallow period. The land is not ploughed and seeds are either broadcast or dibbled into the soil. The output of this system is considered to be higher than that of dryland cultivation and labour and other input requirements are lower than for any other system. There are reports that the productivity of these lands is declining because of factors such as heavy grazing of fallow land, deterioration in the composition of vegetation on fallow land, shortening of the fallow period, erosion, and the depredations of wild animals. Policy makers, however, realize that there is no sustainable alternative to shifting cultivation in the near future. The government appreciates that reverting these lands to forestry without generating alternative sources of livelihood would impoverish the farmers a great deal.

Local Technical Innovations in Agriculture

While no systematic inventory exists of the innovative practices evolved by farmers—men and women—in Bhutan, a very limited listing of practices is presented here to suggest the potential of this knowledge base.

- Before potato tubers are sown they are divided into two sections, strong and weak. The section with more buds is considered strong and the section with less buds is considered weak. The seeds are planted according to the soil fertility in different parts of a field. The possible difference in productivity is also assessed by the difference in the shape. The round tubers are planted on the hillsides and the oval-shaped ones elsewhere. While planting, some people follow the practice of planting four sections of potato seed in a circle with another section in the centre. After sowing, the field is covered with manure and mulch which are burned to control pests and increase soil fertility.
- A rope is taken through the paddy field either to dislodge the dew on the leaves into the soil or to dislodge the eggs of certain insects which may die after falling on to the soil.
- Chillie seedlings are covered with paddy straw and the mulch so formed is burned. The ash is assumed to help in controlling late blight disease.

Local knowledge systems evolve not only about local crops but also about exotic varieties or crops. In some cases, farmers have discovered that hybrid maize plants can withstand strong winds better than the local varieties. Several changes are being brought about in the associated components of the farming system in parts of Tashigang District where strong winds are a problem.

Technological change influences culture just as much as cultural factors influence technology. Further, the teething troubles that are being faced by farmers in the process of technological change indicate that a policy on this topic should be sensitive to the ecological and cultural limits. Several studies pursued by Western scholars in Bhutan have confirmed that most of the ill-designed, rushed efforts for technological change have not been very encouraging. There are studies that have shown that considerable yield potential exists within the local varieties, provided management conditions can be marginally improved.

PART FIVE: TECHNOLOGICAL TRANSITION: EMERGING ISSUES

Bhutan's capabilities for carrying out research, whether in research stations or on farmers' fields, are growing slowly. Part of the problem lies with the way in which technology development and dissemination has been propagated in the recent past. It was hoped that one or two centres, located in the fertile valleys, would be able to generate technology that could be transferred through the training and visit system. Recently, an Australian company, providing assistance for the Tashigang/Mongar Area Development Project, emphatically advocated a case for ecologically balanced, low external chemical input agriculture. It was observed, 'there is no great pest/disease problem in East Bhutan, and your eco-system still seems to be well balanced. Don't fall into the trap of chemical overuse, the only beneficiaries from this are the chemical companies and your balanced eco-system is likely to become unbalanced very quickly' (Thomas 1989).

Agronomic research carried out on several crops is demonstrating that experiments designed without analysing the basis of diversity in the existing cropping systems may not make the best use of time and critical resources—human or material. It was found that seed rates or methods did not make much difference to the yield of mustard. Likewise, line sowing over broadcasting made hardly a difference of 100 kg/ha, which, given the cost and unavailability of labour, was not remunerative.

Most of the new maize species, introduced by the International Wheat and Maize Improvement Centre, have an unacceptably high number of exposed ears and a higher incidence of rot under wet conditions besides the problem of lodging. Only a few species merited further testing. Irrigation research for maize crops was not considered necessary. In the absence of fertilizer, most improved varieties did not outyield the local ones. The technological package tried in Western Bhutan could not be replicated in the eastern or southern parts. The mustard fields not harvested on time face the problem of shattering with the result that the seeds germinate in the next season.

In addition to the technical aspects, the gender aspects deserve serious attention. In most mountain regions, male emigration is almost a rule. Therefore, technologies that build upon the strength of women and help to overcome their limitations need to be carefully screened. Women play a major role in fuel and fodder selection, water collection, handloom activities, and agricultural and livestock management.

Power tillers (about 80 to date) were introduced because labour was thought to be scarce. However, a survey of its use reveals that most of them are used for transportation, being unsuitable for other purposes on terraced fields. It should be possible to involve the companies concerned, supported by international aid agencies, in carrying out research on proper designs, instead of taking the easy option of transferring existing designs. It is natural that given the low-level purchasing power in most mountain regions, the private corporate sector does not have much interest in designing equipment or technologies specifically suited for these regions. This is an issue in which International Centres for Agricultural Research and other donor agencies should become involved. Bhutan's experience shows that, despite the lack of adequate training and skills, appreciable technological changes can, indeed, take place through imaginative policies. For instance, the Royal Government of Bhutan decided that, instead of importing computers and remaining dependent upon the repair services of external consultation, they should try to assemble computers within the country. Today almost all offices in Thimpu have computers asembled in that country at a very low cost and of very high quality. Similar breakthroughs can be achieved in other sectors as well.

Medicinal Herbs: Preserving a Unique Wealth of Knowledge

There has been some concern expressed in the National Assembly on the issue of regulated access to medicinal herbs from the forests and pastures. The mountain regions abound in these herbs. There is a need for an international thrust towards preserving this knowledge reserve. In the absence of a careful strategy aimed at managing the selective collection of herbs, some herbs might disappear because of the ecological principles of succession and dominance. For instance, if elephants are killed in a particular forest, then the grasses and herbs which grow in the micro-environment produced by the trampling of grasses by elephants will not be available. Likewise, in certain forests, livestock browse and when that stops suddenly it also affects the dominance patterns of different species. The result would be excessive growth of some species and total suppression of others. This is a subject that requires very careful monitoring. ICIMOD should collaborate with the IUCN and national governments to form a programme to monitor the changes in ecological succession and dominance patterns due to closure or other major changes in socio-ecological environments.

PART SIX: ENVIRONMENTAL RISKS AND SOCIAL RESPONSE

Farmers, pastoralists, and artisans face several risks which vary over space, season, sector, and social groups. The southern area of Bhutan receives heavier rains and some regions face the problem of floods. In the northern regions, landslides, snowfall, hailstorms, and strong winds cause various types of damage. Damage by wildlife is more serious in the central and lower parts of the country. A repertoire of responses to different types of risk reflects the strengths and weaknesses of indigenous institutions as well as of the modern market and public institutions.

In an informal panel enquiry, we identified nine risks which people had to face; and they are heavy or too little rain, damage by wildlife, hailstorms, snowfall, landslides leading to road blockage, choking of irrigation channels, or dispersal of stones on grazing pastures, strong winds, disease or epidemics, fire, and the drying up of streams. It may be useful to ascertain the composition and order of risks perceived by different social groups and stockholders, such as *lamas*, administrators, and members of the National Assembly in different parts of the country.

Traditional Risk Adjustments: Animal Disease

Yaks are found at approximately 2,500 m above sea level. This would be the lowest height to which yaks descend, even during the winter. They move to pastures at much higher elevations during the summer, giving way to cattle who are driven up after winter grazing in the sub-tropical parts of Bhutan. The migratory cattle naturally mingle with other herds from the South, and, during their grazing, they cross paths with other herds on their route. While doing so they pick up diseases and transmit them to yaks, which are more delicate than cattle. To avoid transmission of any diseases from either side, yaks are moved to the higher pastures a month ahead of the arrival of the cattle. Herders of yaks and cattle do not exchange messages about their movement. They coordinate their behaviour with almost ritualistic precision through ecological signals or other means. The yak-herder expects the cattle-herder to be there on time and the cattle-herder expects the yak-herder to have gone.

Coping with Lean Harvests: Risk Adjustments against Famine and Food Shortages

Within living memory, there has been no famine, on the scale of community or nation, in Bhutan. This may have to do with pure good fortune, with a greater production stability of the farming system, or with the elaborate traditional food security system operated by the State. Probably all of these factors diminished the effects of any incipient famine, but which of these factors played a more dominant role remains to be examined.

Until the 1950s, the State maintained enormous grain silos in each of its massive *dzongs*, i.e., district headquarters. The ultimate use of the kind-taxes, including grain taxes, was to build up reserves to support the State officials and the monasteries, in order to avoid crop failures. The food stock—butter, cereals, meat, mustard oil, etc.—was replaced with new grains by exchanging the old stock with that of the farmers every six years or so. Until the kind-taxes were commuted to cash taxes in the 1950s, farmers suffering from crop failure could get grains either free or on an interest-free loan basis from the State grain silos. There were also remissions of taxes in bad years.

There are localized bad harvests of varying degrees. One hears about insects destroying the crops and hailstorms dislodging them with distressing frequency through the newspaper (*Kuensel*). The households historically maintained a buffer stock from which they could draw during bad periods. This may still be true for households that are more remote. The first course of action is to call on kith and kin for gifts of food on the basis of reciprocity. Grains can also be borrowed from those who have larger food stocks.

Farmers can travel to other places on alms rounds or work for food for villagers in distant places. According to some observers, there always seems to be enough work in exchange for food. It is considered unusual for farmers to take such drastic measures as disposing of their animals, jewels, or other assets to tide them over a period of food insecurity. Strong contingency mechanisms are supposed to prevent them from becoming assetless in the aftermath of a bad harvest. In recent years, the government has begun to build food security stocks equivalent to six weeks of the country's total requirement of essential commodities such as wheat and rice. The actual efficacy of these measures remains to be properly evaluated.

Gathering wild food is an important means of supplementing food shortages in some parts of Bhutan. In some districts of Bhutan, such as Kheng, yields from the forest contribute substantially to the food basket. There are wild yams, wild avocadoes, wild taro, young fronds, young bamboo shoots, orchids, and various kinds of mushrooms, to name just a few edible species. Wild yams, which grow to giant sizes—as long as four feet—are collected even as a normal part of the diet. Since they lie a metre or two below the surface, appropriate extractive tools are used for wild yams. Nowadays, forest food gathering is a growing petty business for some farmers who live within easy reach of the urban markets. Some entrepreneurs have begun to export edible forest products. Farmers are encouraged to scour the forest floors for certain kinds of high-value mushroom. The commercialization of food from the forest, on which the community partially depends, has the danger of excessive depletion.

House Construction and Other Non-monetized Reciprocities

A community is bound by norms of reciprocity and sharing. The members of a community have a network of obligations that includes one's extended family, fellow villagers, and others. Indeed, these social networks of obligations and reciprocity are something of a mixed blessing. Redistribution and equality seem to be a natural outcome of such a social system where the better-off kin must assist the worse off. The levelling effect of such a sharing society is summed up in this ironic Bhutanese saying: 'Relatives will prevent me from becoming destitute; relatives will prevent me from becoming rich.'

There are a large number of norm-oriented customs, among which contribution of labour for the construction of private houses is a good example. The size of houses, especially in the alpine settlements of Bhutan, are in excess of what one could afford to build normally without the community's free labour contribution. The reasons for comparatively larger structures are probably the availability of timber, stones, and mortar, but stronger norms of reciprocity and cooperation possibly grew more easily in clustered settlements in the alpine regions. It remains to be seen whether all the labour contributed for such purposes was voluntary.

Each household usually allocates one person for some part of the construction. Strong norms of reciprocity ensure adequate housing for all. Any family in the village can build a house beyond its means because others help with the work. The average housing conditions in the villages are much better than in places without the norm of reciprocity. It is also a social adjustment to labour shortage and a non-monetized labour market. While the wealthier may choose to build bigger houses and thus benefit more from the norms of reciprocity, on the whole it seems to be a strategy for collective optimization.

The norms of cooperation have been relied upon for a variety of civil works ranging from the upkeep of highways, footpaths, and bridges to the construction of monasteries. In the repair of roads or bridges, the burden of maintenance need not be equally shared by all since the route may not be used equally. In such cases, only special client groups will be bound by the norms of cooperation. It is of course easy to appeal to community labour for the construction of a community temple. There is no apparent differential benefit or vested interest for any section of the village. The benefits are diffused, incomputable, and intangible.

Such solidarity in terms of shared labour would be less spontaneous without a leader to do the organization and without the occasional threat of social sanction.

PART SEVEN: POLICY FRAMEWORK FOR SUSTAINABLE MOUNTAIN DEVELOPMENT

Livestock

The crucial dependence of crop cultivation on livestock because of manure and draft power is well understood. However, livestock are viewed as the most serious threat to the environment today in most parts of the world, particularly in fragile desert and mountain regions. Augmentation in the prices of dairy products should give impetus to the increases in herd size. Improved vaccination and other disease-control programmes also increase the population pressure. At the same time, closure of certain forest areas and pastures (estimated at about 25 per cent) for livestock, burning, and collection of leaves, etc., shifts the pressure on to more marginal grazing lands.

In the short run, the number of livestock is unlikely to go down. Thus, the supply of manure to fields may remain at the existing levels, except on certain migration routes that have been modified due to closure.

The supply of high-yielding exotic breeds such as Jersey and Swiss Brown, together with improvement in the quality of pastures, is making slow progress. The ultimate policy aim of replacing migrant herds of local breeds by sedentary herds of improved breeds does not take the manure problem into account. Nor does it deal with the issue of biodiversity and stability. Whether it is feasible at all in the unique geophysical and cultural setting of Bhutan remains to be seen. We doubt it.

It may be added that cultures that have evolved through movement and exchange of livestock have also provided legitimacy to the efforts of the State in regulating the different resource-use strategies of the people. The ecological basis of this exchange may not be any less vital for the sustainability of society. Politically, pastoralists have suffered in most societies in the process of economic transition. However, the Bhutanese culture and environmentally conscious public administration might resolve this problem provided international aid agencies can invest in the learning and experimentation process. The biodiversity of forests and pastures, influenced by livestock movements, is a global heritage. The cost of maintaining them should not be borne by the poor pastoralists alone. (It should be noted that the majority of herders throughout several regions do not own any pastures.)

Agriculture

Mountain societies generally do not depend upon cereals as the primary source of consumption. Some of the most popular dishes of Bhutan require cheese, potatoes, chillies, maize, *kharang* and buckwheat—ingredients that could continue to be in sufficient supply if the policy shifts in favour of cereals especially rice, are adjusted. Bhutan's cereal food basket has 50 per cent rice, 13 per cent wheat, 24 per cent maize, and 13 per cent other cereals. Rice was never a major staple food. The import of cheap rice has led to this shift.

Self-sufficiency in cereals, particularly rice, as the preferred food, is the main focus of agricultural policy. How the self-sufficiency programme is pursued in practice (and this has not been decided yet) will have ramifications for the land-use patterns in Bhutan. The policy package includes fiscal and institutional measures to change the consumption patterns. Input supplies as well as extension programmes and training of the farmers are being given increased emphasis in every plan period to increase productivity. There is also some possibility that new lands will be brought under cultivation. On the other hand, some agricultural land was lost to urban settlements and industries, before the National Assembly passed a resolution, in 1988, prohibiting construction on wet lands. There is considerable pressure on the forests because of the expansion of orchards. The ecological implications of expanding apple orchards on to steep slopes and lands close to the roads need analysing.

Agricultural research is yet to be reoriented towards the strength of the comparative advantages of the organic, agricultural technological base: less diseases and pest problems, higher yield potential of local varieties through better management, and a very strong institutional base for managing common properties such as irrigation channels, pastures, and forests. Some efforts have been made to identify the scientific basis of eco-adaptive fodder, mulching practices, crop mixtures, etc. Much more remains to be done.

The genetic wealth of the country is unique in several respects. We do not know where else wheat varieties that are suitable as fodder are available. The Consultative Group of International Agricultural Research Institutions has not contributed adequately towards building up dispersed, decentralized germplasm banks. The aid for this must be available as an investment in the global heritage and in the security of genes required in the future. Participatory research approaches do not necessarily require advanced degrees.

Forests

Core policies regarding the forests are embodied in the Forestry Act of 1969 and many other legislations such as the Land Act, the Pasture Grazing Policy, and other National Assembly Resolutions. The National Forestry Policy, 1974, rules that the country should have a minimum forest cover of 60 per cent. Remote-sensing surveys actually show that the country has 64 per cent of forest users. Forest cover and croplands are estimated to cover 8 per cent of the country. About 20 per cent of the country is designated as national parks and reserves, although there are no effective management strategies because of the ever-present constraints of resources and manpower. Bhutan has launched a wildlife and forest conservation programme from a much stronger basis.

The policy of the government is to afforest the degraded areas and harvest the forests close to the road heads on the basis of sustainable yield. A few companies that extract timber for raw materials are required to carry out planting at the same rate as felling. In view of the tendency of private companies to overexploit the forest resources, logging is monopolized by a public sector corporation (Bhutan Logging Corporation).

Debates in the National Assembly between conservationists and revenue maximizers have been quite intense. When foresters demand a further decrease in revenue extraction, the revenue maximizers consider it to be an excuse to mask their alleged inefficiency. On the other hand, the fact remains that timber extraction rates have been considerably slashed.

The conflicts between Western advice for conservation without people and the Bhutanese cultural ethos that holds conservation to be a way of life are becoming stronger. Bureaucratic barriers have seldom proved helpful in insulating natural resources from human greed. Indian and Nepalese experiences are quite instructive in this respect. With the best of technical skills, trained manpower, and coercive authority available with the State, degradation has continued to increase over time in these countries. If a culture can generate indigenous technologies and institutions capable of maintaining forests, water streams, biodiversity, and ecological balance for so long, why should it be found wanting now?

Some people consider the ban on the culling of livestock a rare case of an unsustainable feature of an otherwise sustainable Buddhist ethic. They miss the point altogether. Disease and death balanced, in nature's own way, the population sizes for a long time. However, improved health changed the growth rate. Yet, if consumption patterns and social institutions are not renewed, the imbalance becomes inevitable. The 'principles' and 'rules' must be distinguished. Non-culling is a principle. The sanctity of non-violence remains. The rule is concerned with limiting culling to the degree necessary for survival. There is a need for debate and discussion in monasteries and *dzong*, schools, and pastoral groups about the new rules required to uphold cherished and eternal principles.

CONCLUSIONS

The whole range of institutional norms and social values has regulated the behaviour of people regarding the use of renewable resources without compromising their long-term availability. To some extent, these norms and institutions evolved in recognition of (1) individual behaviour being subordinated to the community's welfare and (2) human beings having rights at par with other components of nature. The homeostasis inherent in the norms of one group was contingent upon respect for the boundaries of norms and institutions of another group. Thus, a very complex network of inter-institutional interactions guided conservation efforts.

A consumerist culture and commercial interests are indeed making sharp inroads into the traditional culture of conservation. Some of the ethnic groups, which compare themselves with other cultures in the plains, perhaps do not appreciate the need for restraint on wants. Questions have been asked about the ability of traditional institutions to withstand the pressure of increasing imbalance in certain parts of the environment. To us, the answer seems to be obvious. We believe that, if policies require careful study of the variability in institutional arrangements and ethical norms across different ecological regions, it should be possible to revitalize the technological and institutional basis of society.

The global concern for sustainable development and the conservation of biodiversity is dominated by the strategies and styles suitable for degraded environments. Since degradation of the environment inevitably is accompanied by the degradation of institutions, these policies presume upon the absence of institutions. Much greater reliance is placed on public interventions which in turn means bureaucratic interventions. In Bhutan's case, the government has already realized the need for limiting the size of the bureaucracy and decentralizing more and more, so that variability in the local socio-ecological systems has been taken into account while designing policies and programmes. However, the problem is one of taking awareness not to the grassroots but from the grassroots. A recent Convention on Environment, organized with the help of the UNDP, identified a long list of measures that were required to maintain environmental variability. However, this was within the context of degradation and consequent cautions. Relatively speaking, the strengths were given much less attention.

How much danger do forests and forest-based agriculture face from livestock? This has to be compared with the danger emanating from private, forest-based industries. While, in the case of other resources, there is some community control available, these industries lie outside the sphere of traditional, regulatory community institutions. It is believed that discouragement to otherwise profitable but extractive enterprises, in areas where they pose a danger to the environment, is necessary if efforts aimed at altering the micro-economic behaviour of small subsistence and deficit-budget farmers and pastoralists have to carry conviction.

The continued functioning and strength of institutions that protect the environment depend on how successfully the future citizens of the country are introduced to the heritage which generates respect for these institutions. The viability of these institutions depends upon the inculcation of these values into the children, especially in schools and urban areas. Environmental studies are now a part of the primary school syllabus in the so-called New Approach to Primary School. Among its many objectives are the understanding of 'the importance of forest and wildlife and appreciating the need for taking proper care of them' and the identification of 'various plants grown in their locality and their classification into food, cash and fodder crops'.

'Education' of the next generation has of course to be accompanied by 'unlearning' by the present generation. Certain paths of development are not sustainable. Administrations and policy makers in several countries are recognizing it belatedly. The need for correspondence between spatial, sectoral, seasonal, and social institutional vectors is slowly being realized.

Culture provides a 'grammar' while technology provides new 'words'. The meaning of a life that is ecologically sustainable and economically just can be discovered only through the blending of both.

REFERENCES

Bates, D.G. 'The Role of the State in Peasant-Nomad Mutualism.' In Yehudi A. Cohen (ed.), *Man in Adaptation: The Cultural Present* (2nd ed.). Chicago: Aldine, pp. 285–296, 1974.

Barth, F. 'Ecological Relationships of Ethnic Groups in Swat, North Pakistan.' *American Anthropologist*, 58:1079–1089, 1956.

Christover, F. 'Sustainable Development in the Kingdom of Bhutan.' Discussion Paper for the National Workshop on Sustainable Development and Environment Thimphu: RGOB, 1990.

Clements, R. (ed.). *Narratives of the Mission of George Boegle to Tibet and of the Journey of Thomas Manning to Lhasa*. New Delhi: Manjusri Publishing House, 1986.

Ekvall, B. *Fields on the Hoof: Nexus of Tibetan Nomadic Pastoralism.* New York: Holt, Rinehart and Winston, 1968. In Vigoda, Religious and Sociocultural Restraints on Environmental Degradation among Tibetan Peoples-Myth or Reality? *The Tibet Journal*, 14(4):17–44, 1989.

Elster, J. Social Norms and Economic Theory. *Journal of Economic Perspective*, 3(4) 99–117, 1989.

Housainou, T. 'Farmers Cope with Risks and Stress in Rural Gambia.' *IDS Bulletin* 20(2) 16–22, 1988.

Harwood, R. *Small Farm Development: Understanding and Improving Farming Systems in the Humid Tropics*. Boulder, Colorado: Westview Press, 1989.

Jodha, N.S. 'Sustainable Agriculture in Fragile Resource Zones: Technological Imperative.' Presented at the International Symposium on Natural Resource Management For Sustainable Agriculture, New Delhi, Feb. 6–10, 1990.

Khalil, E.L. 'Principles, Rules and Ideology.' *Forum for Social Economics*, combined issues 18(2) and 19(1): 41–54, 1989.

Laushe, B.J. Environment and Wildlife Legislation (proposed for Bhutan). Thimpu: WWF and UNDP, 1987.

McKean, M.A. 'Management of Traditional Common Lands (*Iriachi*) In Japan.' Paper prepared for the Fall, 1984, Workshop on Common Property and Environmental Management, Duke University, sponsored by BOSTID, NRC, National Academy of Sciences, USA.

McNeely, J. A. Conserving Cultural Diversity: How the Variety of Human Experience can help Promote Sustainable Forms of Using Natural Resources. Smithsonian, Institution, USA and IUCN, Gland, Geneva 1989.

Ministry of Agriculture. 'Policies and Strategies.' *Accelerated Food Production Programme*, Vol 1. Thimphu: RGB, 1990.

Pradhan, P. *Pattern of Irrigation in Nepal*. Colombo: International Irrigation Management Institute, 1989.

Richards, P. *Indigenous Agricultural Revolution*. London: Hutchinson, 1985.

Sahlins, M. 'Notes on the Original Affluent Society.' In R. Lee and I. Devore (eds.), *Man the Hunter*. Chicago: Aldine, pp. 85–89, 1968.

Sonam Wangmo (n.d.) The Brokpas: A Semi-Nomadic People in Eastern Bhutan. In N.K. Rustomji and Charles Ramble (eds.). *Himalayan Environment and Culture* pp. 140–158.

Stocks, A. 'Resource Management in an Amazon Varzea Lake Ecosystem: The Cocamilla Case.' In Bonnie J. McCay and James M. Acheson (eds.). *The Question of Commons: The Culture and Ecology of Communal Resources*. Tucson: The University of Arizona Press, pp. 108–120, 1987.

Singer, P. 'All Animals Are Equal.' In *Applied Ethics*. Oxford: Oxford University Press, 1979.

Thomas, D. 'Overall Summary of Trial Types, Results, Preliminary Recommendations and Future Programmes.' For Hassal & Associates Pvt. Ltd., Agricultural Consultants, Canberra, Australia: Khangma Research Centre, 1989.

Ura, Karma. 'Economic Policy and Development in Bhutan.' Unpublished M. Phil. Thesis, U.K. University of Edinburgh, 1988.

Vigoda, M. 'Religious and Sociocultural Restraints on Environmental Degradation among Tibetan Peoples—Myth or Reality?' *The Tibet Journal*, 14(4) 17–44, 1989.

24

FARMER-MANAGED IRRIGATION SYSTEMS IN THE MOUNTAINS OF PAKISTAN

E.J. Vander Velde

Dr. Edward J. Vander Velde is associated with the International Irrigation Management Institute (IMMI), Pakistan.

CONTENTS

INTRODUCTION

The small irrigation schemes in the less developed regions, such as Baluchistan, Federally Administered Tribal Areas (FATA), Azad, Jammu and Kashmir (AJ&K), Kohistan, and the Northern Areas (NA) are as important as the integrated development of large irrigation systems in the settled areas of the Indus Basin (The Government of Pakistan [GOP], The Sixth Five-Year Plan).

This paper will focus upon irrigation management: a poorly understood and largely unstudied aspect of agriculture in Pakistan's mountain environments. Given the concentration of national agriculture in the Indus Basin, and the crucial importance of the enormous irrigation infrastructure that has been established to sustain agricultural production, it should not surprise us that most irrigation research in Pakistan has focussed upon the Indus Basin irrigation systems and their problems. After all, Pakistan's mountain environments are found to be peripheral to the 'heart' of the nation; they are not readily accessible and distances are great. Their population is sparse and dispersed as well as contemptuous of, if not hostile to, outside interventions. The cultivated areas are small relative to spatial extent and irrigation systems are decidedly small-scaled, often having more the appearance of meandering watercourses. Altogether, this is the sort of agricultural environment that has attracted comparatively few research professionals and even less research resources—scarcely any of which have focussed upon irrigation systems and their management.

Current signs would seem to indicate that part of the void of past neglect is rather rapidly being filled by ambitious development programmes. National and provincial agencies, supported by both indigenously and internationally mobilized resources, are actively initiating projects designed to establish and/or improve the agricultural infrastructure in the mountain environments of Baluchistan, the North West Frontier Province, the Northern Areas, the Federally Administered Tribal Areas, and Azad Kashmir. Some projects are irrigation development focussed, e.g., the Baluchistan Minor Irrigation Development Project, whereas others are multifaceted, but with a significant irrigation development component, e.g., the Chitral Area Development Project. In virtually all instances, however, detailed, reliable information about the performance of existing irrigation infrastructure, institutions, and irrigation management practices in these mountain environments is absent or seriously deficient.

A potentially undesirable consequence of this condition is that development interventions planned to improve the productivity and sustainability of irrigated agriculture in Pakistan's mountain environments may fail to achieve their objectives in any of several ways. For instance, new irrigation infrastructure may be established using inappropriate design criteria and, thus, fail to deliver water in sufficient amounts at desirable times for planned service areas or cropping patterns. Management requirements for new or rehabilitated systems may not match existing institutional capabilities or practices. Scarce resources may be poorly utilized or wasted as government agencies assume responsibilities for irrigation management, activities previously done by farmers and, perhaps, best done by farmers.

How best to proceed to examine the subject of farmer-managed irrigation systems in Pakistan's mountain environments? First, there will be a brief examination of the general state of our knowledge of mountain irrigation systems in Pakistan. This will be followed by a review of some recently completed research to identify the general characteristics of surface irrigation systems and their management in the mountains of Northern Pakistan as well as changes stimulated in them by recent interventions. Finally, we will conclude with some suggestions as to how continuing gaps in these environments might be expeditiously and effectively filled, in order to provide a more substantial base for development activities that seek to sustain the productivity of irrigated agriculture in Pakistan's mountainous periphery.

EXTENT OF MOUNTAIN IRRIGATION IN PAKISTAN

Small-scale irrigation systems dominate the developed irrigation potential in the mountainous environments of Pakistan, meeting the irrigation requirements of small farmers in the country's least developed and, heretofore, most isolated areas. In a real sense, they constitute the 'veins' whereby the greater productivity potential of irrigated agriculture is made available to the extremities of the national agricultural system. In addition to their comparatively small command or service areas, these systems commonly reflect characteristics that contrast sharply with those characterizing the large-scale irrigation systems of the Indus Plains. Typically, they are farmer-constructed, often have articulated forms of group ownership and management, and usually possess cooperative mechanisms for distributing water and minimizing conflicts.

In the rugged mountains of Pakistan's Himalaya-Karakoram-Hindu Kush, nearly all irrigation is done through *kuhls*, small, often lengthy channels usually constructed and maintained through the collective efforts of farmers and villagers. *Kuhls* carry water directed through a crude intake 'structure' from mountain streams fed by snowmelt, glacial melt, and/or springs for distribution through watercourses to clusters of small, often terraced fields, planted with food grains, vegetables, fodder, orchards, and trees. In basic physical appearance and characteristics, these *kuhl* systems differ little from thousands of others encountered throughout the Indian and Nepalese Himalaya.

As one moves southwest from the Karakoram-Hindu Kush along the mountain periphery of Western Pakistan into Chitral, the Tribal Areas, and the North West Frontier Province (NWFP), elevations decline; the terrain is interrupted by larger valleys drained by such rivers as the Swat, Kabul, and Kurram; and annual precipitation diminishes significantly. Changes in the physical environment and accessibility are mirrored in variations in irrigation development. Larger government-constructed canal systems, such as the Upper Swat Canal and the Warsak Canal, sustain the agricultural economies of the larger intermontane valleys. Smaller 'civil canals'—older systems, usually farmer-constructed and managed but now maintained by public agencies—are also found there and in other lesser valleys. The familiar *kuhl* systems are increasingly restricted to higher elevations and the upper ends of those favourably exposed tributary valleys where small perennial water sources are most likely to exist.

Further south into Baluchistan, the mountains of Pakistan's western borderlands—the Toba Karar, Sulaiman, and Brahui Ranges—continue to decline in elevation and conditions of greater aridity are encountered. Here, the indigenous *karez* systems of irrigation are found. Shafts are sunk in the alluvial fans, linked by galleries to form a tunnel that may tap a spring or, more commonly, collect subsoil water which is then delivered to fields at a lower elevation. Increasingly, tubewell development threatens the continued viability of many traditional *karez* systems here.

How extensive is the irrigated agricultural area in the mountains of Pakistan? Collectively, what is the command area of irrigation systems in this region? The answer to these questions would seem to be anyone's best estimate, considering the fact that formal surveys to assess the cropped, irrigated area or the command area of irrigation systems in this environment have never been done. Irrigation in Pakistan is dominated by the large canal systems prevalent in the Indus Basin, and readily available irrigation and irrigated agricultural statistics mirror this dominance. WAPDA's (Water and Power Development Authority) Irrigation Directories for the provinces give detailed service area data for the Indus Basin systems, but completely omit any reference to irrigation outside that region.[1] The most comprehensive review of irrigated agriculture in Pakistan in the past decade, the Revised Action Programme for Irrigated Agriculture, makes no mention whatsoever of irrigation systems or irrigated agriculture in Pakistan's mountain zones in either the main or supporting reports.[2] Regrettably, the recently published (1988) Report of the National Commission on Agriculture is equally silent on the subject.

The absence of reliable data on irrigated area, system type, and other relevant irrigation statistics for mountain agriculture in Pakistan virtually defines one priority research issue for this workshop, or namely, an accurate inventory survey of irrigation systems and their command areas in the mountain periphery of Pakistan. In the meantime, we must fall back upon the existing, partial data from a variety of sources, supplemented by the estimates of experienced observers, to gain some measure of insight into the extent of irrigated mountain agriculture in Pakistan. We must keep in mind, however, that our data are limited, occasionally contradictory, and subject to unknown error.

For example, in Gilgit District in the Northern Area, nearly 19,000 ha were classified as cultivated area by the 1980 Agricultural Census, virtually all of which can be assumed to be irrigated. More than 9,000 ha of irrigable area apparently has been added to this figure by the irrigation system development activities, covering 166 irrigation schemes supported by the Aga Khan Rural Support Programme (AKRSP) initiatives throughout 1987 (AKRSP 1987). Investigations recently carried out by WAPDA identified another 30 feasible irrigation schemes with the potential of adding a further 4,000 ha to the irrigable area in Gilgit (WAPDA 1988).

In Chitral District, more than 1,000 small, communally owned irrigation channels reportedly irrigate more than 18,000 ha and larger NWFP Irrigation Department Schemes command another 1,500 ha. The proposed irrigation development activities over the next 10 years, through the Chitral Area Development Project, target the addition of 11,000 ha

---

1 See Mohammad Ashraf and Mohd. Asif Khan, *Irrigation Directory: Sind and Baluchistan (1978)*, *Irrigation Directory: NWFP (1981)*, and *Irrigation Directory: Punjab (1984)*. In the case of the NWFP, civil canals that have been linked into larger agency-managed canal systems and their command area are identified.

2 See WAPDA, *Revised Action Programme for Irrigated Agriculture* (1979), for extensive discussions and supporting documentation of Pakistan's water and resources management; irrigation programmes, policies, and projects; and programme recommendations in support of irrigated agriculture for the 1980s, all focussed upon the environment of the plains of the Indus Basin. The recent definitive study on irrigated agriculture in Pakistan by Nazir Ahmad and Ghulam R. Chaudhry contains numerous tables of data and statistics on irrigated systems and irrigated agriculture. However, not a single statistic pertains to irrigation in the mountain regions of Pakistan.

to this total (IFAD 1986). AKRSP is assisting 105 small-scale irrigation projects already underway and their completion will bring about 8,000 ha of cultivated land under irrigation command (AKRSP 1987).

In Baluchistan, the area irrigated by *karez* and springs is reported to be 58,800 ha and wells and tubewells command an additional 113,000 ha (Kahlown et al. 1988). At least 50 per cent of this total irrigated area lies within the mountain environment of the province. If the more than 30 small systems to be developed through Baluchistan Minor Irrigation Development Project in the mountainous districts of Zhob, Loralai, Quetta, and Khuzdar are implemented, over 15,000 ha of new irrigation command area will be created. The total area irrigated by private canals in the North West Frontier Province is reported to be 360,000 ha (GOP 1986a). Again, an assumption that 50 per cent of this area is in the mountain regions of the province would not be unrealistic.

Federally Administered Tribal Areas were reported to have slightly more than 62,000 ha irrigated in 1983, and the Sixth Plan targeted an increase in the irrigated area to nearly 101,000 by 1988 through small surface system and tubewell development (GOP 1986b). The irrigated area in Azad Kashmir was 10,000 ha in 1983, estimated at about 6 per cent of the cultivable area; several new irrigation schemes were to be developed through the Sixth Plan, adding perhaps 8,000 ha to the irrigation command area (GOP 1986b).

To sum up, we can tentatively conclude that the existing irrigated area in Pakistan's mountains is about 380,000 ha. This is probably an underestimation, insofar as we know that the area irrigated by small-scale, farmer-managed systems in other countries in the Himalayan zone is poorly demarcated and surveyed, and there is scant reason to assume a different condition for Pakistan. Irrigation development activities planned or already underway would seem likely to increase this total by at least 55,000 ha in the next few years. True, this extent is dwarfed by the nearly 15 million ha of area commanded by the Indus Basin systems; nevertheless, in absolute terms, the amount of irrigated area in Pakistan's mountain region is not a trifling sum.

MOUNTAIN IRRIGATION SYSTEMS IN NORTHERN PAKISTAN

The literature on mountain irrigation systems in Pakistan, their characteristics, performance, management, and problems, is not extensive. In terms of system 'types', available evidence suggests that small-scale surface irrigation systems, *kuhls* predominate, but in the larger intermontane valleys, large-scale public and private systems are more important. *Karez*, traditional wells, and 'hill torrent' systems are probably less numerous as well as more environmentally specific than *kuhl* systems. Modern tubewell systems, also highly localized, are rapidly growing in number. In the following discussion, the focus will be upon the *kuhl* systems common to the Himalayan-Karakoram environment of Northern Pakistan where a substantial programme of mountain irrigation development has been underway for more than six years.

Water and Agriculture in the Karakorum Mountains

The deep valleys cut by the Gilgit and Hunza rivers and their tributaries, as they drain the Karakorum Mountains, are the locus of permanent settlements in Gilgit District, in villages perched precariously on river terraces or the sides of alluvial fans often threatened by unstable talus just above. The climate is dry continental, characterized by a great range in average temperatures (45°C or more between January and July) and meagre annual precipitation averaging 145 mm/year in Karimabad and 132 mm/year in Gilgit. Moreover, annual variability anywhere in the world is high in regions that are largely in the rain-shadows of the greatest concentration of mountains in excess of 6,000 m. Only at higher altitudes (above 3,000 m), where more precipitation falls and is accumulated as snow, do annual amounts substantially exceed 500 mm.

Throughout Gilgit District—and elsewhere in the Northern Areas and Chitral District—agriculture depends upon irrigation water supplied through small, farmer-constructed, gravity-fed systems. Water in these irrigation systems is derived primarily from snow or glacial melt. Less frequently, they are fed by perennial springs, the scarcest but most reliable water source, or by small rivers.

Generally, glacier-fed irrigation channels show the least year-to-year variability in discharge. However, water from glacial melt often carries large quantities of suspended silt much of which is subsequently deposited in the farmers' fields as a mixed blessing. During the period of seed germination and seedling growth, there is the risk that seeds will become buried too deeply to achieve a satisfactory germination rate or that seedlings will become coated with silt, inhibiting normal metabolism (Saunders 1983). On the other hand, silt is often important in the soil-building process, especially for improving soil structure.

Although channels supplied by springs dependent upon winter-spring recharge reflect some discharge variability, perennial spring water has several advantages over other irrigation sources. It is free of silt, it does not experience great variability, and, as noted by Whiteman (1985), it may be 'up to 5°C warmer, and this has a significant advancing effect on spring growth' of crops. Springs, however, are a scarce source of irrigation water in the region.

The greatest flow variation is found in channels exclusively dependent upon snowmelt, the least reliable irrigation source. Farmers from snowmelt-dependent villages report a severe shortage of irrigation water once every four to five years and general problems of considerable year-to-year streamflow variability.

River-fed irrigation systems are also more vulnerable to annual variations in precipitation and are affected by seasonal fluctuations in river flow as well. A channel intake structure constructed to divert river water for irrigation during crop planting in March may be inundated or washed out when glacial melt subsequently increases river discharges in May-June. Later in the summer, the river diversion may have to be relocated further upstream to sustain irrigation supplies as river discharge diminishes.

Capturing water for irrigation is only part of the task of establishing and sustaining agriculture in Hunza-Gojal. Equally, perhaps even more arduous, is the concomitant and longer process of land development. In bringing land under the cultivation of principal crops for human and animal consumption—grains (wheat and barley), vegetables, potatoes, fruit trees (apricot and apple), fodder (alfalfa), and trees for fuel and fodder (poplar, willow, and Russian olive)—soils have been drastically modified.

In this region, irrigated crops are largely confined to three landform environments and their associated soils. In the valleys of the Hunza River and its tributaries, river terraces and alluvial fans have the greatest agricultural significance. The deeper and better developed soils found on old river terraces are more important than those on terraces of more recent origin. The lower portions of alluvial fans, formed by small streams and hill torrents, are more intensively cultivated than the upper areas, because of the small proportion of coarse soil materials found there. In either instance, better soils are commonly the locus of grain and vegetable crops as well as orchards. Poorer, less developed soils tend to be used for fodder crops and trees for fodder and fuel.

Cone-shaped scree slopes produced by mass wasting of the surrounding barren cliffs and hills below 2,100 m are another locus of irrigated agriculture. However, because of the inherent instability of these slopes, their agricultural development presents special problems and tends to be both more recent and slower. The upper portions where finer materials are concentrated are cultivated first, usually with slope-stabilizing trees and fodder crops.[3]

Village Irrigation Systems in Hunza-Gojal

Initial irrigation system development in Hunza was a highly localized activity, concentrated in locations where water from glacial and snowmelt sources was easily developed by small groups of farmers using locally available technology and resources. Later, traditional chiefs, such as the *Mirs* of Hunza, began to exercise their growing feudal authority to mobilize a larger population for the construction of new *kuhls* in more difficult locations, rehabilitation of older systems, and development of new land. Although a portion of the increased production resulting from enhanced and more reliable water supplies was extracted by a compulsory agricultural tax, the *Mirs* did initiate the irrigation development and modest settlement expansion where smaller, isolated group efforts could not. Following the arrival of the British-supported Dogra administration in Gilgit in the 1980s, a gradual decline in feudal authority began, accompanied by a reduction in the development of irrigation system. This trend continued after the independence of Pakistan until 1974 when the authority of the *Mirs* was formally abolished.[4]

Beginning in 1982, the AKRSP has had an active programme of rural, institutional, and physical development (subsequently expanded to include Baluchistan District, Chitral District, and the North West Frontier Province), which has renewed irrigation development activities in the region. In Gilgit District alone, 166 irrigation projects—improvements on older systems or development of new systems—have been initiated in villages through AKRSP-supported interventions.[5]

---

3 Seasonal agricultural activities also occur at higher elevations, but usually do not involve irrigation. Between 2,100 and 3,300 m, soils have developed in widely scattered locales from physical, chemical, and biological processes acting upon parent material. These locations support trees and grasses and serve as summer pastures for the animals of villages at lower elevations.

4 Kreutzmann (1988:246–250) provides the best historial overview of the origins and development of irrigation systems in Hunza.

5 The primary objective of the AKRSP is to facilitate the development of strong, broad-based village organizations that can continue to undertake a wide range of rural development activities on a permanent, locally sustainable basis. This objective is accomplished through a unique intervention strategy that encourages each village to identify and propose a single 'productive physical infrastructure' (PPI) project which will increase the incomes of most village households. Implementation is then funded by a grant from AKRSP.

Irrigation-focussed PPI projects have proven to be effective foci for the institutional development process. Because agriculture in Gilgit, Baluchistan, and Chitral depends almost entirely upon irrigation,

A recent survey of the Water and Power Development Authority in Gilgit District of 25 localities in the Gilgit River sub-basin and 34 localities in the Hunza-Hispar River sub-basin identified 221 *kuhls* supplying irrigation water to developed agricultural lands (WAPDA 1988).[6] So far, most *kuhls* have perennial flows, but seasonal discharge variations between low and high flows as great as 20 times were reported, reflecting that in more than 85 per cent of the surveyed localities, the water source for *kuhl* systems was a combination of snow or glacial melt and springs.

Kuhls identified in the survey varied greatly in channel length between source of water supply and command area, from a more typical 2 to 3 km to as much as 18 km in the case of *Parri kuhl*.[7] They also varied substantially in size. The discharge range between the smallest *kuhl* and the largest was 7 lps (0.25 cusec) to 425 lps (15 cusec), although more than one-half of the *kuhls* carried discharges between 7 lps and 28 lps (1 cusec).[8].

For systems in the 44 villages covered in the survey, it was possible to calculate the relationship of water supply per unit of cultivated area. In slightly less than one-half of these villages, *kuhl* systems delivered less than 0.75 lps per cultivated hectare (< 6.5 ha/mm/day), reaching as low as 0.13 lps/ha (1.1 ha/mm/day).[9] At such water supply levels, water is apparently scarce relative to land, and one would expect the system and field-level water management practices to reflect such conditions. For seven other villages, the systems supplied between 0.75 lps/ha and 1.25 lps/ha (10.8 ha/mm/day). In general, *kuhl* systems in Hunza-Hispar villages varied more widely in the Gilgit River sub-basin. Also, for roughly one-third of all the villages, *kuhl* systems apparently supply water in relatively abundant amounts. This suggests that in those locations, land suitable for agriculture is more scarce than water.

In 1987, a rapid field survey of irrigation systems was conducted in the upper Hunza River Basin, a part of Gilgit District known as Gojal. The survey was confined to seven villages—Soust, Gircha, Jamalabad, Morkhon, Ghalapan, Khaiber, and Passu—in Gojal.

---

developing and sustaining irrigation systems in this difficult environment continues to require a high degree of organization and collective management. Irrigation systems more frequently meet the AKRSP's criterion of consensus support than any other potential infrastructural development and they comprise about 60% of projects assisted by the AKRSP in Gilgit, Chitral, and Baluchistan. The average cost to the AKRSP for small systems developed in Gilgit has been Rs 139,000 (about US$ 8,175) per project.

6 *Kuhls* also supply water for domestic uses, to meet livestock needs and to deliver water to small-scale hydropower units as well.

7 The total length of channel and all field ditches in these irrigation systems is easily a great deal more. For example, the Hopar community of five village settlements in the Hunza-Hispar River sub-basin is reported to have more than 300 km of irrigation channels supplying melt water to 440 ha (Butz 1987:7).

8 Data reported in the WAPDA survey must be approached with considerable caution, although they may still reasonably reflect the range of irrigation system conditions in Gilgit District. Cultivated area data for villages were taken from tehsil records—a sometimes unreliable source. Unfortunately, it is not known whether or not reported *kuhl* discharge data were based upon actual measurements, and, if it was, where in the system measurements were made. Nor is it possible to be confident that these data are consistently related to either minimum flow or maximum flow conditions. In the discussion that follows, minimum flow condition was assumed to be the case in calculating relationships of water supply to cultivated area which were otherwise unknown

9 By way of perspective, in the Lower Chenab Canal System of the Rechna Doab, Punjab, the sanctioned allocation of irrigation supplies ranges from 0.26 to 0.35 lps/ha, for a water deficit designed system. These figures have been doubled or trebled by supplies from public tubewells in fresh groundwater areas.

Results of this survey provide a more detailed insight into some of the characteristics of *kuhl* systems in Gilgit (Vander Velde and Husain 1988).[10] A total of 25 irrigation channels was identified in these villages, of which 20 predate AKRSP's activities (Table 24.1).[11] Between 1983 and 1987, AKRSP-initiated village organisations (VOs) in these villages completed seven irrigation projects—five new *kuhls* and two improved *kuhls*—adding more than 500 ha of potential agricultural land which will approximately double the previously irrigable area once the time-consuming process of new land development is completed.

Table 24.1. Survey villages, irrigation channels, and water sources

Command area	Channel	Source
Soust [60 hslds]		
Soust (old)		
	1. Aziz Baig's	
	2. Main	Soust *Nullah*
Nazimabad (old)	3. Upper	Glacier + Snowmelt
	4. Main	
Soust (new)	5. New	
Gircha [27 hslds]		Sarteez *Nullah*
Sarteez (old)	6. Sarteez	Snowmelt + Spring
Lower Gircha (imp)	7. Main	Spring
Morkhon [45 hslds]		
Morkhon (old)	8–10. Three left banks	Morkhon *Nullah*
Morkhon (new)	11. High left bank	Glacier, snowmelt
Morkhon (old)	12–13. Two right banks	+ Spring & flow from 7 small *Nullahs*
Ghalapan [10 hslds]		
Ghalapan (old)	15. Jurjurkhon *Nullah*	Snowmelt
Ghalapan (new)	16. Vundergar *Nullah*	Snowmelt (more)
Khaiber [55 hslds]	Khaiber *Nullah*	
Khaiber (old)	17. Lower	Spring
	18. Upper	Glacier + snowmelt
Imamabad (old)	19. Main	Glacier + snowmelt
	20. Small	Snowmelt + spring (less)
Khaiber (new)	21. New	Hunza River
Passu [67 hslds]		
Passu (old)	22. Main	Passu glacier
	23. Nobod	Glacier + snowmelt
	24. Yashvandan	
Passu (new)	25. Batura Glacier	

10 The objective of this survey was to learn more about adaptations, institutional changes, and technical innovations in irrigation management practices in the Gilgit *kuhl* system as a process of managing the expansion of irrigation capacity and realizing its benefits and proceeds. Developments here may have potential for wider dissemination and adoption in other *kuhl* systems elsewhere in the mountains of Northern Pakistan.

11 Of the 20 older systems, at least five *kuhls* were initiated by the former *Mirs* during the last 100 years. Those developments made possible the settlement of about one-third of the total number of households now resident in the villages under study, on the land thereby supplied with irrigation.

Channel Construction

Successful irrigation channel construction in Hunza now involves a combination of local wisdom (knowledge derived from generations of experience) and contemporary engineering technology. Alone, neither the source of knowledge nor skill is any longer sufficient to guarantee success. Instead, the failure to utilize both frequently leads to the construction of poorly performing or failed systems, typically after an expenditure of substantial and scarce resources. The traditional method of determining the slope of a channel was the use of water as a level. Beginning from the source, water flowed along the channel as it was dug on a carefully estimated but unsurveyed line, with the objective of achieving the desired command. The approach 'worked' as long as the scheme was physically possible. Thus, village elders were commonly consulted for advice on past glacial movements, avalanche and mudflow paths, and streamflows from glacial and snowmelt or springs. However, if an impassable outcrop was encountered during construction, or the velocity of water flow dropped so low that command was lost (conditions often discovered only after kilometres of channel had been constructed), the project had to be redesigned or abandoned (Hudson 1983).[12]

Alone, modern engineering science frequently has produced scarcely better results. In the mountainous environment of the Northern Areas, where physical conditions vary greatly within short distances, or from one season to another, the failure of engineering surveys and irrigation designs to draw upon detailed local knowledge greatly increases the probability of failure. The high proportion of unsuccessful irrigation channels in Gilgit District, designed and constructed by the Northern Areas' Public Works Department since 1974, without local consultation or participation, substantiates this conclusion.[13]

In its intervention strategy of assistance to develop farmer-managed irrigation systems, AKRSP purposely links local knowledge with modern engineering skills in the planning, design, and construction of new *kuhls*. Joint surveys of new systems or improved systems sites are done by engineers and knowledgeable farmers from the village and may involve several field visits. During channel construction, frequent consultation continues between farmers and AKRSP engineers to solve unanticipated problems. This collaborative approach has resulted in the successful implementation of several irrigation projects in Gilgit District, previously thought too difficult to implement. Two such projects have been constructed in the surveyed villages of Passu and Soust.[14]

Managing Water Distribution

Warabandi, the practice of irrigation turns, taken according to an established roster, is used in Gojal systems, as it is elsewhere in Gilgit District, to equitably allocate water and ensure irrigation turns during periods of water scarcity in the irrigation system, notably between March and May. When the period of water scarcity is over, or where water

12 For example, Passu villagers reported seven earlier failed attempts at channel construction using such techniques to tap the Batura Glacier melt.

13 Of the 20 schemes undertaken by the Northern Areas PWD, at an average cost of Rs 1.85 million, only one is reportedly still functioning (Hussein et al. 1986:3).

14 Out of the 166 AKRSP-assisted irrigation projects implemented to date in Gilgit District, only one channel is identified as a complete failure.

scarcity is not a problem (e.g., in the older irrigation systems in Passu), water distribution generally follows a relatively informal system of irrigation turns as and when needed. Field observations confirm that the *warabandi* generally remains a durable, not easily changed, irrigation management practice in Gilgit.

Under the *warabandi* system, each household in the *kuhl* command takes its irrigation turn on a specific day, at a specified and equal period of time.[15] Between farmers whose turns are closely proximate to one another, there may be frequent, informal trading or exchange of turns. Generally, food crops are given first priority in water use, followed in order by fodder crops such as alfalfa, and finally by trees. Thus, where night irrigation is practised, it is usually for trees (food and fodder crops are commonly irrigated during the daylight hours). Among food crops, vegetables typically take priority over food grains, even to the point where an operating *warabandi* can be interrupted out of turn should a farmer plead the necessity of water for a vegetable plot.[16]

Managing Channel Maintenance

Maintenance of irrigation systems reflects their common property origins and a continuing collective management basis. Traditionally, the general principle followed for maintenance of the common portion of the irrigation channel was an annual contribution from all farmers served, in the form of labour or produce. The principle continues to be applied, albeit nowadays a farmer may also contribute cash instead of labour or produce. Normally, spring is the time for general annual maintenance, before the first irrigation for the new crop year and when water flows are low or non-existent. On channels where silt loads are heavy, all the farmers may also participate in a one- or two-day mid-season desilting operation. Maintenance of lateral or field channels not common to the system is the responsibility of individual farmers.

Some villages employ a *chowkidar* or watchman during the irrigation season to patrol the common portion of the channel to adjust and clear debris from the channel intake, plug leaks, repair small breaches, and otherwise monitor water supply conditions.[17] In sys-

---

15 For example, in the *warabandi* for the command area of the old Soust *kuhl*, 24 households are divided into two equal groups. One group irrigated between 06:00 and 12:00 hrs, the other from noon to 18:00 hrs. On Nazimabad *kuhl* a four-day rotation is followed with nine households irrigating during each 24-hour period; two groups of four or five farmers each irrigated cereal and vegetable crops daily, one in the morning and the other in the afternoon. The orchards and fodder/fuel trees of each group are irrigated at night. Irrigation turns are longer here because landholdings are somewhat larger on Nazimabad *kuhl* and soils newer and relatively less well formed. The *warabandi* is in force throughout the entire season. All 60 Soust and Nazimabad households have land in two locations in the command of the new Soust *kuhl* which is allocated water on alternate days. On each day, 30 households in one location get their irrigation turn, one-half in the morning and the other half in the afternoon. Thus, in a four-day period, all holdings can be irrigated.

16 (Hussain Wali Khan, personal communication, November, 1987). Interestingly, vegetable plots traditionally are the active focus of cultivation and irrigation for women who otherwise do not share in the common management of irrigation water in Hunza-Gojal.

17 In such cases, the entire water user community on a channel will employ the *chowkidar* with each household making an equal contribution to salary, usually on a seasonal basis. Payment is typically in kind, a combination of foodgrains (wheat) and fodder. At 1987 market prices, the value of such payments ranged from Rs 900 to Rs 1400 per season.

tems where *chowkidars* are not employed, farmers will take regular turns patrolling and maintaining the common channel, usually at the time of their irrigation turn. Whenever a major breach or other maintenance emergency occurs, all the farmers on the channel will participate in its repair.

The rationale for the presence or absence of a *chowkidar* during the irrigation season is somewhat uncertain. There are several possible explanations for the practice, e.g., channel length, the amount of silt load in the channel, and whether or not night irrigation is done. An examination of these conditions for the surveyed systems (Table 24.2) suggests that channel length is the most common variable in the use of *chowkidars*. Insofar as channel length reflects both the quantitative nature of the likely maintenance requirement and the cost of walking to the head to regulate the discharge at various times during the day as well, this is not surprising.[18] For new *kuhl* systems, whether or not there previously was a *chowkidar* in the village also appears to be significant. Conditions of other systems seem to be somewhat less important.

Table 24.2. Variables in the presence or absence of *chowkidar* for irrigation channel maintenance

Irrigation command	*Chowkidar* present	Lengthy channel	Irrigation at night	Significant silt load
Soust (old)	Yes	No	No	Yes
Nazimabad	Yes	Yes	Yes	Yes
Soust (new)	Planned	Yes	No	Yes
Sarteez	No	No	Yes	No
Lower Gircha	No	No	?	No
Morkhon (old)	No	No	No	No
Morkhon (new)	No	Yes	No	No
Jamalabad	Yes	Yes	Yes	No
Ghalapan (old)	No	No	No	No
Ghalapan (new)	Planned	Yes	?	No
Khaiber (old)	Yes	Yes	No	No
Imamabad	Yes	Yes	No	No
Khaiber (new)	Yes	Yes	No	No
Passu (old)	No	No	No	Yes
Passu (new)	No	Yes	No	Yes

Other Physical Infrastructures

The physical infrastructures of other systems are generally straightforward and not elaborate. Sets of flat(ish) stones are often used as channel and field ditch-drop structures. Rudimentary, but functional, turn-outs generally are constructed from selected rocks; occasionally carefully fitted wooden turn-out gates or small pine outlets are encountered along channels. Sedimentation tanks or stilling basins have been built at the head of

18 A weir at the head of a *kuhl* system usually requires modification when the melting rate of snow or glacier increases the discharge at the source (early afternoon) or when the irrigation demand is minimal (after sunset).

the main channels in *kuhl* systems in Soust and Passu to reduce heavy silt loads carried by the glacial-origin water. The tanks on Soust *kuhl* are meant to trap rock debris carried in the glacial melt as well, and they must be desilted from time to time during the irrigation season—a task done collectively by the irrigation community. Several farmers in Soust have dug shallow stills close to their fields. Here, removed silts are mixed with animal manure and spread on the fields to improve both soil structure and fertility.

In two older *kuhl* systems in Soust and Nazimabad, overnight storage tanks have been built. These permit the augmentation of channel flows during day-time irrigation. However, such infrastructure is not as widespread among Hunza *kuhls* as might be anticipated, perhaps because inexpensive construction of tanks that will not leak, and at the same time will be of sufficient size for irrigation water, is difficult (Hudson 1983).[19]

CHANGE IN MOUNTAIN IRRIGATION SYSTEMS

After a generation or more of comparative quiescence, the 1980s have heralded a period of renewed activity and change for farmer-managed irrigation systems in Northern Pakistan, primarily in response to the AKRSP's rural development programme. Although there is evidence from systems surveyed in Hunza-Gojal of the continued vitality of proven ways to solve problems, and of carefully adjusting new systems to fit environmental conditions, it is also clear that there has been both institutional innovation and considerable farmer-initiated experimentation to modify the previous irrigation management practices or techniques. In the following discussion, a few examples that substantiate these observations will be described.

Innovation in System Design Parameters

From the perspective of agency intervention in, and technical assistance to, small-scale mountain irrigation systems in Pakistan, the AKRSP's success in using a strongly participatory strategy that draws upon local knowledge and experience is possibly the innovation with the greatest long-run significance. The design criteria for irrigation channels assisted by AKRSP illustrates the value of such an approach. Following a survey and measurement of bed slope and conditions in older farmer-managed *kuhl* systems in Gilgit District, empirically based parameters were adopted and used as basic design criteria for new systems (AKRSP 1987).[20] The stimulus for this change, at least in part, was the visible evidence of the failure of previous government-constructed irrigation systems in the Northern Areas where engineers had used textbook standards that were more appropriate to environments outside this region. The general farmers' satisfaction with and apparent absence of failure in AKRSP-assisted systems is firm evidence that studying existing *kuhl* systems and skillfully drawing upon farmers' past irrigation experiences

19 Night irrigation, of course, is the commonly practised alternative to overnight storage during periods of water scarcity.

20 To obtain a discharge of about 28 lps per 40 ha of command area, a bed slope of 1:300–400 was found to be appropriate.

are valuable complements to modern engineering science in developing new mountain irrigation systems.

Adjusting New Systems

Along with the substantial increase in command area created by renewed *kuhl* development, considerable attention has been focussed upon the process of bringing new land into actual production. Some observers have thought the process inefficient and perhaps too slow (World Bank 1987). The new system in Passu illustrates the situation. A *kuhl* completed in 1985 commands at least 273 ha, an area sufficient to increase fivefold the average landholding in Passu. However, more than three years after channel construction, less than 10 per cent of the command was actually developed. Northern Area farmers know from experience that, regardless of how well-designed and constructed a new irrigation system is, it will not immediately fit its environment. An initial period of adjustment is commonly required and farmers in the new system at Passu are engaged in that process. Since the construction of the new *kuhl*, they have relocated the intake to compensate for glacial retreat and to improve the bed slope condition in the upper reach of the channel; because of substantial leakage, which affected the Karakoram Highway, a 100 m reach of new channel had to be lined with cement and stone. Lastly, the new channel required stabilizing or 'hardening', a process encountered elsewhere in new systems in Hunza-Gojal.

Each year, as silt from glacial melt fills soil interstices along a longer reach of the bottom and sides of the main channel, discharge at the system head is gradually increased. Stabilizing or hardening the *kuhl* in this way reduces the likely occurrence of major breaches that would be difficult and costly to repair.[21] Clearly, new land cannot be brought under irrigated agriculture ahead of adequate water supplies, and Passu farmers estimate that it will be another four or five years before they can confidently operate their new system at full supply levels.

Irrigation Specialists

The *chowkidar* is a traditional and familiar figure in *kuhl* irrigation systems in Northern Pakistan. Over time, a *chowkidar* develops a highly detailed knowledge of the irrigation system in which he works. He is accountable to the farmers and they willingly pay him because he provides economies of specialization for an essential service. However, should he fail in his duties, they also are likely to replace him quickly. In three Gojal villages—Soust, Morkhon, and Passu—land not immediately adjacent to the already developed command area is supplied water through lengthy new channels. Rather surprisingly, none of these villages yet has a *chowkidar* for its channels, although farmers on the Soust *kuhl* say they plan to hire a *chowkidar*. Older systems in Morkhon or Passu do not have *chowkidars*, thus their absence on the new *kuhls* may be related to this fact.

On the new channel in Khaiber, however, a variation of the *chowkidar* system has emerged in conjunction with another innovation: the VO's decision to collectively develop

[21] Another element in the process of hardening a channel is the planting of saplings of fast-growing willow and poplar along the embankments to establish reinforcing root systems.

the command area for at least the first five years of operation. The new Khaiber *kuhl* commands land located 2 to 3 km from the village. Because new land remains in collective ownership, there are no specific individual responsibilities for irrigation, and this situation required Khaiber farmers to devise a new approach to manage irrigation of the new land. Of the two obvious possible solutions, namely, irrigation done by small groups of farmers on a rotational basis or a modification in the traditional patrolling and maintenance responsibilities of hired *chowkidars*, Khaiber farmers chose the latter. At a monthly salary somewhat equal to the local wage labour rate, three men were hired for the four-month agricultural season to do daily field irrigation activities in the new command area. These 'specialist *chowkidars*' also continue to perform other tasks traditionally associated with them.

Changes in Water Allocation

It was noted earlier that *warabandi* is one of the most durable water management techniques in farmer-managed irrigation systems in Gilgit. Although in its details the *warabandi* often varies from system to system and is poorly understood, it also can be rather flexible. Both characteristics are revealed in the *warabandi* adopted for the new Soust *kuhl*. Here, within the same time of the channel's operation, a different water allocation procedure is followed for the smaller, terraced fields of fruit trees and intercropped fodder on steep slopes than for the larger, more level fields below planted with annual grains, although both are irrigated simultaneously! In allocating water for the command area of the new Soust *kuhl*, there also was an unusual revision in water rights of the two older systems in Soust and Nazimabad. Formerly, in periods of water scarcity, these two *kuhls* received water on alternate days; now, when water is scarce, each command area receives its turn every third day.

Field-level Water Management Experimentation

In general, and in contrast to land-use patterns in older, irrigated areas, most of the developed land commanded by new *kuhl* systems in Gojal has been planted with trees and fodder. This situation undoubtedly reflects the present physiologic environment of many newly commanded areas (e.g., steep slopes or newer river terraces) that have poor soil structure with high infiltration rates and low fertility, the conditions of which can only be changed rather slowly. Plantation of fuel trees, fodder crops, and orchards on the scale now underway is also an unusual phenomenon in Hunza-Gojal, and there is interesting evidence of farmer innovation and experimentation on field-level water management practices that fit the current conditions better.

In Morkhon and Jamalabad, farmers have adopted a different technique for the irrigated cultivation of crops planted on the scree slopes in the new *kuhl* command. Instead of constructing the usual and costly stone-walled terraces and using basin irrigation practices, they constructed field ditches along the contour to deliver irrigation water to trees and alfalfa planted on shallow reverse slope terraces. This appears to be an adaptation of the furrow irrigation practices already used for potato cultivation on less steeply sloping fields. In several locations in the command area of the new Soust *kuhl*, another modification of furrow irrigation is evident on steep slopes. There, in a few individual holdings, field irrigation ditches have been made as a series of linked S's down slope, and small drops have been fabricated from stone and polyethylene to reduce soil erosion as water is carried from one terrace level to the next one below.

CONCLUSIONS

To the casual observer or non-specialist, the foregoing discussion may appear to have focussed unduly upon modest developments. However, it is clear that the changes described therein reflect the significant initiative and willingness on the part of farmers in this mountain region to improve or fine tune their use-efficiency as well as management of scarce water and land resources. It also demonstrates the resilience and flexibility of some traditional institutions in adapting to an environment of rapid social and economic change. This leads us to one important, initial conclusion, perhaps even a principle: the existing irrigation systems and farmers who managed them are a critical resource to be used in any effort to develop irrigation potential in Pakistan's mountain environments.[22] The AKRSP's programme in the Northern Areas already has demonstrated the utility of grasping this point. Unfortunately, in the case of Pakistan's formal irrigation bureaucracy, one can not be confident that it is yet even perceived, much less understood and heeded.

On the other hand, it is clear that many gaps remain in our knowledge and understanding of small-scale surface irrigation systems in Northern Pakistan. For example, although recently several surveys have been made of irrigation systems in Gilgit District, there continues to be an absence of systematic information about how well the older *kuhl* systems perform, either physically or institutionally, in sustaining productive irrigated agriculture.[23] The inference is that these systems perform reasonably well within their environmental contexts, but that does not leave us with much insight as to their potential for sustaining more productive irrigated agriculture. Nor do we know anything about how long it may take before such a system begins to perform reasonably well—a matter of considerable importance in the context of new system development activities in Northern Pakistan.[24]

In general, we can conclude that there are three priority needs that must be filled in the near future if successful irrigation development strategies that fit the mountain ecosystems of Pakistan are to be designed and implemented.[25] One of these was alluded to much earlier in this paper, namely, an inventory of mountain irrigation systems that would provide reliable information on system type and service area. Rapid appraisal

---

22 Rapid appraisal methodologies adapted to gathering such information for irrigation systems are readily available and already tested in mountain environments (e.g., see Yoder and Martin 1985).

23 The recent thesis study by Butz (1987), judging from an abstract, may very well have begun to fill this gap for Gilgit systems. Unfortunately, it is unpublished as yet and copies are not easily available in Pakistan.

24 In collaboration with the AKRSP, The International Irrigation Management Institute (IIMI), Pakistan, will begin a research project designed to measure the performance of farmer-managed irrigation systems in Hunza-Gojal in March, 1989. Both the old and the new systems will be selected for a comparative study of their irrigation efficiencies.

25 I have drawn here considerably upon a thoughtful general overview of research issues in farmer-managed irrigation systems in Martin, Yoder, and Groenfeldt (1986) and on a recent working paper by Coward, Johnson, and Walter (1988) that focussed upon ways to improve government policies and programmes for small-scale irrigation systems in Asia.

reconnaissance studies, combined with careful interpretation of aerial photographs and, perhaps, remote sensing imagery analysis, provide a ready and fairly rapid means of filling this knowledge gap.

The second need is for a modest set of studies that would systematically examine how farmers manage their irrigation systems in different mountain environments in Pakistan. Some of the studies should be comparative examinations of each type of irrigation system, focussing upon three sets of management activities, those for water (e.g., allocation, distribution), those for physical structures (e.g., design, construction, maintenance), and those for organization (e.g., resource mobilization, communication, conflict management). Of course, there is interaction between these sets of activities and the processes for those interactions need to be clarified as well.

Finally, there is a need for a more intense body of knowledge on how well irrigation systems are managed in the mountains of Pakistan. Research that ascertains system performance and system constraints is needed if we are to accurately identify potential foci for improvement. These studies of different mountain systems should include measurements of water flows, determination of irrigation efficiencies, assessments of crop yields, and evaluation of institutional arrangements. Such studies will demand a set of multidisciplinary competencies in both research design and implementation that, in itself, will break new ground in irrigation management research in Pakistan.

REFERENCES

The Aga Khan Rural Support Programme. *Fifth Annual Review, 1987*. Gilgit, Pakistan: AKRSP, 1987.

Ahmad, N. and G.R. Chaudhry. *Irrigated Agriculture of Pakistan*. Lahore, Pakistan: Shahzad Nazir, 1988.

Ashraf, M. and M.A. Khan. *Irrigation Directory; Sind & Baluchistan*. Lahore, Pakistan: WAPDA (Survey & Research Organisation, Planning Division), 1978.

Ashraf, M. and M.A. Khan. *Irrigation Directory: Punjab*. Lahore, Pakistan: WAPDA (Survey & Research Organisation, Planning Division), 1984.

Asian Development Bank. *Appraisal of the Chitral Area Development Project in Pakistan* (SRC Draft). Manila: ADB, 1987.

Butz, D.A.O. 'Irrigation Agriculture in High Mountain Communities: The Example of Hopar Villages, Pakistan.' Unpublished MA thesis. Wilfrid Laurier University, Waterloo, Ontario, 1987.

Coward, E.W., R.L. Johnson and M.F. Walter. Other Channels; Improving Public Policies and Programs for Small-Scale Irrigation Development in Asia. Working Paper for the Irrigation Studies Group, Cornell University. Ithaca, N.Y.: Cornell University, 1988.

Government of Pakistan. *Agricultural Statistics of Pakistan, 1985*. Islamabad: Ministry of Food, Agriculture and Cooperatives (Food and Agriculture Division [Planning Unit]), 1986a.

Government of Pakistan. *The Sixth Five Year Plan, 1983–88*. Islamabad: Planning Commission, 1986b.

Government of Pakistan. *National Commission on Agriculture Report*. Islamabad: Ministry of Food and Agriculture, 1988.

Hudson N.W. *Report on Irrigation Consultancy for the Integrated Rural Development Project, Gilgit UNDP/FAO Pak/80/009*. National College of Agricultural Engineering, U.K., 1983.

Hussein, Maliha H., W. Khan, Z. Alam and T. Husain. *An Evaluation of Irrigation Projects Undertaken by AKRSP in the Gilgit District of Northern Pakistan*, AKRSP Programme Evaluation Report No. 5). Gilgit, Pakistan: AKRSP, 1986.

International Fund for Agricultural Development. 'Pakistan Chitral Area Development Project: Preparation Report' (Draft). Rome: IFAD (Asia Division, Project Management Department), 1986.

Kahlown, M.A., M. Khalil and M. Munir. *Karez Irrigation in Baluchistan (Bench Mark Study)*. (Irrigation Systems Management Research Project, GOP-USAID). Lahore, Pakistan: WAPDA, Planning Directorate (South), P & I, 1988.

Kreutzmann, H.J. 'Oases of the Karakorum: Evolution of Irrigation and Social Organisation in Hunza, North Pakistan.' In Nigel J.R. Allan, Gregory W. Knapp and Christoph Stadel (eds.). *Human Impact on Mountains*. Totowa, N.J.: Rowman & Littlefield, 1988.

Martin, E., R. Yoder and D. Croenfeldt. *Farmer-Managed Irrigation: Research and Issues*. ODI-IIMI Irrigation Management Network Paper, 86/3c. London: Overseas Development Institute, 1986.

Saunders, F. *Karakoram Villages: An Agrarian Study of 22 Villages in the Hunza, Ishkoman and Yasin Valleys of Gilgit District*. Islamabad, Pakistan: UNDP/FAO (IRDP, Gilgit, PAK/80/009), 1983.

Spate, O.H.K. and A.T.A. Learmonth. *India and Pakistan; A General and Regional Geography*. London: Methuen & Co. Ltd, 1967.

Vander Velde, E.J. and T. Husain. 'Intervention and Innovation in Farmer-Managed Irrigation Systems in Northern Pakistan.' Paper presented at the Seventh World Congress for Rural Sociology, 26 June–2 July, Bologna, Italy, 1988.

Water and Power Development Authority. *Northern Areas Regional Development Plan Reconnaissance Report, Gilgit District* (Main Report and Appendix I) Lahore, Pakistan: WAPDA (Regional Planning Directorate, Planning Division, Water Resources Planning), 1988.

Whiteman, Peter T.S. *Mountain Oases: A Technical Report of Agricultural Studies (1982-1984) in Gilgit District, Northern Areas, Pakistan*. Islamabad, Pakistan: UNDP/FAO (IRDP, Gilgit, PAK/80/009), 1985.

World Bank. *The Aga Khan Rural Support Program in Pakistan: An Interim Evaluation*. Washington D.C.: The World Bank, Operation Evaluation Department, 1987.

Yoder, R. and E. Martin. *Identification and Utilization of Farmer Resources in Irrigation Development: A Guide for Rapid Appraisal*. ODI Irrigation Management Network Paper 12C. London: Overseas Development Institute, 1985.

PART 5

Transformation of Mountain Areas: Some Innovative Approaches

25

ANTI-POVERTY FOCUSSED PROGRAMMES IN THE MOUNTAINS: EXPERIENCES IN CHINA

Gao Hongbin and Ye Xingquing

Gao Hongbin and Ye Xingquing are the Deputy Director and Secretary, respectively, of the Leading Group Office for the Economic Development of Poor Areas under the State Council of the People's Republic of China.

CONTENTS

A ROUGH SKETCH OF POVERTY IN CHINA AND ITS BASIC PROBLEMS

Characteristics of Poverty in Contemporary China

China is still a developing country with many minority nationalities. In comparison to ancient China, where scenes of devastation and accumulating poverty prevailed, we can claim with great confidence that the living standards of the Chinese people have improved gradually. By 1986, China had solved the problem of providing enough to feed and clothe its people. There has also been a fundamental change in the overall situation of poverty and backwardness.

It is necessary to explain that, in China, despite the overall increase in per capita annual income, proportionate development results in discrepancies in the economic status of different regions. Despite the regular increase in average figures, low income continues to be a problem. Therefore, the overall development of the rural economy throughout the Chinese mainland does not mean that China has rid itself of poverty and backwardness. Yet, fundamental changes have taken place in the characteristics of poverty and its implications. By 1986, domestic poverty could be described by referring to three specific types of situations.

(1) Up to 1986, nearly 30 million people were suffering from hunger and were scantily and poorly clothed, living in semi-open adobes or even in huts with dried corn stems as walls.

(2) It is believed that 15 per cent of the total population in rural areas earn half or below half of the average income of the people living in urban areas. Statistics show some 80 million peasants with a per capita net income below RMB 200 yuan (4.73 yuan = US$ 1).

(3) A large number of underdeveloped areas with poor socioeconomic situations were revolutionary bases in the past or are remote hilly areas where minority nationalities live. These areas once played an important role and made great contributions to the establishment of the People's Republic of China. Therefore, successful implementation of development activities focussed on these areas has special significance.

It is believed that the current problem of poverty and backwardness in China can be divided into situations with three types of characteristics.

(1) Absolute poverty, in which situation a labourer can barely earn sufficient income for survival. Farm households are not only production but also consumption units in which a labourer cannot expand the household capital stock, and this determines the poverty status.

(2) Comparative poverty, a historic concept undergoing slow change. Normally it is relative to the bottom 10 to 15 per cent of the population in rural areas.

(3) The poverty-stricken and backward areas, with concentrations of comparatively and absolute poor people. The importance of eliminating poverty at present stems not only from the large number of absolute poverty-stricken people or from the fact that comparative poverty has affected the uniform economic development of the country, but also from the fact that this has affected the social stability and unity of the nation.

The Distribution of Poverty-stricken Areas

Problems relating to poverty are highly concentrated in the mountainous areas of China. Currently, the poverty-stricken and low income generating populations are concentrated in 18 poverty-stricken contiguous mountain areas. Out of the 664 counties in these areas, 425 are sustained by funds from the central and other levels of government (see Table 25.1). With such a distribution, the problems that need immediate attention are those dealing with poverty in these mountainous areas. The poor areas fall into five categories: (1) old revolutionary bases; (2) minority nationality areas; (3) remote areas; (4) mountainous areas; and (5) backward and underdeveloped areas. As a matter of fact, mountains are the elementary geographical feature of the above five areas.

Table 25.1. The 18 poverty-stricken contiguous areas

Sr. No.	Areas	Provinces	Number of poor counties
1.	Daba and Qinling Mountain area	Sichuan, Shaanxi, Hubei, Henan	68
2.	Wuling Mountain area	Sichuan, Hunan, Hubei, Guizhou	40
3.	Wumeng Mountain area	Sichuan, Yunnan, Guizhou	32
4.	Dabie Mountain area	Hubei, Henan, Anhui	27
5.	Southeast Yunnan Mountain area	Yunnan	19
6.	Hengduanshan Mountain area	Yunnan	13
7.	Taihangshan Mountain area	Shaanxi, Hebei	25
8.	Luliangshan Mountain area	Shaanxi	21
9.	Northwest Guangxi Mountain area	Guangxi	29
10.	Jiuwandashan Mountain area	Guangxi, Guizhou	17
11.	Nuluerhu Mountain area	Liaoning, Inner Mongolia, Hebei	10
12.	Xihaigu Mountain area	Ningxia	8
13.	Dingxi Mountain area	Gansu	27
14.	Tibet		
15.	North Shaanxi Mountain area	Shaanxi, Gansu	27
16.	Southwest-Northeast Fujian Mountain area	Fujian, Guangdong	23
17.	Jinggangshan Mountain area	Jiangxi, Hunan	30
18.	Yimeng Mountain area	Shandong	9

Source: Study team.

OPERATIONAL CIRCUMSTANCES IN THE 18 POVERTY-STRICKEN MOUNTAIN AREAS

Constraints

In terms of social and economic development, the 18 contiguous poverty-stricken areas share the following common constraints.

Poor Infrastructure

These areas are all plagued by infrastructural problems such as poor transportation facilities, ineffective communication systems, severe shortage of technical personnel, poor skills of farmers, simple and crude medical equipment, and rampant endemic diseases. The 1986 statistics show that 16.1 per cent of the townships in the 664 poor counties could not be reached by highways; 22.2 per cent of the townships did not have electricity; and the average per capita consumption of electricity in the rural areas stood at 30.3 kWh only—55.8 per cent lower than the country's average. A comparative study indicates that, in 1985, there were approximately 68 technical personnel for every 10,000 people in 11 provinces (or autonomous regions) that reported a relatively concentrated poor population (about 23.2% less than the figure of 89 in 10 relatively developed provinces or municipalities); there were 5,000 people with primary school education among every 10,000 people in the 11 poor provinces (18.4% lower than the national average); there were 2,943 illiterate persons among every 10,000 people living in the poor provinces (24.6% higher than the country's average); there were only 21.7 hospital beds for every 10,000 people (9% less than the nation's average); and there were 30.3 medical workers among every 10,000 people (7% below the country's average level). According to a survey of the 109 poorest counties in China, 94 counties or 86.3 per cent of the sampled counties reported the existence of the main endemic diseases such as snail fever, *Keshan*, *Kaschin-Beck*, and fluorine poisoning.

Grave Ecological Imbalances

Many of the poor regions suffer from adverse agricultural conditions. For instance, the deep mountain areas suffer from serious soil erosion, the arid areas in Northwest China are frequently afflicted by drought, the karst region in Southwest China lacks both soil and water, and the red soil region in areas south of the Yangtze River and the poor regions in North China almost every year experience natural disasters such as droughts, floods, and severe salinization. Such adverse factors have directly led to a vicious circle of decline in agricultural production due to the ecological degradation experienced.

Excessive Population Growth

Studies show that the phenomenon of 'the poorer the people are, the more babies they have' has become widespread in many poor regions. The population growth rate in such regions often exceeds 2–3 per cent, almost 40–100 per cent higher than the nation's average growth rate. This is to a certain extent due to the fact that minority nationalities in China are exempted from the one-child policy.

Low Level of Social Development

The low level of social development in poverty-stricken areas, especially in areas where minority nationalities live, results in slow social development. Before 1949, in some of the poverty-stricken areas of China, social life still bore the characteristics of primitive communes. Social customs, lack of commercial mentality, ways of behaviour, etc., which had primitive communistic characteristics, imposed social restrictions on the development of a commercialized economy. External factors, too, have affected the productive capacity of these areas since 1949.

Potentials

In spite of the fact that the 18 areas face severe disadvantages, they have some comparative advantages over the plains, such as being specific habitats for special medicinal plants, fruits, and mineral resources. Generally speaking, the 18 areas have the following common potentials for economic activities.

Abundant Agricultural Resources

The 18 areas can be divided into six categories in terms of agricultural conditions. Each of them has its own special agricultural resources.

The first category is the border region between Eastern and Western China, including Qingba, Wuling, and Taihang mountain areas. These areas have the advantages of both the mountains and the plains. From south to north, mid-subtropical, north-subtropical, temperate, and central plains conditions prevail and from east to west, this area extends through humid, semi-humid, semi-arid, and arid regions. The diversification of natural conditions as well as the topographies, climates, lands, and resources make these areas suitable for developing forestry, agriculture, and animal husbandry, as well as for the diversification of their economies. To be more specific, in Qingba Mountain area there are 5,000 species of wild plants, including 1,216 varieties of medicinal herbs. There are also 627 species of wild animals, including 42 rare animals. In addition, this area grows and produces mulberry, tea, *Tung* tree, sealing wax, edible fungus, and medicinal herbs. Wuling Mountain area is known for its *Tung* trees, oranges, sealing wax, flue-cured tobacco, and ramie.

The second category is the karst region in Southwest China, including the Wumeng Mountains, the Southeast Yunnan Mountains, the Hengduanshan Mountains, the Northwest Guangxi Mountains, and the Jiuwandashan Mountains. The waterpower potential in this region is the highest in China, accounting for two-thirds of the country's total. However, so far only 2 per cent of the waterpower potential has been tapped. Situated in the humid subtropical zone, it has favourable conditions in terms of water supply and temperature and has mild winters and plant growing seasons. Subtropical perennial plants can be grown widely. This region is also one of the most important bases of forestry and livestock products as well as economic forests. Various native products and medicinal herbs, with the areas they are well suited to, are listed below:

- the Wumeng Mountains—forestry and animal husbandry;
- the Wumeng Mountains of Sichuan Province—flue-cured tobacco, oil-seed crops, fruits, tea, and forests;
- the Wumeng Mountains of Yunnan Province—tobacco, fruits, oil-seed crops, and phosphorus ore;

- the South-east of Yunnan Province—sugarcane, rubber, shellac, tea, coffee, spice, oil-seed crops, hides, Southern Chinese medicinal herbs and fruits; and
- the Hengduanshan Mountains—sugarcane, sealing wax, tea, *Tung* tree, walnuts, and fruits.

The third category is the arid region in Inner Mongolia, that is the Nuluerhu Mountain Area. This region enjoys abundant solar energy resources. It has a large amount of solar radiation due to long hours of sunshine. Besides, this region has a lot of untapped resources such as uncultivated land, grasslands, and woods.

The fourth category is East China's hilly areas, including Dabieshan, Southwest and Northeast Guangdong Province, and the Jinggangshan and Yimengshan Mountain areas. This region abounds in woods such as rainforests on the fringe of the tropical zone, evergreen broad-leaved forests and coniferous forests in the subtropical zone, and deciduous broad-leaved forests in the temperate zone. This region is one of the important forest bases of China. There are plenty of special economic forest products in these mountain areas. In addition, it has an ample labour force. The typical monsoon climate is favourable for agricultural production. This region can exploit and utilize its agricultural, forestry, animal husbandry, and mineral resources. The details are as follows:

- the Dabieshan Mountains—fruits, cattle, rabbits, sheep, geese, bamboo, mulberries, tea, Chinese medicinal herbs;
- the Jinggangshan Mountains—bamboo, tea, citrus fruits, etc.; and
- the Yimengshan Mountains—hawthorn, fruits, etc.

The fifth cateory is the hilly/ravine area on the loess plateau, including Luliangshan, Xihaigu, Dingxi, and the Northern Shanxi Province. This region has vast lands of which 70 per cent is loess, loose, fertile, and cultivable. However, the topography is rather complicated and land here is of various types. Through development, this region can become a mixed farming area for agriculture, forestry, and animal husbandry, as follows:

- the Xihaigu and Dingxi areas—commercial crops such as common flax, sallow thorn, Chinese medicinal herbs, and leather; and
- the North Shanxi Province—abundant flue-cured tobacco and fruits (has a great potential for developing animal husbandry).

The last category is the high mountain region in Tibet. The insolation here amounts to 2,600 to 3,200 hours per year. The percentage of insolation is 60 to 70 per cent. This region has plenty of sunshine. The extensive grasslands with quality grasses provide excellent pastures for animal husbandry. Yaks, Tibetan sheep, and goats are the three main types of domestic animal in this region.

Mineral Resources Suitable for Exploitation

There are plenty of mineral resources suitable for exploitation in China's poor mountain areas. For example, in Qinling and Daba, the key mineral resources suitable for exploitation are gold, silver, precious stones, coal, phosporous, cement rock, plaster stone, marble rock, and clay. There are 298 small proven mineral mines and deposits in the area. In the Hengduanshan Mountain area, there are abundant resources of gold, tin, mercury, stibine, asbestos, serpentine, coal, cement rock, cement burden, lead, and zinc.

Cheap Labour Force

Wage rates in the poor mountainous areas are much lower than in the plains areas. This is a very important advantage for the poor mountain areas in developing labour-intensive manufacturing, mining, and infrastructure. The government can mobilize agricultural surplus labour to participate in the construction of public projects such as roads and water conservancy. In recent years, the labour cost in coastal areas has increased very sharply; some labour-intensive production has tranferred from there to the less developed mountain areas. On the one hand the low wage rate is the result of low development, but, on the other hand, the low wage rate is an advantage for development.

DEVELOPMENT INTERVENTIONS: ANTI-POVERTY PROGRAMMES

Targets

In 1986, the Leading Group for the Economic Development of Poor Areas under the State Council had stressed that economic development in the 18 poor mountain areas was a long task that would have to be undertaken in two stages. These have been outlined below.

The First Stage (Near and Short-term Targets)

Efforts should be made to guarantee that there is enough food and clothing in the poor areas within a relatively short period of time; in other words, before 1990. During normal harvest years, 90 per cent of the people living in the poor areas should have enough food and clothing (except for households living on social welfare and the so-called 'five-guaranteed households', meaning the childless, infirm, and old people who are guaranteed food, clothing, medical care, housing, and burial expenses by the government). All the concerned provinces and autonomous regions should work out their respective standards and set schedules to ensure sufficient supplies of food and clothing in their poor areas according to the above-mentioned basic targets and the status quo of local economic development.

The Second Stage (Long-term Development Targets)

When the problem of guaranteeing enough food and clothing to the poor areas has been basically solved, efforts should then be shifted to regional economic development to help those poor areas to establish production bases that rely on local resources, operate on a fairly large scale, and are in conformity with social services. Morever, such areas should develop a series of products and key enterprises, by utilizing local resources and boosting the processing industry, so that they can bring the income level of the local farmer households closer to the average of the local provinces or autonomous regions in order to eradicate poverty as soon as possible.

MAIN PROGRAMMES

In order to achieve the above targets, the Chinese government has gradually implemented a series of programmes.

Financial Programmes

Funding for the 'Three Xi' Special Agricultural Construction

At the end of 1982, the Chinese government selected Xihaigu area of Ningxia and the adjacent Dingxi area and Hexi area in the middle region of Gansu as key areas to be given financial support. The government developed a 'Three *Xi*'—Ten-Year Development Plan from 1983 to 1992, and the central exchequer gave 2 billion yuan to support the programme. The purpose of this programme is to improve the infrastructure and the agricultural production conditions.

Fund to Support Underdeveloped Areas

The Ministry of Finance has been engaged in this programme since 1980, and it will last until the year 2000. The scale of funding is 800 million yuan per year, and its purpose is to promote social and economic development in poor areas.

Food and Clothing for Labour Services

From 1985 to 1987, the State offered 2.7 billion yuan of grains, cotton, and cloth as payment for labour services to help poor areas improve transport and water conservation facilities. During the three years, the poor areas in China built and revamped a total of 120,000 km of highways, tractor paths, and post roads (including the construction of 46,000 km of new State-class highways); built 7,200 bridges and 65 wharves; increased irrigated farmland by 172,600 ha and improved 7,033 million ha; repaired and consolidated 558 reservoirs and increased the capacity of generators at small hydropower stations by 1.56 million kW; and solved the water supply problem for 14.5 million people and 9.71 million domestic animals.

Middle and Low-grade Industrial Products

From 1989 to 1991 and from 1990 to 1992, the State has allocated middle and low-grade industrial products valued at 600 million and 1.5 billion yuan respectively to further strengthen the transport and water conservancy programmes in order to increase productivity and improve living conditions.

Credit Programmes

Special Interest Discount Loans for Poverty Alleviation

The People's Bank of China entrusted the Agricultural Bank of China to carry out this credit programme from 1986 to 2000. The scale of funding is 1 billion yuan per year. The input principle of this programme is less investment, quick results, easy business for every household, economic benefit to every household, and a solution to the clothing and food problem. Therefore, the priority is crop raising, animal and fish raising, and product processing.

Special Interest Discount Loans for Poverty Alleviation in Pastoral Areas

The source and purpose of this programme is the same as the one above. However, the term is from 1988 to 2000 and the scale of funding is 50 million yuan per year.

County-owned Industrial Loans

In order to improve the industrial structure, to increase agricultural product processing and the capability to use resources, and to raise the self-support ability of local financial institutions, the State-owned banks have introduced special credit programmes. These include the People's Bank of China, which allocates 400 million yuan annually (1988–2000), the Industrial and Commercial Bank of China, which allocates 300 million yuan annually (1988–2000), and the People's Construction Bank of China which allocates 100 million yuan annually.

Loans for Development

From 1984 until 2000, the People's Bank of China and the Agricultural Bank of China have agreed to respectively allocate 1 billion and 300 million yuan annually to boost the local economy in the poor mountain areas.

Preferential Policy Programmes

In recent years, the State and provinces (in autonomous regions) have each adopted some new policies aimed at alleviating the burden of the poor areas. The State has decided to exempt the poor areas from agricultural tax for three to five years, starting from 1985. Beginning in 1987, the State has also exempted poor counties that receive key State aid from taxes for energy and transport construction funds, and has lowered the proportion of bank reserves in those areas. At the same time, each province (or autonomous region) has formulated its own preferential policies to offer more benefits to poor areas. For instance, Guizhou Province has decided to reduce or even write off the quotas for grain purchasing contracts between the government and the poor counties; to offer income, product, and operation tax waivers for township enterprises for three to five years in such counties; and to lower the proportion of cash needed by poor farmers who apply for bank loans. In addition, the province has decided to exempt pupils studying at primary schools, at the township level or below, from tuition fees and textbook expenses and to allocate an additional 20 million yuan each year out of the local government reserved funds for economic development in the province's poor counties.

Mobilization of the Whole Society to Help the Poor Areas

Since 1986, the State Council has convened two conferences during two consecutive years to hear reports by its ministers and commissions on their efforts to develop the poor areas. The conferences called on the participants to play a leading role in supporting construction and economic development in the country's poor areas and to make greater contributions to this end. Twenty-eight central government departments were cited for their outstanding performance in this field at the conferences. So far, there are 36 State institutions that have established direct relationships with designated poor areas to support the latter's economic development. Generally speaking, all central government departments have taken active, prompt, and practical steps and have achieved initial results that are quite promising.

Meanwhile, economic development in the country's poor areas has won support from all social circles, including democratic parties and relatively advanced areas, large and medium-sized cities, scientific research institutions, colleges, and universities as well as the People's Liberation Army. Every year, a large number of volunteers, including teachers, doctors, researchers, skilled farmers and workers, and entrepreneurs come to the poor areas from other parts of the country to help formulate local development programmes, train local personnel, diffuse new technologies, and help solve local economic development problems.

EFFECTS OF ANTI-POVERTY PROGRAMMES

In order to relieve poverty in the 18 mountain areas, the Chinese government has been implementing the above-mentioned anti-poverty programmes since 1985. These have not only proved to be efficient in lowering the incidence of poverty but have also resulted in marked effects on the social and economic development of the poor mountain areas. Poverty is closely connected with the social and economic conditions in these regions, and is an integral part of the social and economic systems. Eliminating poverty will inevitably lead to social and economic changes, more significant than relieving poverty itself when one considers the restriction placed on the development of poverty-stricken mountain areas by their social and economic systems. Therefore, systematic development of poor areas requires a comprehensive plan and steady progress. China has gained noticeable achievements in this field over the past few years. However, due to incomplete statistics, the analysis here had to be based on cases.

Improved Infrastructure

The infrastructure of poverty-stricken areas is experiencing or will experience big changes. During 1985 to 1987, the Chinese government implemented a programme of Food and Cloth for Labour in association with the building of highways and water conservation facilities in poor mountain areas. This programme has promoted infrastructural construction there. Actually, quite a number of local government departments obtained the necessary materials such as steel, cement, and blasting caps by exchanging the allocated grain, cotton, and cloth. The peasant farmers worked even without pay. Moreover, a large portion of the funds for supporting underdeveloped areas in the State financial budget were used to better the infrastructure, especially between 1985 and 1987, when the funds were jointly used with the grain, cotton, and cloth mentioned above as a complete package in many areas.

For example, from 1985 to 1987, the following work was completed in the Yimengshan Mountain area in Shandong Province.

- The nine poor counties in the area completed 1,497 schemes for water facilities. Among them were 380 ordinary wells, 212 motor-pumped wells, 99 pumping stations, 710 cisterns, 90 diversion works and supportive facilities, and a 138.5 km long pipeline. A total of 238,000 people in 469 villages, or 19.5 and 21.5 per cent of the total population and villages, respectively, have benefited from these projects.

- The counties built 15,576 new schoolhouses and added facilities to existing ones. Compared with 85 per cent in 1984, 90 per cent of all school-aged children are now in school.
- More than 1,400 km of highways connecting county towns and smaller towns and 11,000 km of roads among the villages were built. Buses now have access to nearly half (1,903 villages) of the total number of villages previously not open to traffic.
- Newly erected power transmission lines transversed 183 km. They have enabled 1,058 villages to use electricity (30% of the total unelectrified villages in 1984).
- Also erected were 1,885 km of broadcasting line which made it possible for 801 villages (19.4% of those without) to receive broadcasts.

Government departments concerned with infrastructural construction have contributed a great deal to the improvement of living conditions in the poor areas. The Ministry of Communications, for example, drafted its Seventh Five-Year Plan (1986–1990) to improve communications in these areas. The plan included 18 large or medium-sized highways (8,500 km) and an investment of 190 million yuan in highway and waterway traffic below county level. The Ministry also allocated 10 million yuan for railways, 2,000 traffic management personnel and technicians for 201 needy counties in 12 provinces and autonomous regions, and 50 million yuan to improve local traffic conditions. The former Water Conservancy and Electric Power Ministry's plan to assist poverty-stricken areas covers the years up to 2000. In 1987, the Ministry set aside 13 million yuan to help poor counties across the country to build one or two small projects each. Another 2 million yuan for five needy counties in the Three Gorges Area went to construct drinking water facilities for 40,000 people and 30,000 domestic animals. The Ministry also provided 12 billion kilowatt-hours of electricity to people situated far away from reservoirs and 70 million killowatt-hours to old revolutionary bases, remote areas, and minority areas. A total of 166 million yuan was allocated by the Ministry to construct more than 30 power generating projects. Another 32 million yuan was provided for 36 counties in the country which were experimenting with electrification. Ten other poverty-stricken counties in Sichuan and Jiangxi provinces and Ningxia Hui Autonomous Region received more than 60 million yuan to help build power stations.

Closed and Semi-closed Conditions Undergoing Changes

Most of the poverty-stricken mountain areas are closed or semi-closed to the outside world and live within traditional social and economic systems. Their few contacts with other parts of the country and the world outside have strong administrative characteristics. Previously, the connections of these areas with the outside world were limited to the transfer of public employees, raw materials, and consumer and relief goods. All these aspects are constrained under their economic and social systems. For reasons such as these and inaccessibility these areas became more isolated, until recently, when the government strengthened its aid to poverty-stricken mountain areas, resulting in a gradual increase of the social and economic contacts and a break down in the closed or semi-closed conditions.

One way to alleviate such conditions in poor areas is to provide advanced areas with surplus labour from the poor ones. The Jinggangshan Mountain Economic Development Working Group of the Ministry of Civil Affairs helped send more than 5,000 people from the region to work in other parts of the country, and this stimulated local labour emigration. Through these labourers, local people changed their ideas and increased their income, broadened their perspectives, and trained skilled personnel.

Another method of breaking the closed and semi-closed conditions is to introduce personnel, technologies, information, and management experiences from the advanced areas to the impoverished areas. A noticeable trend in advanced regions is that some local crafts and industries, such as straw weaving and carpet weaving, are now gradually moving to poor areas due to rising production costs, especially rising labour costs. The transfer of such industries to different regions is due to economic pressures. Administrative forces play no role here. The poverty-stricken areas were involved in the labour division system in the national economy. Technical personnel go to poor areas to establish township enterprises through contracts and promote technology in agricultural production, playing an important part in opening up these areas.

The breaking up of closed and semi-closed conditions has given a great impetus to the development of the market economy in poverty-stricken areas. To receive government assistance a product must be able to earn profits, since the aid has changed from grants to loans. As the government has its assistance, the previously autarkic economy in poor areas has started to integrate with the market economy. The development of the market economy, in turn, further breaks down the closed and semi-closed condition of the impoverished areas. The development of the market economy and the breaking down of closed conditions complement each other.

The Economic Behaviour of Poor Families

The economic behaviour of poor families has changed greatly due to the government's policy adjustment on assistance. They are gradually adjusting to the standards required by the market economy. Linqu County in the Yimengshan Mountain area, for example, reclaimed 20,600 ha of undeveloped hills, slopes, sands, and water during 1985 and 1986. They included 3,000 ha of timber forests, 234 ha of water to raise fish and lotus roots, and 7.07 million fruit trees of various kinds. The County also built 2,520 km of mountain roads and completed 1,801 new water conservation projects. The total man-hours spent on such projects by local farmers was 336 million or 1,016 for every farmer on an average. The farmers were enthusiastic and their work was of unprecedented speed and quality. This time, instead of external political pressure for self-supply, it was the internal economic impetus and impulse for expansion that governed the farmers' enthusiasm. In the late 1950s and 1970s, the people's political responsibility directed their economic behaviour and they scarcely cared whether what they were doing would benefit them or not. The change in policy brought out changes in the farmers' economic behaviour. Improvements in the external economic environment, such as the service system, marketing channels, and so on, resulted in a rise in the farmers' marginal propensity to invest. Farmers became more and more interested in self-development.

According to our investigation, the 'minimum critical level of consumption' is 150 yuan of per capita annual consumption. Below that income level, farmers use outside funds to satisfy their most elementary consumption needs. If their per capita annual income is more than 150 yuan, they use the surplus money either on improving living conditions or on investment.

Scope for New Economic Growth and Training Pioneer Entrepreneurs

The government's policy of helping poor areas to set up economic concerns and, through them, providing poor farmers with enough to feed and clothe themselves, has helped to widen the scope for new economic growth and to train pioneer entrepreneurs. The poverty-relief organizations promoted local economic improvement through investment. Our investigation into 27 poverty-relief organizations in the Luliang Mountain area in Shanxi Province gave us a general idea of how such activities are promoting regional economic development.

At present, we cannot exactly estimate the revenue change in these counties, but increases in investment are affecting revenue in two ways: (1) they offer new markets to local products, thus upgrading the regional economy through demand increase, and (2) the investments provide more productive materials for local areas, thus pushing the regional economy ahead. We can analyse the effect of aided businesses, through the integration with the local market, on the economy in poor areas. Table 25.2 shows those businesses that obtained 68 to 98 per cent of their inputs locally, and sold 47 to 66 per cent of the output in local markets. The poverty-relief organizations become growth points of the regional economy.

Table 25.2. Sources of input for and purchasers of output from enterprises developed with poverty-relief funds

County	Number of organizations sampled	Sources of raw materials consumed by organizations(%)		Purchasers of output (%)	
	Within county	Outside county	Within county	Outside county	Within county
Linxian	13	68	32	63	37
Xingxian	9	84	16	47	53
Kelan	5	98	2	66	34
Total	27	79	21	59	41

At present the government pays special attention to establishing two types of enterprise, processing enterprises which can purchase raw materials from nearby poor families, such as farm produce and raw minerals, and enterprises that employ families for processing, such as carpet weaving. The government also trained a group of entrepreneurs for needy areas by establishing the organizations needed. The managers of these poverty-relief organizations were village cadres or soldiers who had been to places outside the counties. Those who had matured through years of managing experience were the elite among the local people. The establishment of an entrepreneur stratum has undoubtedly had a positive effect on local economic development.

Traditional Culture: Confronted with Forceful Challenges

Traditional culture in poverty-stricken areas faces strong challenges as the market economy develops and outside contacts increase. Many cultural ideas and social customs unfit for the development of a market economy have gradually changed. Economic development promotes social progress.

For example, while the agricultural surplus labour in coastal areas was turning to industry and commerce, people in poor areas resisted labour emigration, because they regard working outside as the 'selling of labour and being exploited' and being children's nurses in urban families as 'being slaves'. However, years of effort from the cadres sent by the government to aid the impoverished areas, and success stories from the first group of labourers who left the areas, have led to obvious changes in the local people's ideas. Many of them have now begun to be active in registering for work outside, but previously they believed that all businessmen were treacherous. When they occasionally sold some goods at country fairs, they would avoid forming relationships with outsiders. Mostly the surplus goods were given to relatives or neighbours. Such things are rarely seen nowadays. People are gradually basing their principles on the market economy and selling more goods instead of sending them out as gifts. We discovered an interesting 'toilet revolution' in the Luodian area of the mountains of Southern Guizhou Province, inhabited by the *Buyi* and *Miao* minority nationalities. People here regard human waste and domestic animal waste as too dirty to be used for farming and for quite a long time have had neither toilets nor stys. But vegetable experts in the Scientific Institute of Agriculture of the Province helped local peasants grow early maturing vegetables, suited to the county's high temperatures in spring and winter. The vegetables fill a seasonal vacancy in the markets. Fertilizer application then became a must since vegetable growing there was intensive. As a result, every household started to build toilets and stys in pens, leading to the toilet revolution. The local government had long advised people to do this, but without success. The experience was enlightening. Only through the demands of the market economy can there be any fundamental change in the traditional concepts of value, thinking habits, and modes of behaviour which gives rise to social evolution.

CONCLUSIONS AND SUGGESTIONS

Conclusions

From the Chinese government's anti-poverty programmes mentioned above, we can sum up the main successful experiences.

Anti-Poverty through Economic Development

In the past three decades, the State spent a lot of money and effort in helping development in poor areas and in solving the key problems of poverty. Though there were some achievements, overall, the results were not satisfactory. The main problem was that most of the money was used for temporary relief work instead of for investment in efforts to help poor areas and poor farmer households to build the capability for self-development. Therefore, such aid not only failed to solve the basic problems of the efficiency of fund use and the augmentation of local production, but also encouraged some areas and farmer households to be dependent on the government. After realizing the drawbacks in its old policies, the State Council decided to make substantial adjustments in its policies for poor areas by shifting the policy focus to development and reform. The new economic development policy for poor areas calls for efforts to arouse the initiative of the cadres and the masses in the poor areas; bring into play the spirit of self-reliance; stimulate such areas' internal vitality for economic development; conduct development production and construction by tapping abundant local natural resources with necessary State support; and gradually help poor areas to form the capability for self-development in introducing a commodity economy, solving the problems of shortage of food and clothing, and eradicating poverty.

Increase of Investment Intensity

Considering the economic characteristics of the poor households, the government holds that poverty-relief funds should be used according to the nature and priorities of the national policies, and the 'super-stabilized structure' of poverty should be broken down. If the funds are equally distributed, the 'minimum consumption trap' cannot be avoided. And, as a result, the economic development approach may not be effective.

As an alternative to consumption subsidy, the government, through village and township institutions, buys some productive inputs and distributes them to poor households instead of giving them cash. This enables the poor to expand reproduction under a condition of low-level consumption. If specific restraints are not placed on the use of poverty-relief funds, the minimum consumption trap may not be overcome. Such tendencies are contrary to the purpose of an economic development approach.

Introduction of the 'Project Management' Method

Project management is one of the major contributions made by the World Bank to global economic development. The World Bank has accumulated a lot of experience in project management. The Chinese government introduced this method in its poverty-relief efforts and issued 'Provisional Regulations on Project Management in Poor Areas'. Project management is an effective way to approach poverty relief. Distributing funds according to project priorities not only improves the efficiency of fund use, but also avoids misuse and embezzlement of poverty-relief funds.

Strengthening the Organization of Economic Development

Since China has long practised centralized planning, even in remote poverty-stricken areas, administrative institutions are well-established. With the economic reforms that have taken place in the past few years, this extensive government network has an important role to play in organizing poor households to participate in economic development by helping them with planning, project selection, and implementation. The State cannot do a good job in the area of poverty relief without the participation of local organizations.

Suggestions

Currently, China is undertaking economic and political reforms, and, although the Chinese government has made many adjustments to its anti-poverty programmes, further improvements are still necessary and urgent.

Adjustments in the Industrial Structure

To develop the economy in poverty-stricken areas, appropriate industrial policies are needed. The poverty-stricken areas have devoted most of their resources to grain production and their natural ecological balance has been destroyed. This leads to low grain production and slow development of non-agricultural undertakings and rural industries. Recently, with the adjustment in the industrial structure, the poverty-stricken areas have begun to pay attention to the development of non-agricultural undertakings and rural industries. However, some areas seem to go too far. A tendency to neglect grain production is emerging. Industrial policies with flaws mean failure in economic development. Therefore, we must be clear about the basic aspects of industrial policies.

Ensuring Food Sufficiency by Increasing Supplies and Cutting Demands

The grain production situation all over the country cannot be viewed optimistically. It is very difficult for the government to allot grain in large quantities to the poor areas, even when the country on the whole enjoys a large supply of grain, because of transportation difficulties and price differences. So self-sufficiency remains the basic target of the country's food policy for poverty-stricken areas. Measures to secure food sufficiency include: (1) employment of cheap labour to undertake farmland capital construction and gradually improve agricultural production conditions and (2) transformation of traditional production with high technology and sophisticated equipment and extension of quality strains and intensive farming to increase per unit yield. Measures to cut demands for grain include appropriate control of industries, while greatly increasing pastured livestock, poultry, and fruit-based wine making.

Developing Rural Industries and Household Agricultural Undertakings

Efforts should be made gradually to transfer the labour-intensive and less profitable industries of the developed areas to poverty-stricken areas. Obviously, poor areas have a large, cheap labour force. We should take advantage of this in the national division of labour.

Promoting Labour Emigration

Unemployed and under-employed labour resources constitute a big waste in poor mountainous areas. Emigration of labour not only provides employment and income but also eases the population pressure on natural resources. Labour emigration is an industry with economic, social, and ecological benefits. While forming industrial policies, priority should be given to it.

Restraining the Government's Economic Behaviour

Local government departments, especially those with county-relief funds, currently independently design projects. The local officials tend to initiate projects that will bring large profits rather than those that will benefit the poor. To understand the local officials' preference, we must analyse the budgets and expenditures of the local governments.

At present, many counties have deficits due to excessive expenditure and low revenue. Some counties cannot balance their budgets, even with their superior government subsidies. Budget limits are often exceeded. Administrative, educational, and operational funds account for the high expenditure. We can see that government officials and the non-agricultural population are closely linked to the budget expenditures. Budget deficits affect the welfare of this group. This explains why government officials prefer to initiate projects that can generate tax revenue.

Having understood the features and motives of local officials, we suggest that the following two measures be adopted to guide them. On the one hand, the examination and supervision of projects should be enhanced and the poverty-relief funds from the central and provincial government invested in projects that are developed to solve food and clothing problems for poor households. In the process of analysis, special attention should be paid to the distribution of project benefits. On the other hand, the special conditions in poverty-stricken areas should be taken into account by the central and provincial governments while macro-economic policies are being formulated, and the problems, despite difficult conditions, should be solved. For instance, it is irrational for the central government to offer uniform loans to the localities. Regulations should differ among different areas. If the central and provincial govenments extend some more financial allowances and special care to the counties needing relief in mountainous areas, the unfairness of the government officials using funds to enrich the county will be eliminated and more funds will be used to solve the problem of basic needs. Poverty relief in mountainous areas will thus be greatly accelerated.

The thrust for county enrichment is understandable, but it should be guided by the future strategy in economic development. Future plans might be based on the attainment of basic needs but the emphasis on poverty-relief work should gradually change from the support of poor households to regional economic development on a large scale. Transition of strategies can only be realized step by step. The implementation of the strategy should be guided according to the differences between regions.

More Emphasis on Training in Anti-poverty Programmes

In order to raise the executive efficiency of anti-poverty programmes, the Chinese government pays due attention to personnel training. However, the training activities conducted in the past few years were mainly for leaders. Farmers' training has been neglected to a great extent. We believe that training priority should be given to farmers who are directly involved in anti-poverty programmes. The basic principle should be to train those who implement programmes and to train those needed for the anti-poverty programme.

26

THE EXPERIENCES OF AN AREA-BASED DEVELOPMENT STRATEGY IN HIMACHAL PRADESH, INDIA

L.R. Verma and T. Partap

L.R. Verma and Tej Partap are working as professional staff with the Mountain Farming Systems Division of ICIMOD

Oxford & IBH Publishing Co. Pvt. Ltd., New Delhi, India.

CONTENTS

INTRODUCTION

Himachal Pradesh is a small Indian state situated in the heart of the Western Himalayas. Its entire territory is mountainous with altitudes ranging from 350 m to 7,000 m. It has a total population of 4.28 million (0.63% of the total Indian population) settled throughout the valleys and on the mountain slopes and ridges in 16,916 villages and 39 towns. Tables 26.1 and 26.2 give the population figures of specific areas of the Indian Himalayas and a demographic profile of Himachal respectively. There are no indications of plentiful arable land holdings and rich farmers. The data clearly show that, in terms of fertile land, resource scarcity, the size of land holdings (Table 26.3), and irrigation, Himachal is no different from other mountain areas of the Hindu Kush-Himalaya region.

Table 26.1. The Indian Himalaya: area, population, villages, and political units (states)

Political units/states	Geographic area (sq km)	Population in millions (1981)	Population density per sq km (1981)	Number of villages (1981)
Western Himalayan zone				
Jammu and Kashmir	222,236	5.987	43.09	6,503
Himachal	55,673	4.281	76.89	16,916
Hill areas of Uttar Pradesh	51,125	4.835	94.59	15,010
Eastern Himalayan zone				
Sikkim	7,299	0.313	42.88	405
Meghalaya	22,489	1.336	59.41	4,583
Tripura	10,477	2.053	195.95	4,727
Manipur	22,356	1.421	63.56	1,946
Nagaland	16,527	0.775	46.85	960
Mizoram	21,087	0.494	23.42	229
Arunachal Pradesh	83,578	0.637	7.62	297
Hill areas of Assam	15,212	6.3	7.62	1,931
Darjeeling Hills of West Bengal	2,417	0.507	209.76	507
Total for Indian Himalaya	530,476	28.939		54,014
Total for India	3,287,263	685.185	208.4	575,000

Source: Compiled from *Agroclimatic Zones: Profiles and Issues* ARPU Working Paper No. 2 Planning Commission, India, ARPU, Ahmedabad, 1989 and other sources.

But if the human development process, in terms of available choices and access to resources essential for a decent standard of living, is any gauge (UNDP 1990), then Himachal as a state seems to have achieved better results than many other mountain areas within the HKH region (Tables 26.4 and 26.5). Appropriate development planning enabled the state to accomplish a rapid transformation within a short period of two decades. Today, by and large, people living in the state have access to essential comforts and therefore a better standard of living. Singh (1989), while reporting on the progress made by Himachal, stated that it belies the myth that mountain regions are condemned to be poverty stricken because of their geography. Development observers consider the State to be an example of a success story in terms of development planning for mountain areas, because Himachal provides its people with higher incomes, better employment

Table 26.2. Himachal at a glance

		Year 1984/85
Total area		55,673 sq km
(Survey or General of India)		
(By village revenue records)		3,215,375 ha
Forests		874,331 ha
Total cropped area		988,059 ha
Net area sown		580,025 ha
Gross irrigated area		169,007 ha
Percentage of the gross cropped area		17.2
Net area irrigated		95,044
Percentage of net area sown		16
Intensity of cropping		170.4
Total output of food crops		1,061,034 MT
Total output of non-food crops		4,772 MT
Population	1991	1981
Persons	5,111,079	4,280,818
Male	2,560,894	2,169,931
Female	2,550,185	2,110,887
Density of population per km^2	92	77
Decennial growth rate (1981–91)	19.39%	23.71%
Sex ratio (number of females per one thousand males)	996	973
Literacy rate	53.29%	42.48%
Percentage of rural population	92.0 %	92.39%
Percentage of scheduled castes		24.62%
Percentage of scheduled tribes		4.61%
Number of districts		12
Number of households		783,974
Number of villages 1981 census only		
Total		18,721
Inhabited		16,807
Uninhabited		1,914
Total livestock population (based on Thirteenth Livestock Census, 1982)		4,988,540

Source: Annual Season and Crop Report for Himachal Pradesh, 1984/85. Directorate of Land Records, Himachal Pradesh.

opportunities, superior social services to meet local development needs, a responsive planning and administrative apparatus, and an enthusiastic political culture. The State has created a new concept in the development of hill economies through the transformations brought about in agriculture, horticulture, and animal husbandry. Development of an extensive infrastructure and the harnessing of the natural resources of the mountains, along with the necessary precautions against negative side-effects, are the noteworthy aspects of these development initiatives. The significant point is that this success has been achieved without using the conventional strategies which focus mainly on industrialization. Finally, whereas the dominant picture characterizing most of the mountain areas in the HKH region reflects a state of stagnation, Himachal's achievements appear unique.

Table 26.3. Status of farmers and farming

Total area of operational holdings (Agricultural Census, 1981)	980,425 ha
Average size of land holding of the family (1984/85)	1.5 ha
Average number of *parcels* per land holding (1984/85)	6
Average area per *parcel* (1984/85)	0.26 ha
Marginal farmers (below 1 ha) (1984/85)	52.2%
Small farmers (1–2 ha) (1984/85)	22.0%
Semi-medium farmers (2–4 ha) (1984/85)	15.1%
Medium farmers (4–10 ha) (1984/85)	6.6%
Larger farmers (above 10 ha) (1984/85)	1.1%
Gross irrigated area (1987/88)	147,900 ha
Net Irrigated area (1987/88)	98,026 ha
Total output of food crops (1987/88)	892,700 MT

Source: Annual Season and Crop Report for Himachal Pradesh, 1987/88. Directorate of Land Records, Himachal Pradesh.

Table 26.4. Basic indicators of growth in Himachal Pradesh

Indicators	1967/68	1972/73	1982/83
Population (millions)	3.22	3.57	4.28
% of population living in rural areas	N.A.	94.97	92.38
Population density/km^2	57.8	64.5	77.0
Net State Domestic Product (Rs in millions)	1,830	2,391	2,969
Per capita income (Rs/yr)			
(a)at current prices	528	769	1,658
(b)at 1970/71 prices	568	669	686
Literacy rate	21.27	31.96	42.48
No. of doctors per million of population	21.8	203	296
No. of hospital beds per million of population	1,440	1,270	1,355
No. of hospitals and dispensaries	480	590	831
% of villages electrified	6.15	24.83	75.63
Per capita domestic consumption of electricity (kWh)	3.1	5.6	75.63
Electricity generated (million kWh)	3.7	162.6	540.5
Mileage of roads (km)	4,308	7,609	13,600
(a) Per 100 sq km of area	8.72	16.85	24.44
(b) Per thousand of population	1.51	2.61	3.18

Source: Statistical Outline of Himachal Pradesh (various issues). Shimla: Directorate of Economics and Statistics, Himachal Pradesh Government, 1985.

Table 26.5. Selected socioeconomic indicators for hilly regions of the Indian Himalaya in comparison to Himachal

Regions/ states	No. of villages (1981)	Surface road length 1,000 sq km. (1980–81)	Electricity consumption per capita kWh (1977–78) capita	Villages supplied with electricity % (1984)	Post offices 1,000 pop. (1984)	Educational institutions pop. 1,000 (1981'–82)		Public health facilities 1,000 pop. (1983)		
						Primary	High School	Hospitals	Dispensaries	PHC
West Himalayan zone										
Jammu and Kashmir	6,503	5.2	65.0	85.2	0.22	1.25	0.48	0.06	0.11	0.02
Himachal Pradesh	16,916	35.4	54.2	83.7	0.55	1.46	0.48	0.01	0.26	0.02
		38.2[1]	147[2]	100[3]	0.50[3]	1.45[3]	0.41[3]	0.01[3]	0.13[3]	0.03[3]
Uttar Pradesh hills	15,010	24.3	82.5	35.4	N.A.	0.02	0.04	0.10	0.26	0.02
West Himalayas	38,429	21.6	67.3	68.1	0.4	0.9	0.3	0.03	0.21	0.02
East Himalayan zone	575,540	21.8	18.5	33.0	0.2	1.48	0.53	0.01	0.12	0.04
Sikkim	405	14.5	30.2	38.0	N.A.	1.49	0.28	0.02	0.15	0.09
Meghalaya	1,931	8.4	24.8	58.8	0.01	N.A.	N.A.	0.01	0.08	0.06
Assam Hills	1,946	26.1	4.5	20.3	0.28	2.01	0.51	0.01	0.15	0.09
Tripura	4,727	74.7	9.3	17.0	0.25	1.36	0.25	0.01	0.15	0.01
Nagaland	960	35.3	23.3	93.7	0.21	1.46	0.54	0.04	0.15	0.02
Mizoram	229	5.4	5.3	13.3	0.82	1.57	0.92	0.02	0.31	0.04
Arunachal Pradesh	297	15.7	38.1	21.4	0.38	1.47	1.02	0.03	0.10	0.07
Darjeeling W.B.	507	2.9	6.5	11.4	0.24	1.98	0.30	0.01	N.A.	0.02
Indian Himalayas	613,969	21.1	42.3	50.5	0.3	1.19	0.4	0.02	0.16	0.03

[1] For the year 1984/85

[2] For the Year 1986/87

[3] For the year 1988

Source: Compiled from Agroclimatic Zones: Profiles and Issues. ARPU Working Paper No. 2. Planning Commission, India, ARPU Ahmedabad 1989 and other sources.

THE VISIBILITY OF SUCCESS

The indicators of success in Himachal Pradesh are numerous and varied, depending upon the interests and capabilities of the evaluators. For instance, the availability of one horticultural product, i.e., apple juice, at practically every railway station or major bus stand and airport in India is a symbol of success for the State and for the common man. The electrification of villages, the elimination of landlessness in rural areas, the extensive network of all-weather roads, the transformation of extensive wastelands into productive orchards, the increased marketability of hill products, fair prices for producers, generally improved incomes, and access to communication, education, and health facilities are some of the major indicators of transformation of interest to those researchers who like to define development in terms of qualitative changes. Singh (1989) reported that signs of Himachal's development were not simply statistical but concrete. He described this visibility of success in the following ways. 'Grocery stalls in the interior towns are full of easily affordable fresh produce. The government ration shops dispense basic cereals at cheap prices. The highways are wide and roomy by hill standards and are equipped with large night-time reflectors that are more commonly found on European motorways. While cooking gas is rarely found in most Himalayan urban centres, red natural gas canisters can be seen everywhere even in the most remote areas of Himachal. Even the street vendors have discarded kerosene and firewood in favour of the clean blue flame. For statisticians and economists, the State provides time series and other data on almost every aspect of economic research. Statistics show clearly that the people have a development record unequalled throughout the Himalaya. They had a literacy rate of 43 per cent compared to that of 40 per cent for the affluent Punjab and an Indian national average of 35 per cent in 1981. Safe drinking water is provided to 93 per cent of the population. For planners from Nepal, Bhutan, Tibet, and the other hilly states of India, these statistics are staggering.' The quantitative dimensions of change will be presented later while discussing the factors and processes behind Himachal's success.

FRAMEWORK FOR ASSESSING DEVELOPMENT

Depending upon the nature of the enquiry and the sectoral affiliations of commentators, a large inventory of factors contributing to the achievements of Himachal can be found. However, to derive meaningful lessons and replicable experiences from Himachal, the whole process of change has to be analysed within an integrated framework. This framework, as outlined by Jodha (Chapter 2), consists of a development approach based on the mountain perspective. By mountain perspective, one means that one applies explicit or implicit considerations of specific mountain characteristics and conditions while designing and implementing development options. These mountain characteristics are: inaccessibility, fragility, marginality, diversity, 'niche', and people's adaptation mechanisms. As listed by Jodha (Chapter 2), these characteristics generate a number of objective circumstances. The latter act as sources of constraints and opportunities in mountain areas. The appropriate handling of constraints and the harnessing of opportunities are the basic factors behind successful development in mountain areas. Himachal's development performance has been summarized in an ICIMOD Workshop Report (1990). In the following discussion, we will examine Himachal's development experience within the context of the above framework. In other words, we will consider how mountain specificities were understood and applied, while planning and implementing the programmes in Himachal, in order to make a successful case of mountain area development. We use 'area development' in the sense that Himachal is completely a mountain area and that, despite its internal heterogeneity, an integrated perspective was followed while designing its development strategies.

For further enquiries into a mountain perspective-based development approach the more detailed hypothetical queries would be as follows:

(1) To what extent were mountain constaints, such as inaccessibility, marginality, and fragility, overcome through development interventions?
(2) How were mountain characteristics, such as agro-ecological diversity and 'niche', which offer several unique opportunities and have comparative advantages over the plains, harnessed by the State?
(3) In what way has exploitation of natural resources through development interventions affected sustainability and the quality of life issues in order to make Himachal a successful case of mountain area development?

The development scenario in Himachal Pradesh shows that the constraints and opportunities provided by mountain specificities are not uniform. Their scales would differ from area to area. It is these scales of impact of mountain specificities which decide the priorities for appropriate development in a particular area. This paper explains how Himachal successfully identified the right development priorities leading to the desired impacts.

DIMENSIONS OF SOCIOECONOMIC MARGINALITY AND APPROACHES

The marginality of mountain populations from the mainstream sociopolitical culture is widely accepted. It is one of the basic reasons for the general neglect of local, need-based development in mountain areas. Remedial measures warrant an understanding of the adaptations to changing circumstances. Himachal Pradesh is no exception to this mountain characteristic. It was recognized as a major constraint in the progress of the State as soon as Himachal came into existence (Singh 1988).

The Struggle for Statehood (1947–1971)

From 1947 to 1971, the hill communities of this mountain area fought a peaceful struggle against sociopolitical marginality (Singh 1988). The aim of this struggle was to receive recognition from the Indian Union as an autonomous state capable of managing its own governance and development planning affairs. From 1948 to 1951, while India was trying to organize itself as an emerging democratic nation, the small princely states of Himachal were merged into a single hill state, as a consequence of demands from the local mountain communities. It led to the direct administration by the Government of India, which meant a negligible role for the local people in development planning. The administrative system was changed in 1951 when Himachal became a 'part C' state, with an elected government under the control of the Lieutenant Governor. It had the opportunity to launch its First Five Year Plan (1951–1956). Priority was given to the development of roads, education, agriculture, and public health. Because of the limited resources and powers of the state government, the implementation of programmes was not satisfactory. Eventually, political autonomy and a free hand in development planning were achieved when Himachal Pradesh was declared a Union Territory in 1956. A territorial council was established to comply with the people's demand for representation in the development and planning process. The second five-year plan was launched during this period without any significant achievements. In 1963, the status of Himachal Pradesh was again upgraded. It was given a legislative assembly and a popular government. The third five-year plan was launched during that period and it continued the trend of giving priority to roads, followed by horticulture and hydropower development.

Until 1966, a major part of the present-day Himachal was under the administrative control of the Punjab, a prosperous state in the northern Indian plains. The people of the hilly area of the Punjab and the then Himachal made joint efforts to put forward a case for the merger of these areas into one mountain state. In 1966, this demand was accepted by the Government of India. The hill districts of the Punjab were merged with those of Himachal Pradesh in 1966, doubling its population and its size, which increased from 27,000 sq km to 55,673 sq km. The final step in the removal of sociopolitical marginality was accomplished when Himachal Pradesh became the 18th full-fledged state of the Indian Union and was granted autonomy in political, administrative, and development planning processes.

Post-1971 Focus on Underprivileged Sections of Mountain Society

The issue of socioeconomic marginality within Himachal Pradesh is also equally important. It has a sizeable population of tribal people, backward classes, and scheduled castes (Joshi 1984). The majority of these three categories of people lived below the poverty line and were not included in the mainstream. For the sake of convenience, each of these marginal communities is discussed separately.

Tribal Areas

The tribal areas of the state constitute 42.49 per cent of its geographical area and 3.13 per cent of its total population (Table 26.6). Physical inaccessibility, remoteness, conservatism, poor mobility, severe climates, sparse populations, and centuries of alien rule had resulted in the marginalization of these areas from the mainstream. Glaciers, high altitudes, and extremely rugged terrain, transversed by fast-flowing rivers and their tributaries, are the striking features of the tribal belt. Some of these areas remained inaccessible for more than six months a year, and this has far-reaching effects on the socioeconomic conditions of these people. The economy of tribal areas was predominantly agro-pastoral.

Initial efforts for the development of the tribal areas started in 1955 in the form of 'Special Multipurpose Tribal Development Blocks'. However, such programmes were run in an *ad hoc* manner, mainly in the form of welfare schemes, without considering the mountain specificities of these areas. As such, there was no visible impact on the quality of life of the people belonging to scheduled tribes up to the period of the fourth five-year plan. A new strategy for the development of these tribal areas was implemented from the beginning of the Fifth Five Year Plan. The salient features of this strategy were:

— Division of all tribal areas into two categories, i.e., areas with more than a 50 per

Table 26.6. Statistical profile of tribal area

Item	Unit	Ref.	Tribal	H.P.
Area	Sq km	1981	23,654	55,673
			(42.49)	(100.00)
Population	No.	1981	133,432	4,280,818
			(3,13)	(100.00)
Density of population per km^2	No.	1981	6	77
Scheduled tribe population	%	1981	77.52	4.61
Decennial growth rate	%	1971–1981	17.14	23.71
Workers	%	1981	57.91	42.38
Agricultural Workers	%	1981	63.99	70.81
Literacy	%	1981	30.73	42.48
Cultivated area agricultural workers	ha	1980–81	0.45	9.59
Area under fruits	ha	1982–83	4,939	108,676
Area under Forests	ha	1982–83	444,625	2,114,200
Average size of operational holdings	ha	1976–77	1.4	1.5
Gross irrigated to net area sown	ha	1980–81	45.23	16.47
Livestock population	No.	1977	304,874	4,795,226
			(6.36)	
No. of blocks	No.	1983–84	7	69

Source: Statistical Outline of Himachal Pradesh, 1985.

cent tribal concentration and those with dispersed tribal populations and initiation of development efforts, in the first instance, in the former category of area,

— Funds for the tribal sub-plan were allocated from the state's general plan rather than from the funds for welfare of backward classes, as was formerly practised,

— Special financial assistance from the Central government to supplement the state sector sub-plan for the development of tribal areas, and

— Identification of thrust areas for tribal development in the following sectors:

- infrastructure: minor irrigation facilities, soil and water conservation, cooperation, rural roads, and land reforms,
- social services: drinking water supplies, general education, technical education, and health, and
- production: agriculture, animal husbandry, dairy development, fisheries, forests, and small village and cottage industries.

Although only 3.11 per cent of the population lives in tribal areas (Table 26.6) the State plan investment reached the level of 9 per cent during the Seventh Plan (1985–1990) from 3.65 per cent in 1974/75. In order to make working conditions in tribal areas more attractive, a scheme of enhanced compensatory allowances, ranging from 120 to 180 per cent of the total pay, was made available to government employees, and this continues even today. Special winter allowances and housing facilities are also provided. Government officials working in tribal areas are given special financial, technical, and administrative powers so that development activities do not suffer due to bureaucratic procedures. Provisions also exist for the non-diversion of tribal plan budgets and such budgets are not tied to the otherwise prevalent system of fixed expiry dates.

Backward Areas

Generally, hill areas have been categorized as underdeveloped, because of the disparity of economic development between them and the lowlands. However, even within these mountains, interregional disparities exist in the overall context of socioeconomic development. For example, in Himachal, the apple belt and off-season vegetable areas can be categorized as affluent by mountain standards, but there are a number of less developed areas that suffer because of either environmental conditions or untapped potentials. These areas are economically marginalized and some people may be living at subsistence level. Such areas were categorized as backward areas, although all the people living there may not be tribals or from underprivileged classes. The following economic indicators were used to identify such areas in Himachal.

(1) Remoteness and inaccessibility.
(2) Demographic indicators: 25 per cent of the population belonged to scheduled castes or tribes, at least 90 per cent of the population engaged in primary occupations, such as agriculture or animal husbandry, and less than 20 per cent of the children between the ages of 6 and 14 years attended school.
(3) Infrastructural indicators: water scarcity in the villages and lack of social and community services such as electricity, health, sanitation, banking, and marketing facilities.
(4) Agricultural indicators: average land holding of one ha or less, of which only 50 per cent was under cash or cereal crops.

Each of the above economic indicators was allotted 25 points. Areas scoring 60 per cent or above were declared backward.

Based on the above criteria, more than 243 village *panchayats* were registered as backward. Ten per cent of the total outlay for education, health, and water supply was spent in these backward areas. Similarly, these criteria were also applied to other selected sectors such as animal husbandry, food and supplies, forests, roads, rural electrification, and small-scale industries.

As a result of concentrated efforts, many villages are now integrated into the mainstream and are developing rapidly.

The Scheduled Castes

The underprivileged sections of Himachali society were identified by two important characteristics. They were either geographically isolated (scheduled tribes) or they were culturally marginalized (scheduled castes) from the mainstream. The latter suffered the disabilities of severe economic exploitation and social discrimination. They constituted 22.2 per cent of the total population of the state (Khosla 1985). Keeping in mind these facts, a special component plan for uplifting scheduled castes was formulated. Since then, this component plan has been consistently maintained with sufficient financial resources, for example, the proposed outlay for this component plan in the overall seventh five-year plan was Rs 1,447 million (Anonymous 1984).

The salient features of the new strategy for development in the special component plan included heavy subsidies to these classes for agricultural inputs, rural electrification in such areas as well as in individual houses, subsidized education, monthly stipends, free uniforms, the opening of free coaching centres for admission to technical courses, and free legal aid. Seats are also reserved in professional courses and in the universities. These component schemes have helped the scheduled castes improve their economic conditions and have brought them into the mainstream.

Tenant Farmers and the Landless

One of the important causes of low agricultural productivity in Himachal Pradesh was that few of those who actually cultivated the land had any proprietary interests in it. They were either tenants at will or share-croppers only. So reform of the agrarian structure, to protect the marginalized and vulnerable farming communities, was essential for building up an efficient agricultural economy. With strong political will and a sense of equity, the state took bold initiative in implementing land reforms which abolished intermediary tenancy, such as *zamindari*, *jagiris*, and *inams*. Historians and development analysts rate it as a silent revolution. Implemented with sincerity, it helped improve the social and economic position of tenant families. The tenants became the owners of the land they worked. In 1972, a land ceiling was fixed and excess land was distributed among the landless. Final efforts were made to rehabilitate the remainder of the landless poor by settling them on village common lands, which were converted into farmlands for the landless. In addition, incentives based on economic devlopment programmes were initiated for these mountain communities.

As a result of all these measures, the erstwhile depressed farming classes became socioeconomically well off. Today they are a part of the mainstream life of the State, making significant contributions to overall development.

INACCESSIBILITY AS A TOP PRIORITY OF DEVELOPMENT STRATEGY

Besides sociopolitical efforts to achieve autonomy in development planning and programming, a look at the planning goals and the data on resource allocation for the various plans (Table 26.7) will reveal that there has been definite mountain orientation in the priorities of the five-year plans of Himachal Pradesh. The highest priority was given to transport and communications up to the fourth plan. Once they had reached a satisfactory stage, the highest priority was given to water and power development in the fifth, sixth, and seventh plans. By and large, the second position was given to the agriculture and allied services sector in the various plans of the state. Social and community services were accorded third position in the first three plans and fourth position later on. Industries and minerals ranked fifth in priority.

Economic development in mountainous regions can hardly be visualized without effectively reducing the impact of inaccessibility in mountain areas. Before 1949, when Himachal Pradesh was divided into small feudal states, inaccessibility was the most serious constraint. Pack track and human backs were the only modes of transport. Stray motor roads (the total length in 1949 was 500 km) and bridle paths, usable only in good weather, connected the larger towns and the capitals of some princely states. During winter and the rainy season, many areas remained cut off from the rest of the country for weeks and months as roads were badly damaged due to heavy rains and snow. This isolation and remoteness was the normal way of life, and it delayed the economic growth of the hinterlands. Isolation from urban influence resulted in economic, social, cultural, and political backwardness. Rural populations were not able to take advantage of the opportunities offered by natural resources. The potentials of cash crops, such as fruits and vegetables, for which ideal ecological conditions existed, could not be harnessed as these were not remunerative due to the exorbitant cost and difficulty of transport. At that time, the means of transport in larger interior areas were mules, sheep, and

Table 26.7. Percentage allocation of expenditure of different sectors in the five-year plans of Himachal Pradesh

(Percentage)

Percentage annual Sector	First Plan 1951-56	Second Plan 1956-61	Third Plan 1961-66	Annual Plans 1966-69	Fourth Plan 1969-74	Fifth Plan 1974-78	Annual Plans 1978-80	Sixth Plan 1980-85	Seventh Plan Targets 1985-90	Growth of Expenditure 1951-66	Growth of Expenditure 1966-85
1. Agriculture and allied services	14.1	15.7	22.7	18.6	24.1	26.2	26.6	25.1	28.97	62	113
2. Cooperation	14.1	14.9	10.3	4.5	3.2	1.1	1.7	1.1	0.77	25	6
3. Water and power development	4.1	9.4	7.1	27.9	21.7	26.7	22.2	26.2	37.41	67	392
4. Industries and minerals	1.7	2.5	2.5	3.2	3.6	3.3	3.0	3.2	2.92	56	132
5. Transport and communications	46.2	37.1	35.2	34.0	29.1	24.1	22.9	17.7	18.13	26	48
6. Social and community services	19.8	19.2	21.2	11.5	18.0	16.3	21.3	24.2	10.25	39	117
7. Economic and general services	0.1	1.2	1.0	0.3	0.3	2.3	2.3	2.5	1.55	666	270
Total	100.0	100.0	100.0	100.0	100.0	100.0	100.0	100.0	100.0	42	97
Rs million	(52.7)	(160.3)	(338.4)	(397.8)	(1,134.3)	(1,614.8)	(1,475.6)	(6,556.6)	(9,070.0)		

Source: Draft Seventh Plan, 1985–1990, Shimla: Planning Department, Government of Himachal Pradesh, 1984.

goats. Thus, in its overall development strategies, Himachal gave top priority to transport and communications in the initial development phases (1954—1974). Road development throughout the state received a great deal of attention during the first four five-year plans (Table 26.7). From the inception of the first five-year plan, about 25 to 40 per cent of the total state expenditure was incurred annually on the construction and improvement of roads. Road development across the Himalayan ranges, connecting most of the remote mountain areas of the State, was further facilitated by the national focus on good road networks in the mountains, as a result of the India-China War in 1962. Himachal benefited from two national highways passing through its most remote, high mountain areas, and these have been built without direct costs to the State. Thus, its resources were invested in building locally important state and district roads. Today, the State maintains four categories of road:

(1) primary roads which provide inter-district and inter-state connections such as national and state highways,
(2) secondary roads which connect district towns with their sub-divisions, generally called district roads,
(3) tertiary roads connecting towns with the villages to facilitate the movement of goods and human transport; having fulfilled requirements under (1) and (2) above, village roads are currently receiving attention in all agro-ecological zones of the State, and
(4) quarternary roads; wherever village roads in category (3) have not been provided villages are linked to motorable roads by mule tracks and bridle paths.

Except for one or two cases there is no area of Himachal Pradesh which cannot be reached from the motor road by a few hours' walk. The network of roads grew from 500 km in 1949 to 16,589 km in 1980 and 21,585 km in 1989. Further, to improve transport facilities, the unsatisfactory private transport system was abandoned, because it was poorly managed, insufficient, and ill-suited to hill roads. Transport services were thus nationalized and State transport facilities extended to all possible motorable roads.

The efforts made by Himachal Pradesh to build a network of roads throughout the State, especially in remote rural areas, have facilitated development activities in all key sectors. The barriers of isolation and remoteness have been broken down, adding to the attractiveness of rural life. This has a positive influence on net domestic output and agricultural production, because transportation of both people and goods has been facilitated and the expansion of rural settlements and the functional growth of towns and service centres have been promoted with a consequential increase in the area of influence of urban centres. Easier accessibility has provided vital linkages between producing and consuming centres and also facilitated the provision of jobs to landless labourers and marginal farmers. Seasonal migration from high altitude areas has decreased. The boost given to the tourism industry has grown into an important income-generating activity. Immediately after Independence, there were hardly any manufacturing units in the State, but progress was made, especially in the small-scale cottage industry sector, following the creation of transport facilities.

By reducing inaccessibility, the process of agricultural transformation from traditional cereal crops to commercial cash crops, such as off-season vegetables and fruits, has taken place. It has remarkably improved the socioeconomic status of farmers. The road network not only solved the problem of marketing agricultural produce but it also facilitated the availability and distribution of essential agricultural inputs and the introduction of new farm technologies in rural areas. Even the remote villages now have direct access to markets in the major cosmopolitan cities of India for marketing their agricultural and horticultural produce, e.g., apples and potatoes. Some of these items are being exported to neighbouring countries.

Access has also helped strengthen the animal husbandry sector. There are extensive networks of milk producers' cooperatives and milk chilling plants which collect milk from accessible rural areas and supply milk to almost all the towns in the State. The increased efficiency of transportation has decreased the cost of cultivation, brought down the prices of all commodities, and reduced the pressure on land because of shift from agricultural to non-agricultural enterprises. Thus, reduced inaccessibility has helped to solve economic under-development to a great extent.

On the more fragile, steep mountain slopes, the state has plans to install cableways or ropeways as an alternative means of reaching remote villages. An allocation of Rs 100 million was made for cableways and ropeways in the 7th five-year plan.

AGRICULTURAL DEVELOPMENT STRATEGY

As in any other mountain area, agriculture is the major occupation of the people of Himachal Pradesh. It provides employment to about 71 per cent of the total working population. Farmers do not have large land holdings and in fact small and marginal farmers, with less than 3 ha of land, constitute the majority of the farming class (Table 26.3). Even the per capita availability of arable land (0.14 ha) is an indicator of the predominance of small and marginal farmers.

Such conditions can be observed throughout these mountain areas. What is noteworthy in the case of Himachal is the positive approach to the transformation of subsistence systems into economically promising commercial agricultural systems. In two decades agriculture has been commercialized to the extent of bringing in an annual revenue of Rs 600–700 million (US$ 32–37 million) to mountain farming communities.

Harnessing Diversity for Agricultural Development

Resource use and production in the agricultural sector (which for the purposes of our discussion includes crops, horticulture, livestock, pastures, and forestry) is largely influenced by the geographical and environmental diversity prevailing in different areas (Joshi 1984). The wide range of altitudes and variations in other agro-climatic parameters, such as rainfall and temperature, broadly classify Himachal into four major agro-climatic zones. In the following discussion it will be shown how biophysical diversity was harnessed to bring improvements to the agricultural systems. In planning the cultivation of cash crops, the comparative advantages of the mountain climate and the crops that had better scope for exchange and marketing in the hinterlands were taken into consideration.

Below are zone-specific descriptions of the resource base and mountain specificities of the different zones explaining the modifications and adaptations made to change the existing constraints to potential benefits.

The Low Hills and Valleys Near the Plains

This area has a sub-tropical climate. It rises up to an elevation of 900 m and receives an approximate rainfall of between 600 and 1,000 mm. The districts of Una, Hamirpur, Bilaspur, and parts of Kangra and Solan fall into this zone, which occupies about 35 per cent of the geographical area and 33 per cent of the cultivated area of Himachal Pradesh. It has a high population density of 234 persons per km^2 compared to the State average of 77 persons per km^2. Crops usually face moisture stress outside the monsoon season. Irrigation is practised by means of wells, tubewells, ponds, lakes, and traditional irrigation channels. Agriculture is dominated by annual crops, mostly foodgrains. Due to the heavy population and livestock pressure, forests and grazing lands have been drastically reduced to provide more agricultural land. Over the years, the livestock pattern has demonstrated a gradual predominance of buffaloes over all other stock.
The agricultural development strategy of this zone is based on:

(1) promotion of food crops in the fertile valleys,
(2) improvements in livestock quality and reduction in the number of small and marginal farmers depending on milch cattle to supplement their farm income through selling milk,
(3) industrial development as a source of employment to remove pressure from agricultural lands, and
(4) horticulture, i.e., fruit crops, promoted on a secondary level on less fertile and slopy lands.

The Middle Hills and Valleys with Sub-humid Climates

This zone consists of areas of the Lesser Himalayas, lying between 600 m and 1,800 m. It covers the districts of Sirmour, Solan, Mandi, and parts of Chamba. Precipitation is between 1,500 and 3,000 mm, 70 per cent of which is received during the monsoons. The zone occupies about 32 per cent of the geographical area and 53 per cent of the cultivated area of the State. The population density is 143 persons per km^2. Small valleys with good irrigation facilities are found along the banks of streams and rivulets. A major portion of the uncultivated area is composed of grasslands. Forests are limited to the northern and northeastern slopes only. The zone is a producer of grain crops, off-season vegetables, and stone fruits. Livestock husbandry is an integral part of the farming system. Both buffaloes and cattle are kept. Forests and pastures are under pressure from both agricultural expansion and from the livestock population.

Agricultural development was earlier concentrated on food crops, livestock, and stone fruits. In recent years, the zone emerged as the vegetable bulwark of the State. The demand for vegetables in the adjacent plains (in the cities of the Punjab and in Delhi) and an appropriate cool climate for cultivating vegetables when they are out of season in the plains has given rise to the phenomenon of off-season vegetable cultivation.

Himachal is now being identified as the 'vegetable state' (Singh 1989). Cabbages, cauliflowers, peas, hill capsicum, tomatoes, and beans are produced during periods when they cannot be grown in the adjoining plains, and they are in great demand. For this reason, hill vegetables fetch higher prices and are sold at a premium. A favourable climate and easy accessibility to developed markets in the neighbouring states are the main reason for such a spectacular success.

Earlier, off-season vegetable growing was not at all a common practice, despite the fact that the State had vast potential for the production of off-season vegetables on a commercial scale. This was because of the non-availability of critical inputs, unawareness among the farmers, lack of irrigation and marketing facilities, and the non-availability of disease-free, high-yielding germplasm. In view of the economic potential of vegetable cultivation (after the inception of the State) several Vegetable Research and Extension Farms were established in four of the different agro-climatic zones. Following the founding of a university of horticulture and forestry, off-season vegetables received a lot of attention in both research and extension activities.

As a result of these efforts, marginal farmers are bringing more areas under these crops because the per hectare net returns from off-season vegetables are certainly higher than the returns for cereal crops. If apples helped boost the economies of families with relatively large, slopy, marginal land holdings in the mountain zone, then vegetable cultivation has been equally successful in improving the economies of small and marginal farmers who have sufficent family labour and land holdings. Vegetable production is also a source of income for people not involved in horticultural development, whether within or outside the horticultural zone (Tewari 1990).

Complementary Diversification Efforts

Beekeeping and mushroom cultivation are activities with comparative advantages in mountain agriculture. They do not need large areas of land, and beekeeping provides essential ecological services such as the cross-pollination of crops. Agricultural strategies in Himachal Pradesh incorporated these diversified activities into mountain agricultural development programmes. Although they are not zone-specific, the climate in this zone is especially suitable for them.

Beekeeping has traditionally been an economically marginal farming occupation, because it is a low investment activity unless operated on a commercial scale. In addition, it is flexible enough to adapt to any scale of operation, and, hence, ideally suited for small farmers. Inaccessibility plays a less constraining role because hive products are characterized by low weight, high value, non-perishability, high storage capacity, and easy transportation. Further, the role of honeybees as pollinators assumed priority in horticultural development strategies. The government maintains a substantial number of bee colonies which are distributed to orchardists during the apple flowering season to enhance productivity.

High Mountains and Valleys with a Temperate Climate

This area includes the districts of Kullu, Shimla, and parts of Mandi, Chamba, and Kinnaur. Characterized by a humid temperate climate and alpine pastures, the zone commences at an elevation of 1,600 m. It contains 25 per cent of the geographical area and 11 per cent of the cultivated area of the state. The population density is only 63 persons per km^2. The annual precipitation varies from 1,000 mm to 1,500 mm, 60 per cent of which is received during the monsoon season. Most of the cultivated areas are located in the higher reaches where there is a lack of irrigation facilities. On a limited scale, perennial streams and springs are used for irrigation, and the zone is a major producer of temperate fruits such as apples, pears, plums, and off-season vegetables. Most of the valuable timber forests are also located here. Thus, it also benefits from forest resources, both major and minor, to a considerable degree. Crops include a mixture of the major food crops, wheat and maize, as well as mountain crops. Cattle, sheep, and goats are kept. Due to the favourable climate, tourism also flourishes. The major resorts are Shimla, Kullu, Manali, and Dalhousie in Chamba. Currently, because of the abundance of resources, moderate population pressure, and appropriate development initiatives, the zone has experienced considerable prosperity. The extent of human migration, both inter-state and intra-State, as well as seasonal and permanent, is greater in this zone.

The agricultural development strategy for the zone focussed on the promotion of fruit cultivation (Azad et al. 1987) because of the suitability of the climate for apples and other pome fruits. Most of the horticultural development efforts of the State have been concentrated here. In addition to some agricultural land, the approach has been to convert slopy non-agricultural land which was suitable for crop cultivation into fruit orchards (Table 26.8).

Table 26.8. Land-use shifts occurring because of horticultural development in Himachal

Category	Land-use pattern before orchards			Existing land-use Pattern			Previous use of orchard land		Shift of area to orchards from	
	Land holding (ha)	Area under field crops	Area put to other uses	Area under field crops	Area under orchards	Area under other uses	Field crops (ha)	Barren and other land (ha)	Field crops (Col. 8 as % of Col. 3)	Barren and other (Col. 9 as % of Col. 4)
1	2	3	4	5	6	7	8	9	10	11
					Temperate zone (Prome fruits)					
Marginal	0.71	84.4	15.6	52.2	44.3	3.6	0.225	0.084	38.2	77.2
Small	1.142	79.0	21.0	52.8	39.7	7.5	0.365	0.192	33.1	64.2
Medium	2.34	79.9	20.1	51.2	37.9	10.9	0.689	0.216	35.9	45.8
Large	5.81	61.3	38.7	33.6	41.7	24.7	1.619	0.816	45.2	36.2
Overall	2.57	69.6	30.4	41.6	40.7	17.7	0.721	0.327	40.3	41.8
					Sub-tropical zone (Stone fruits)					
Marginal	0.66	86.0	14.0	65.2	28.4	6.4	0.138	0.051	24.2	54.3
Small	1.54	86.2	13.8	72.9	18.3	8.8	0.204	0.045	15.4	36.3
Medium	2.46	81.7	18.3	69.7	16.4	13.9	0.295	0.108	14.7	23.9
Large	2.58	53.8	46.2	46.7	8.7	44.6	0.517	0.108	13.0	3.1
Overall	3.01	65.4	34.6	55.9	12.4	31.7	0.288	0.086	14.7	8.2
					Himachal Pradesh					
Marginal	0.68	85.2	14.8	58.6	36.5	4.5	0.182	0.067	31.2	66.7
Small	1.48	82.7	17.3	63.2	28.6	8.2	0.288	0.135	23.5	52.7
Medium	0.24	80.8	19.2	60.7	26.3	12.5	0.482	0.184	24.8	35.2
Large	6.61	57.2	42.8	41.0	23.1	35.9	1.067	0.461	28.3	16.3
Overall	2.79	67.3	32.7	49.2	25.5	25.3	0.505	0.206	26.8	22.6

Note: Figures in Col. 3 to Col. 7 are percentage of total land holding (Col. 2).

Source: K.C. Azad, et al., *Horticultural Development in Hilly Areas*, Table 4.10. Delhi: Mittal Publication, 1987.

The horticultural development strategy of the state has brought about a number of positive changes in the form of increases in the incomes of the hill farmers, better use of land that is otherwise not suitable for agriculture, the checking of soil erosion on hill slopes, and the development of horticulture-based industries (Nadda 1988, Azad et al. 1987, Swarup 1990). Marketing, transportation, and the provision of cold storage are handled by the Horticultural Products Marketing Corporation, a specialized (state sector) agency established for this purpose (Rana 1990).

Before the formation of Himachal Pradesh in 1948, hardly 1,000 ha of land was under fruit tree cultivation. The area increased to 44,329 ha in 1970/71, and 149,284 ha in 1988/89. The total fruit production was 1,485,800 MT in 1970/71 and it increased to 3,086,900 MT in 1988/89. Apples are the principal crop and they account for more than 80 per cent of the total fruit production. In 1987, apple exports from the State amounted to 207,421 MT (Anonymous 1989). The priority given to horticultural development can be assessed from the five-year plan outlays for this sector (Table 26.7). While the total outlay for horticulture in the sixth plan (1980–1985) was Rs 115 million, it more than doubled (284.3 millions) during the seventh plan (1985–90). As a consequence of horticultural development, the economic status of farmers from this zone is much better than in other zones (Bhati et al. 1990).

Further, horticulture flourishes in the State because of the unlimited demands for these products in the adjoining lowlands. The large towns and cities of northern India provide a huge market with favourable terms of trade.

Besides fruits, medicinal plants, mushrooms, and other non-wood forest products contribute substantially to the economy of this zone. It is rich in forest resources and has an optimum population. Under the changing trends of horticulture-dominated farming, livestock composition patterns are also changing slowly. Rabbits for wool, Jersey cows to meet milk needs, and some fine quality sheep for wool are replacing conventional livestock in apple growing areas. This is partly as a response to declining common lands and partly because of the growing scarcity of cheap human labour in this zone. Within the transformed economy, these activities are held to be less productive and not very promising economically.

Trans-Himalayan Cold Dry Zone

This area includes the high mountain areas of Kinnaur and the Pangi areas of Chamba and Lahaul-Spiti. Elevations range from 2,000 to 3,500 m. It occupies about 8 per cent of the geographical area and 3 per cent of the cultivated area of the State. The population density ranges from nine persons per km^2 in Kinnaur and Pangi to two persons per km^2 in the Spiti area. Precipitation is less than 200 mm during summer but heavy precipitation occurs in the form of snow during winter. The dry temperate climate has resulted in scanty vegetation. Cultivation is wholly dependent on irrigation, provided by harnessing perennial streams and springs. Soils are prone to heavy erosion from glaciers and avalanches. Man-made vegetation belts are developed on water courses for use as fuel and fodder. Agriculture is concentrated on food crops such as barley, buckwheat, wheat, and maize. During the last decade, a great deal of effort has been made to transform farming in this zone. The production of potato seeds, chicory, hops, and vegetable seeds was promoted on a large scale for cash crop farming. Medicinal plants, as cash crops, are also produced in parts of the zone. A decade ago, there was hunger in this zone but great strides have been made through the commercialization of agriculture and the transformation of the farming economy. Cold desert improvement programmes, focussing on afforestation and soil conservation, have proved beneficial in terms of both farming needs and environmental conservation.

The agricultural development strategy for this zone focussed on the following aspects.

(1) Taking advantage of road development, the potential of the niche of the cold, dry climate was used through promotion of disease-free potato seed cultivation on a large scale. The Northeastern Himalayan areas, which were earlier source areas for potato seed, had been hit by an epidemic of potato disease. Thus, to fulfill the demands for potato seeds from many Indian States in the plains this zone was chosen as the most suitable area. Backed by ensured demand and favourable terms of trade, this zone became an agricultural area of great potential within two decades. Today, apart from apples, seed potato is another major product of Himachal Pradesh. Potato production has nearly quadrupled in the last 35 years. Although the crop is grown in all districts, this area is the major producer. Potato seed export brings in about Rs 10 crore (US$ 5.4 million) annually.

(2) Apart from the commercial cropping strategy, special efforts have been made through an anti-desertification programme to improve the well-being of both people and the environment.

The Anti-desertification Programme

This programme, funded by the central government, is based on the recommendations of the National Commission on Agriculture concerning the development of the desert areas of Lahaul and Spiti. The programme emphasizes irrigation, afforestation, community plantation, fodder development, soil conservation, crop farming, animal husbandry, and fisheries and is independent of the above-mentioned special tribal development programmes.

The major achievements (Negi 1985) so far include the development of irrigation facilities—for both agriculture and afforestation purposes through a network of channels—from melting glaciers, lift irrigation schemes, and the construction of water storage tanks. However, such irrigation schemes in this cold desert area are facing the problems of damages caused by snow and avalanches, high costs, non-availability of repairing facilities, short working seasons, and labour shortages. Keeping these problems in mind alternative irrigation methods are being recommended. These include the use of windmills, hydraulic dams, collapsible pipes, drip irrigation, and lift irrigation.

Under the anti-desertification project, the afforestation, agroforestry, and agricultural sectors are being given top priority. In agriculture the major emphasis is on the diversification of crops. Earlier, barley was the major staple crop. Now wheat, potatoes, peas, mustard, onions, beans, cabbages, and tomatoes are successfully grown up to an altitude of 3,810 masl. Farmers receive subsidized agricultural inputs such as improved seeds, fertilizers, and agricultural implements. There are also incentives for bringing barren lands under cultivation.

Under this project, improved livestock such as yaks from Ladakh, Jersey cows, crossbred Merino sheep, and Angora goats are distributed to the farmers on a subsidized basis. A network of veterinary dispensaries and an increase in the number of fodder trees and fodder grasses help promote livestock production.

The trans-Himalayan cold dry zone provides an ideal 'niche' for harnessing non-conventional sources of energy such as the sun and wind. For example, weather data from the cold desert of the Spiti Valley reveal that there are as many as 306 dry, clear days in a year when solar energy can be harnessed. In this context, some photovoltaic lighting systems have already been installed in public places. Solar photovoltaic pumps are used to dry fruit, vegetables, and seeds. Similarly, wind (velocity of 45 km/hour) is also harnessed by installing aero-generators and windmills; water is drawn from the rivers and tributaries for power generation.

The impact of these efforts has resulted in 50 per cent of the 11,815 target group families moving above the officially defined poverty line by the end of seventh plan period (1985–1990). All the villages in the tribal areas receive electricity and are provided with piped drinking water. The literacy rate is now approximately 50 per cent, compared to 42 per cent in the State as a whole. The increases in apple production in Kinnaur District and in potato seed cultivation in Lahaul-Spiti District have changed the subsistence economy into a commercial one (DES, GOHP 1989).

The multi-pronged strategy to develop this underprivileged, resource-scarce zone had a definite impact. Potatoes, hops, and other cash crops, together with several socioeconomic incentives made available through the special tribal development programmes, have brought comparable affluence to mountain farming families. Today, by and large, people from this zone are considered to be better off than people from the first two zones in education, employment, and in living standards.

Livestock

Almost 90 per cent of farming families maintain bovines. Animal husbandry is in fact an integral part of the farming system and provides manure, draft power, milk, wool, and meat. The state has a livestock population of 5 million including cattle, buffaloes, sheep, goats, horses, and yaks at an average density of 99 livestock units per km^2. Improvements in the quality of cattle were undertaken to facilitate promotion of a network of dairy development programmes and provide cash income to families having little agricultural land and few alternative sources of income.

The contribution of animal husbandry to State income is 12 per cent in Himachal Pradesh as compared to a national average of 6 per cent. This signifies the importance of livestock in the State's agricultural systems. In view of this, livestock received appreciable attention in the development planning process. The principal livestock development programmes include: (1) animal health and disease control through a network of 216 veterinary hospitals and more than 500 dispensaries, (2) cattle development through key village schemes, hill cattle development programmes, and intensive cattle development projects and breeding facilities through hospitals, dispensaries, bull centres, and artificial insemination centres, (3) sheep-breeding and the production of wool through breeding farms, (4) poultry development, (5) feed and fodder development schemes, (6) development of dairy and milk supply schemes, and (7) a special component plan for animal husbandry for economically marginal farmers from tribal areas or from scheduled castes or scheduled tribes (Negi 1990). As a result of such programmes, there is now great diversity in the livestock resources and products in the State. Cattle, buffaloes, yaks, sheep, goats, poultry, and rabbits are the major types of livestock reared. These animals provide milk, mutton, wool, hides, motive power, transport, and manure. The projected target is that each person in the State will have access to 300 gm of milk, 100 gm of meat, and an egg a day (Negi 1990) in the near future.

HARNESSING MOUNTAIN 'NICHE'

Water Resources

Water resources occupy a vital position among the natural resources of Himachal Pradesh. The hill topography, the steep fall of rivers down the slopes into mountain catchments, and the perennial water supplies from the streams favour the development of hydroelectricity in the State (Joshi 1984). The five major river systems originating from the Western Himalayas, namely, the Chenab, Ravi, Beas, Sutlej, and Yamuna, pass through Himachal Pradesh. A conservative estimate of the total hydropower potential of Himachal Pradesh comes to 20,000 MW or even more. Out of this, 12,700 MW can be generated from the five river basins cited above (Anonymous 1985). From the Sixth Five-year Plan onwards, exploitation of the hydroelectric power potential in the State has received top priority. The installed capacity of hydropower has increased 75 times since 1950/51, and 3,286 MW have already been harnessed out of a total potential of 12,700 MW. More hydel projects with power potentials of 126.52 MW have been completed while others are under way (DES, GOHP 1989).

There are a number of benefits from hydropower to the state. Electricity for household consumption relieves the pressure on fuelwood and kerosene. The establishment of industrialization based on a renewable energy source is possible and industrial houses and entrepreneurs are drawn to Himachal Pradesh because of the uninterrupted electricity supply. This is a significant advantage to industrialists because of the general shortage of electricity in the adjoining states. Selling the excess electricity to adjoining states brings in huge revenues. It is this latter benefit that planners are most interested in. Harnessing the water resources' 'niche' helps the State generate more revenue so that it can invest in development to further improve the quality of social services. Fortunately for Himachal Pradesh, there is a prevailing power shortage throughout the whole northern region of India. This affects industries, agriculture, and the overall economy. However, as small hill state such as Himachal Pradesh may not be able to provide the outlay for big hydro-projects from its own yearly or five-year plan budgets, alternative arrangements for financial support from the Government of India and from neighbouring states have been made. A plan to exploit an additional 4,700 MW at a cost of Rs 50,000 million has been formulated for the coming decade.

The visible impacts of hydropower development are that several industrial complexes have been established in the State, and electricity is provided in all of its rural areas. Pressure on the forests, as a source of fuelwood, is decreasing and people are finding electricity an ideal substitute for wood. This enables the State government to implement other schemes such as water supply schemes and lift irrigation schemes. With the completion of major hydel projects, electricity is becoming a major source of income.

Tourism

Tourism is one of the few off-farm employment activities of the State. The major focus of tourism development has been on harnessing the scenic, cultural, and recreational resources. While the state has been looking after the promotional aspects, people have benefited from the increased employment opportunities in off-farm professions such as hotel management and private transport services. Taxi services and private hotels have become a big business. To help unemployed, educated youths to enter into these professions, very attractive financial policies have been adopted by the State.

INDUSTRIAL DEVELOPMENT AS A TERTIARY SECTOR FOR REDUCING PRESSURE ON LAND

To reduce the pressure on land, it was essential to develop the potentials of the industrial sector (especially small and cottage-scale industries). In the late 1960s, manufacturing in Himachal Pradesh was still in the cottage industry stage. Since then, the creation of infrastructural facilities and an aggressive government policy of incentives and subsidies have resulted in the emergence of a relatively sophisticated industrial framework. A large number of industrial firms, 97 in the large and medium-scale sector and about 6,400 in the small-scale sector, employ about a hundred thousand workers. However, it is in the cottage industry sector, with more than 60,000 units, that employment has been impressive. The State produces diverse goods such as cement, engineering products, electronic goods, watches, and fertilizers.

In terms of resource allocation, licensing and the provision of materials the industrial development strategy of the State has the following priorities.

(1) Industries that produce items having considerable value added (top priority).
(2) Units based mainly on local resources (top priority).
(3) Units catering mainly to local requirements.
(4) Units based on scarce and controlled raw materials and competitive with outside units.
(5) Industries based on outside resources and demand.

Categories (4) and (5) receive the lowest priority under the industrial policy, because the approach is to discourage such industries in the State.

Thus, the industrial development policy aims at ensuring the optimum utilization of local, raw materials and stimulating employment. Industries with greater employment potential and labour intensive units are accorded preferential treatment. Emphasis on the growth of small-scale industries is in tune with mountain perspectives and specificities, as these have the comparative advantages of low investment, high potential for employment generation, and easy dispersal, especially in rural and semi-urban areas.

A future approach for the expansion of industrial development has already been identified.

- Plentiful natural resources, especially hydroelectricity, ensuring cheap and sufficient power.
- Cool environment, suitable for engineering precision and electronic industries.
- Major and minor forest products which provide raw materials.
- Large mineral deposits, especially limestone.
- Development of horticulture, providing raw materials for agro-based industries.
- Political stability and a peaceful environment.

DIMENSIONS OF FRAGILITY AND APPROACHES

The pressure on mountain ecosystems arises from *natural catastrophies* as well as from the high use intensity of natural resources. The former include snowstorms, cloud bursts, avalanches, forest fires, and landslides. They cause disruption in communications and the loss of life and property. Their sudden impact on large volumes of stored water is often disastrous. Like other mountain regions of the world, Himachal Pradesh is subject to natural hazards that are often outside the direct control of man.

Pressures also arise from *man-made catastrophies*. Mountain resources decrease and deteriorate because of population pressure, expanding industrialization and urbanization, and the emergence of new needs connected with the development of communications and trade. The population growth is 2.3 per cent per year in Himachal Pradesh, and this provides a serious set-back to economic growth, which is 2.4 per cent per year. It also places enormous pressure on natural resources. Similarly, forests, pastures, medicinal plants, and wildlife are important natural resources and fall under the fragile resources category for development activities.

The salient features of land use in Himachal Pradesh include a low percentage of areas suitable for agriculture, intensive land cultivation beyond its carrying capacity, low percentage of areas lying fallow, extension of cultivation to marginal and sub-marginal land, and a high percentage of areas under village pastures and grazing lands. Similarly, forest and grassland ecosystems are becoming more fragile, as a result of destruction of the natural habitat, overexploitation, and degradation caused by grazing beyond the carrying capacity and encroachment on forest and grassland areas for the cultivation of agricultural and horticultural crops. Harmful exploitation of economically important plant and animal species and the practice of monoculture are having serious ecological consequences. As the management of fragile environments was considered to be primarily related to forests, related approaches and programmes have been launched under the forestry programme.

A comprehensive State policy to manage the fragile mountain environment was formulated in 1980. Its major goals are:

(1) to cover 50 per cent of the total geographic area with forests by 2000 A.D., to minimize the diversion of forest areas into other farming sectors such as agriculture or horticulture,
(2) to raise multiple-use tree species to meet requirements for timber, fuel, fodder, and manure as well as soil and water conservation needs,
(3) to create biosphere reserves,
(4) to control grazing,
(5) to develop suitable substitutes for wooden fruit-packing cases, and
(6) to reduce annual tree felling by 10 per cent.

As a result of the implementation of this plan, the prescribed annual extraction of wood decreased from 628,000 m^3 to 407,000 m^3 in 1984. The felling of trees for commercial purposes has gradually decreased. The system of contracting out timber extraction work to private contractors was abolished. Now, the State forest corporation plans commercial timber extraction on the basis of the productivity potentials of a forest area. A substantial increase in resource allocation for afforestation and soil conservation activities during the 1980s is now producing encouraging results (Table 26.9).

Table 26.9. Changes in land use in Himachal Pradesh from 1970/71 to 1985/86

Land use	1970/71 Area ('000 ha)	1970/71 Percent to total	1985/86 Area ('000 ha)	1985/86 Percent to total	% Change from 1970/71 to 1985/86
Total geographic area:					
a) By professional survey	5,565.8		5,567.3		0.03
b) Reporting area for land use purposes	2,932.5	100	3,227.1	100	10.05
Forests	638.2	21.8	884.8	27.4	38.64
Barren and uncultivable	118.8	4.0	159.6	7.0	34.34
Land put to non-agricultural uses	172.1	5.9	220.1	6.8	28.89
Cultivable waste	167.7	5.7	128.0	4.0	−23.67
Permanent pastures and other grazing lands	1,188.0	40.5	1,153.63	5.7	−2.90
Land under misc. tree crops not included in area sown	40.8	1.4	40.8	1.4	0.00
Current fallows	58.3		43.9		
		2.1		1.8	−4.95
Other fallow land	2.3		13.7		
Net area sown	546.4	18.6	582.6	18.1	6.63
(Cropping intensity)	(166.9)	—	(167.1)	—	(0.12)

Source: Directorate of Land Records, Himachal Pradesh, Shimla.

CONCLUSIONS

In terms of both (1) conventional yardsticks, e.g., productivity and income growth, infrastructural facilities, and modern input use, and (2) quality of life indicators such as reduced landlessness, education, housing, health, and nutrition, Himachal Pradesh is a success story in area-based development. The rapid transformation of Himachal Pradesh could be attributed to several factors and could be explained in terms of the sensitivity of its development interventions to different mountain specificities. The replicability and applicability of these development interventions to other mountain environments will vary according to the prevailing conditions of the environments in question.

Himachal passed through a major political and socioeconomic transformation process within a period of four decades. The State has been instrumental in removing interregional disparities within its jurisdiction. This has been achieved through a specified strategy for the socioeconomic development of underprivileged classes. In its plans there has been an emphasis on infrastructure, social services, and production through liberal economic assistance and subsidies. As a result of such efforts, marginalized classes have joined the mainstream. Agricultural development was planned according to agro-ecological zones, emphasizing different lead farming activities in each zone. It was a strategy for harnessing the environmental diversity of the State, and its success transformed the subsistence farm economy in a most promising way.

Further, Himachal's success primarily lay in the ability of its planners to consider mountain specificities in development approaches (Table 26.10). It identified and harnessed the variety of 'niche' it had. Horticulture was accepted as a lead sector in its development strategy. Political will and the people's traditional experiences played an important role in promoting this land-extensive activity, even on fragile slopes. Concomitant, and in some cases sequential, emphasis on related activities, such as animal husbandry, beekeeping, and vegetable crops, with R&D and marketing support, played an important role in Himachals transformation. Simultaneous attention to the production and marketing aspects also led to an improvement in and utilization of infrastructural facilities. The availability of an extensive hinterland, i.e., the non-mountain areas of India, to absorb the products, e.g., fruits, for which Himachal has a comparative advantage, not only encouraged their production but also facilitated the harnessing of other major 'niche' (the irrigation and hydropower potential) by the State without having to bear the huge overhead costs involved in the process. The states that were potential users of irrigation and power provided the investments for their development. After Himachal became a full-fledged state of India, it was in a better position to promote these projects.

Table 26.10. Consideration/incorporation of mountain perspective in public policies and programmes in selected areas of HKH region

Levels of consideration/ incorporation of mountain specificities in public policies and programmes	West Sichuan	Himachal Pradesh	Nepal	Pakistan NWFP
1. Understanding of broad ecological zonation	*	*	*	*
2. Presence of (1) at micro-level	P	*	P	P
3. Macro-level policies and planning based on (1) and (2)	P	*	P	P
4. Micro-level projects and resource allocation based on (1), (2) and (3)				
a. Wider coverage	x	x	x	x
b. Limited coverage	*	*	*	*
5. Farm-level initiatives	P	P	P	P
6. Implementing agencies perspectives	P	P	P	P
7. Consideration of mountain specificities as explicit constraints	P	*	*	P
8. Persistent information gap				
a. Macro-level	P	P	P	P
b. Farm-level	*	*	*	*

Note: * = yes; x = no; P = partial or limited.

Source: MFS-ICIMOD, August, 1990. *Progress Report of the Project on Strategies for Sustainable Mountain Agriculture*.

Efforts made by Himachal Pradesh to remove the constraints caused by inaccessibility, by building a road network throughout the State, especially in remote rural areas, have facilitated development activities in all sectors. Roads have broken the barriers of isolation and remoteness and added attractiveness to rural life, promoting the expansion of rural settlements and the functional growth of towns and service centres with an increase in the area of influence of urban centres. By reducing inaccessibility, the process of agricultural transformation, from traditional cereal crops to commercial cash crops was expedited and resulted in remarkable improvements in the socioeconomic status of the farmers in the state.

Regarding the explicit or implicit consideration of mountain specificities, such as diversity 'niche', marginality, and adaptation mechanisms, Himachal provides us with a number of useful lessons for other mountain areas to follow. However, since the extensive hinterland played an important facilitative role in the harnessing of 'niche', Himachal's experience may not be applicable to mountain areas having limited linkages with the plains. Nevertheless, the State's integrated approach (which takes into consideration the potential of specific ecological zones), evident in the synchronization of various related activities, the creation of institutional structures like the Horticultural Products Marketing Corporation, and the creation of infrastructural support, is not dependent upon having an extensive hinterland. Such activities are worth documenting in order to facilitate their replication.

REFERENCES

Anonymous. *Draft Seventh Plan* 1985–90 and *Draft Annual Plan* 1985–1986. Himachal Pradesh: Planning Department, Govt. of Himachal Pradesh, 1984.

Anonymous. *Development Profile of Himachal Pradesh.* Shimla: Directorate of Economics and Statistics, Himachal Pradesh, 1985.

Anonymous. *Statistical Outline of Himachal Pradesh.* Shimla: Economics and Statistics Department, Himachal Pradesh, 1989.

Azad, K.C., R. Swarup and B.K. Sikka. *Horticultural Development in Hill Areas.* Delhi: Mittal Publications, 1987.

Bhati et al. See Chapter 18.

DES-GOHP. *Statistical Outline of Himachal Pradesh.* Department of Economics and Statistics, Govt. of Himachal Pradesh, 1989.

ICIMOD. *Agricultural Development Experiences in Himachal Pradesh, India* (Workshop Report). Kathmandu: ICIMOD, 1990.

Jodha, N.S. 'Mountain Perspective and Sustainability: A Framework for Development Strategies.' Chapter 2, Vol. I, this poublication.

Joshi, K.H. *Geography of Himachal Pradesh.* New Delhi: National Book Trust, 1984.

Khosla, R. 'Scheduled Caste Development Component Plan Schemes and Their Impact.' In *Development Profile of Himachal Pradesh.* Himachal Pradesh: Directorate of Economics and Statistics, Govt. of Himachal Pradesh, 1985.

Nadda, A.L. *Economics of Apple.* New Delhi: B.R. Publishing Corporation, 1988.

Negi, G.C. *Livestock Development in Himachal Pradesh: Performance, Policy, and Prospects.* ICIMOD/MFS Discussion Paper No. 7. Kathmandu: ICIMOD, 1990.

Negi, J.P. 'Approach to the Problem of Arid and Semi-arid Areas in Himachal Pradesh.' From Regional Workshop on Development of Wasteland in Himachal Pradesh and North-west Himalayan Region, pp. 1–16. Solan, H.P., India, 1985.

Rana, R.S. *The Horticultural Produce Marketing and Processing Corporation of Himachal Pradesh.* A Critical Evaluation. ICIMOD/MFS Discussion Paper No. 9. Kathmandu: ICIMOD, 1990.

Sharma, K.K. 'Socio-economic Spectrum of Tribals in Himachal Pradesh.' In *Develop-*

ment Profile of Himachal Pradesh. Himachal Pradesh: Directorate of Economics and Statistics, Govt. of Himachal Pradesh, pp 45–119, 1985.

Singh, M. *Social, Cultural, and Economic Survey of Himachal Pradesh*. Shimla: Minerva Book House, 1985.

Singh, M. *Himachal Pradesh: History Culture and Economy*. Shimla: Minerva Book House, 1988.

Singh, M.M. 'Himachal, Worries in a Prosperous State.' *Himal*, March/April, 2(2):8–11, 1989.

Swarup, R. 'The Role of Commercial Crops in the Development of Himachal Pradesh.' In ICIMOD-MFS Workshop Report No. 1. Kathmandu: ICIMOD, pp. 49–50, 1990.

Tewari, S.C. *The Role of Off-season Vegetables in the Development of Hill Agriculture* ICIMOD/MFS Discussion Paper No. 8. Kathmandu: ICIMOD, 1990.

UNDP. *Human Development Report*. New York: Oxford University Press, 1990.

Verma, L.R. *Beekeeping in Integrated Mountain Development: Economic and Scientific Perspectives*. New Delhi: Oxford and IBH Publishing Co. Pvt. Ltd., 1990.

27

A LOCAL RESOURCE-CENTRED APPROACH TO RURAL TRANSFORMATION: AGRO-BASED COTTAGE INDUSTRIES IN WESTERN SICHUAN, CHINA

Liu Zhaoguang and Wu Ning

Liu Zhaoguang and Wu Ning are both associated with the Chengdu Institute of Biology, Academia Sinica, Chengdu, Sichuan, China.

CONTENTS

INTRODUCTION

Agro-based cottage industry involves the use of house and courtyard space for plant and animal husbandry; in addition, it involves the manufacture of handicrafts. Concurrently, peasant farmers forage for wild plants and hunt for wild animals and carry out the rough processing of their products. This activity is not new, as it has long been a subsidiary occupation in China.

This paper covers the Ganzi Tibetan Autonomous Prefecture, the Aba Tibetan and Qiangs Autonomous Prefecture, and Lianashan Yi Autonomous Prefecture in Sichuan. The total area of all three is over 300,000 sq km and their population is 4.8 million. About four million people (approximately 82.06%) are involved in agriculture and 0.4 million (8.3%) in animal husbandry. The agriculturalists are distributed throughout the mountain areas and river valleys, and those involved in animal husbandry mainly inhabit the plateau of Northwestern Sichuan. Agro-based cottage industries are mainly concentrated in the agricultural areas.

The paper illustrates approaches to harnessing the 'niche' and diversity characteristics of mountain areas (Jodha, Chapter 2) and thereby exhibits the mountain perspective of agro-based cottage industries in selected areas of China.

POSITIVE FACTORS IN THE DEVELOPMENT OF AGRO-BASED COTTAGE INDUSTRIES

Abundance and Diversity of Natural Resources

The region is one of the richest in China for a variety of animals, plants, and edible and medicinal fungi. There are over 800 species of animals, 6,000 species of plants, and over 200 species of fungi. Among them are a number of well-known and rare species such as Chinese caterpillar fungus (*Cordyceps sinensis*), glossy ganoderma (*Ganoderma lucidum*), black edible fungus (*Auricularia auricula*), white edible fungus (*Tremella fuciformis*), golden edible fungus (*Tremellia mesenterica*), fragrant mushroom (*Lentinus edodes*), pine mushroom (*Tricholoma matsutake*), hedgehog hydnum (*Hericium erinceus*), morel (*Morchella esculenta*), musk, bear's gallbladder, young pilous antler pangolin (*Manis pentadactyla*), salamander (*Batrachuperus pinchonii*), squirrel (*Tupaia glis chinensis*), the bulb of fritillary (*Fritillaria cirrhosa*), Forbes' notopterygium (*Notopterygium forbesii*), large-leaved gentian (*Gentiana macrophylla*), the root of membranous milk vetch (*Astragalus membranaceus*), the root of herbaceous peony (*Paeonia veitchii*), Auchandia (*Aucklandia lappa*), Chinese Nardostachys (*Nardostachys chinensis*), and Chinese Ligusticum (*Ligusticum sinense*). They are of high yield and good quality.

Sericulture is an example of a profitable cottage industry that could be developed throughout the river valleys of the sub-tropical zone. The mulberry plant flourishes there, sustaining fresh leaf growth throughout the winter and shedding only in the dry season. As many as four or five generations of silkworm can be bred in one year, and this is double the number raised in the Yangtze River Basin. The quality of silk is very fine, because of the natural resources and the conditions prevailing throughout West Sichuan.

Climate

There is an obvious bio-climatic variation throughout the low to the high elevations. At low elevation, conditions are sub-tropical and these change in the frigid temperate belt, culminating in a permanent snow belt. Inter- and intra-climatic belts have a wide variety of topography and flora and fauna: 'four seasons on one mountain and different climates within a five kilometre radius.' This range of topographies and climates is advantageous for the growth and abundance of living things and natural resources.

There are generally 1,500–2,000 hours of strong sunshine per year and the heat distribution decreases from the low river valleys to the high mountains. The daily range of temperature is great and the annual range small, and this creates beneficial conditions for photosynthesis. The small annual range promotes the stability of stenothermic species. Most of the land is cultivable and, in general, conditions are excellent for the development of a variety of agro-based cottage industries throughout the various belts.

THE DIVERSITY OF POSSIBLE PRODUCT DEVELOPMENT

Given the natural environment, the products that can be developed are wide-ranging. They fall into three main categories and these are described in the following passages.

Plant-based Cottage Industries

The cultivation of flowers, fruits, herbal medicines, and some plants, which are not only useful for domestic decoration purposes but also for income-generating activities, is possible in Western Sichuan where many species and varieties of plants are of economic value. They include the *bailan (Michelia alba)*, mango (*Mongifera indica*), tamarind (*Tamarindus indica*), and grape (*Vitis vinifera*), cultivated in the sub-tropical valley of the Jinsha River and the Anning River; a variety of oranges, cottage bamboo (*Sinocalamus affinis*), peaches, and cherries, cultivated in the Luding area of the Dadu River Valley and the low and middle belts of the hills along the Anning River Valley; and apples, walnuts, and wild peppers (*Zanthoxylum simulans*), cultivated in the middle or upper reaches of the Dadu River and the Minjiang River.

Animal-based Cottage Industries

In addition to raising domestic fowl and animals, peasant farmers are also engaged in apiculture, sericulture, and breeding animals such as roe-deer, deer, bears, and ermine. A limited number of farmers breed bull frogs, but not yet on any substantial commercial scale. However, pheasant breeding has already become a profitable industry.

Agro-cottage Industries Based on Handicrafts and on Plant Collection

Handicrafts include weaving, braiding, embroidery, woodwork, and ironwork based on the traditional crafts of the minority nationalities. The collection of plants for herbal medicines and the processing of wild fruits, vegetables, and edible fungi, many for indigenous use, is also carried out as a subsidiary occupation.

According to the rough statistics available, there are approximately 300–400 products that can be produced in this region. This provides a substantial basis for the development of agro-based cottage industries.

Flexibility in Implementation

Production is carried out within the family unit. Scale of production is small with short lead times for finished products. The fixed assets and amount of investment involved is very little and, since most of the work is not physically exacting, cottage industries create job opportunities for the old and the handicapped also. Since products can be changed easily, the risks involved are minimal. When there is a glut of one product on the market, production can be diverted to another product without causing significant loss. Both the benefits and the advantages of cottage industries are obvious.

Existing Foundation for Cottage Industries

The level of cottage industry development is indicative of the level of the commodity economics in the region. For a long time, this region remained a self-sufficient subsistence economy area and commercial transactions were carried out through barter exchange. However, most of the produce was for self-consumption. During the last decade, many changes have taken place and changes in productivity have resulted in a transition from a self-sufficient subsistence economy to one that is based on commodity production and marketing exchange. This has stimulated the development of agro-based cottage industries (Table 27.1).

Table 27.1. The development of subsidiary occupations in Western Sichuan

Year	Ganzi			Aba			Liangshan		
	TOVA	OVSO	%	TOVA	OVSO	%	TOVA	OVSO	%
1950	10,160	747	7.35	5,581	1,035	18.55	16,671	1,409	8.45
1957	11,070	850	7.68	8,240	1,573	19.09	25,666	2,104	8.20
1965	12,938	1,186	9.17	11,938	1,769	14.82	35,601	3,357	9.43
1970	15,349	1,259	8.20	12,515	2,029	16.21	37,508	2,877	7.67
1975	18,613	1,748	9.39	15,801	2,312	14.63	45,229	4,129	9.13
1980	19,713	1,501	7.61	20,903	2,448	11.71	59,168	6,110	10.33
1985	29,931	4,424	14.78	29,111	4,625	15.89	88,099	13,050	14.81

Notes:

The numbers are calculated on 1980 values.
The unit of output value is ten thousand yuan.
TOVA = the total output value of agriculture.
OVSO = the output value of subsidiary occupations.
% = OVSO as a percentage of TOVA.
There are 4.70 yuan to one U.S. dollar.

Source: Ao Chepu et al. 1988

Throughout the river valleys in the southern part of this region, almost every farming family raises bananas. The average annual income per family is 200–300 yuan annually, but in some families it is as much as 1,000 yuan.

Plantations of the flower *bailan* (*Michelia alba*) are a new feature throughout the river valleys of the south. In Ningnan County, in particular, in two villages, every family has planted three trees or more. After three years of growth the trees yield more than 10,000 flowers a year and they bring in an income of over 200 yuan. After five years of growth, each tree will yield 20,000 flowers annually, thus realising an average income of 400 yuan.

In Luding and Mao County, wild pepper is sold. On an average each family earns over 400 yuan but as much as 700–800 yuan can be earned. Table 27.2 shows the production of wild pepper in Western Sichuan from 1965 to 1985.

Table 27.2. The production of wild pepper in Western Sichuan, 1965–1985

(in kg)

Year	Ganzi	Aba	Liangshan
1965	63,550	101,350	156,200
1970	91,600	172,050	48,750
1975	83,650	151,900	68,200
1980	99,100	125,850	271,600
1983	157,850	161,850	459,500
1985	208,350	270,950	844,500

Source: Ao Chepu et al. 1988

Apart from in Mao County, there was no apple production on a large scale before the 1970s (see Fig. 27.1). A few farmers planted trees to meet the needs of the indigenous inhabitants. After the 1970s, families began to plant apple trees on a large scale and not only are local requirements met but large amounts of apples are also sold in Chengdu and Zhongging (Table 27.3).

Table 27.3. The production of apples since the 1970s

(in tons)

Year	Butuo	Puge	Xide	Yuexi	Yanyuan
1970	0.1	0.4	0.0	0.6	15.0
1975	177.5	5.2	8.4	212.5	174.9
1980	9.5	28.8	152.5	477.9	560.3
1983	475.8	154.9	278.2	1,323.0	1,805.3
1985	423.6	298.0	788.2	2,215.8	2,252.9

Source: Ao Chepu et al. 1988

Apples are now a significant source of income for many peasants. In addition, peasant will always produce a number of other products, as can be seen in the following examples of two families in Jinzhou Village of Mao County in 1989.

These two families represent the higher and lower income groups in the village. In the 1970s, their income from grain was twice as much as it is now, but there were very few other sources of income.

At present, the use of gardens to cultivate strawberries, plums, and cherries has been considerably successful in various districts. Gardens have also been used to breed silkworms and sericulture has rapidly developed (see Table 27.4).

The production of cocoons in Ningnan County reached 800 tons in 1989, and the

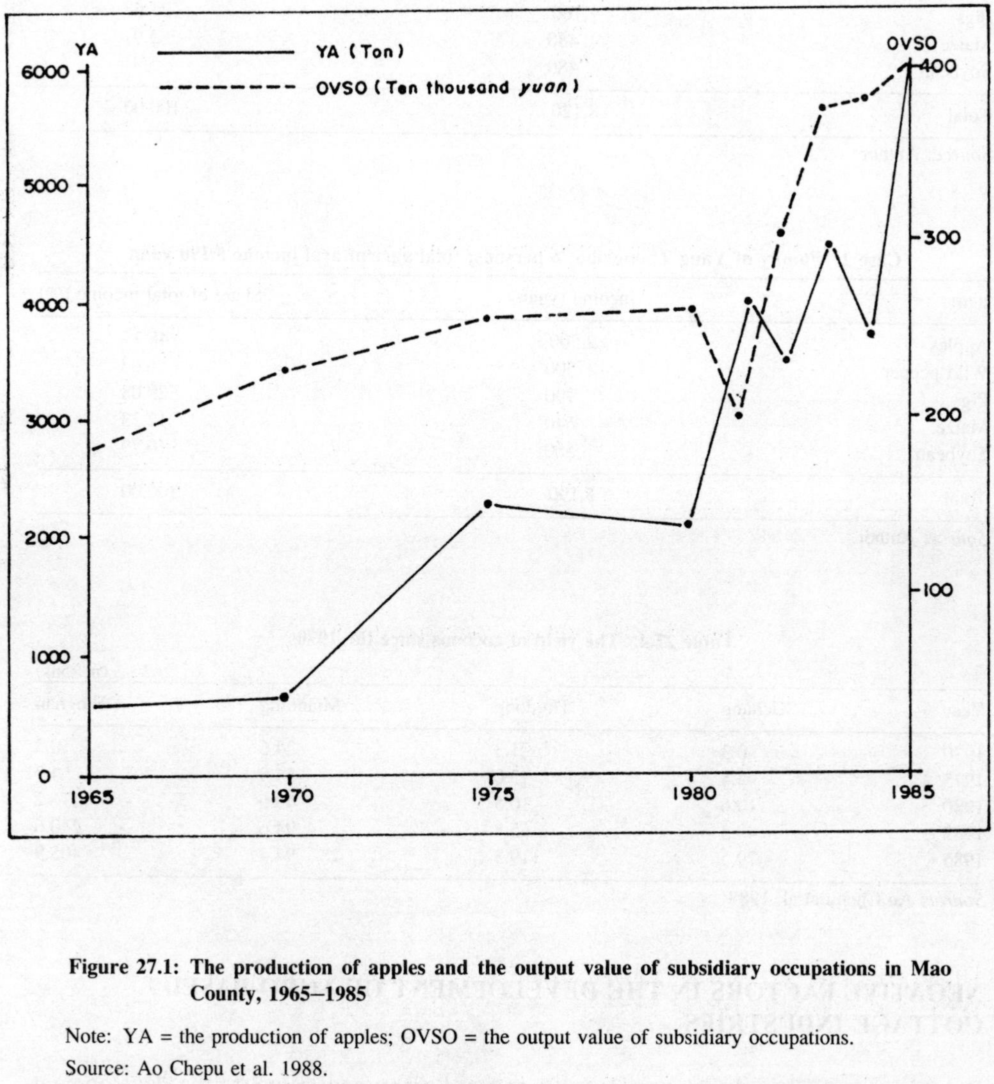

Figure 27.1: The production of apples and the output value of subsidiary occupations in Mao County, 1965–1985

Note: YA = the production of apples; OVSO = the output value of subsidiary occupations.

Source: Ao Chepu et al. 1988.

average income per family increased to 500 yuan. With the development of sericulture, the silk filature industry also developed. A silk filature mill with an output of 70 tons per year has been built, and it employs 600 people. This has led to the opening up of many more job opportunities.

Other cottage industries involve the planting of medicinal herbs; in addition, the rough processing of edible fungus and fodder and grain are new cottage industries that have emerged within the last decade.

Case 1. Family of Chen Mingxin: 5 persons; total agricultural income 8,120 yuan

Item	Income (yuan)	Share of total income (%)
Apple	4,800	59.11
Wild pepper	600	7.39
Pigs	1,760	21.68
Maize	480	5.91
Soybean	480	5.91
Total	8,120	100.00

Source: Author.

Case 2. Family of Yang Zhongxiao: 6 persons; total agricultural income 5,190 yuan

Item	Income (yuan)	Share of total income (%)
Apples	2,500	48.17
Wild pepper	500	6.63
Pigs	990	29.08
Maize	640	12.33
Soybean	560	10.79
Total	5,190	100.00

Source: Author.

Table 27.4. The yield of cocoons since the 1970s

(in tons)

Year	Xichang	Huidong	Mianning	Ningnan
1970	0.3	1.3	24.6	0.3
1975	3.5	3.0	34.8	4.7
1980	10.6	10.5	43.6	76.2
1983	40.5	65.8	94.6	240.6
1985	79.5	119.3	94.4	405.9

Source: Ao Chepu et al. 1988

NEGATIVE FACTORS IN THE DEVELOPMENT OF AGRO-BASED COTTAGE INDUSTRIES

In reviewing the last decade, we have discovered many constraints to the development of agro-based cottage industries, and these have been summed up in the following passages.

Distance from Market

Most of the cottage industry products are small items for daily use or edible consumer goods such as fresh edible fungus, fresh vegetables, and other mountain products. Goods that are difficult or expensive to transport are not suitable for production in this environment because the major markets are too far away.

Lack of Access to Market Information and Transport Facilities

Because of the mountainous terrain, the dissemination of market information is slow and difficult. For example, in recent years, dahurian angelica (*Angelica dahurice*), and pilose asiabell (*Codonopsis pilosula*) saturated the herbal medicinal market and sales began to slacken. However, this information did not reach the mountainous regions, resulting in overstocking. In Ganzi, over 2,000 tons of fresh songrong mushroom (*Tricholoma matsutake*) is produced each year, but sales in the region only dispose of 800 tons a year. The rest is difficult to transport and thus the selling price of fresh songrong mushroom has to be reduced. A transportation network is vital for the development of agro-based cottage industries.

Absorptive Capacity and Market Risk

For a long time the economic structure of a farming society has depended upon the production of sufficient grain to meet its daily needs, with some surplus for barter in exchange for household necessities and small luxuries. A few fowl and livestock were also kept on the same basis. They have few possibilities to increase their income. If any problems occur and their products are not saleable, they have little to fall back upon. Thus, should a selected product fail to be marketable it will adversely affect daily life and lead to increasing poverty.

Low Educational Level and Lack of Exposure to Science and Technology

Agricultural production in this region is underdeveloped and basically stagnant. The potential for increasing output through scientific methods is quite considerable, but, because the proportion of the population that is illiterate and semi-illiterate is high, it is difficult to introduce and popularize advanced technology. This also hinders the development of agro-based cottage industries.

APPROACHES AND STRATEGIES FOR THE DEVELOPMENT OF AGRO-BASED COTTAGE INDUSTRIES

The Principles for Selecting the Type of Industry

Comparative Advantage and Market Demand

The predominant resources in each area should be selected as development items, given the variety of resources, topographies, climates, flora, and fauna throughout the region. In this region, edible fungi, such as wild fragrant mushroom (*Lentinus edodes*) and songrong mushroom, are the predominant resources throughout the three prefectures. In 1988, songrong mushroom brought in a revenue of 80 million yuan in Ganzi Autonomous Prefecture, and this averaged out at 100 yuan per capita. Earnings in Xiangcheng, Daucheng, Yajiang, and Jioulong counties were as high as from 10,000 to

20,000 yuan. This is due to the abundance of oak (*Quercus* spp.) and Yunnan pine forests (*Pinus yunnanensis*). Both species are symbiotic with songrong mushroom.

The correct techniques for processing this mushroom are easily mastered by peasant farmers in comparison to those for processing common mushroom (*Agaricus bisporus*), common mushroom (*A. campestris*), and umbrella mushroom (*Pleurotus ostreatus*), which are difficult to process and store. The latter are not a good option for this area because they are easily damaged by micro-organisms. In order to promote agro-based cottage industries, items must be selected carefully according to the availability of natural resources and market demand. The latter two are interlinked. Without market demand, abundant natural resources alone will not create a dynamic economy.

Local Specialization

Because of limited transportation and communication facilities and the lack of science and technology, products from this region cannot compete with like products from other areas. Therefore, those products for which this region has a comparative advantage should be developed. Within this category are Chinese medicinal herbs such as Forbes' notopterygium (*Notopterygium forbesii*), tendril-leaved fritillary (*Fritillaria cirrhosa*), and large-leaf gentian (*Gentiana macrophylla*). These are famous products from this area. Their processing should be developed into a basic pharmaceutical industry, since there is not a lot of competition for these products.

Sericulture, which is a principal industry in the Jinsha River Valley, as well as throughout its hinterland, is another example of an industry for which the area has comparative advantages. Silkworms can be bred there at the rate of four or five generations a year in comparison to the lower regions of the Changjiang River where only one or two generations can be bred yearly and where the quality of the silk is not so good. In the valleys of Ningnan and Miyi County, the insolation is excellent, making it possible to grow *bailan* (*Michelia alba*) and mangoes (*Mongifera indica*) in cottage gardens. Earnings, on an average, for each family are as much as 600 to 1,000 yuan annually, and this is 25 to 40 per cent of the total earnings from agriculture. In developing agro-based cottage industries it is always important to build on existing strengths and circumvent weaknesses.

Economies of Scale

One of the principal aims in promoting cottage industries is to foster self-sufficiency. This is quite difficult, especially in terms of sufficient productivity and market share, when each family develops its own subsidiary occupation or cottage industry. The breeding of bull frogs is a case in point. This is only pursued in a few families and the amounts produced are not sufficient to have any significant impact in terms of market share. To develop agro-based cottage industries, it is essential that inter-linkages among the families involved are established. At present, farmers have a number of integrated activities that are initiated by them in order to achieve a better scale of production.

Cultural Standards: Acceptance and Levels of Technology

The culture of the area is traditional and conservative; as a result, educational standards are low. Consequently, technology is primitive and the acceptance of new technology quite low. In order to ameliorate this situation, it is essential to raise the standards of the farmers through training and through increasing the number of technologically skilled personnel in the region. In this way, natural resources will be used for all-round development.

Improvement of the Agro-ecological Environment

Two main factors should always be considered in establishing agro-based cottage industries. First, pollution-causing industries should never be permitted, and, second, the gathering of wild plants and the hunting of wild animals should be kept within rational limits. The continuity of natural resources must be a principal concern of all parties involved. Sustainability of both the natural resources and the dependent activities are interlinked.

The Major Alternatives Available

Within the context of the above principles, four development factors are discussed for consideration.

Cultivation of Plants

Fruits. In the southern part of the region, tropical and sub-tropical **fruits**, such as guavas, can be the main crops, and oranges, peaches, and plums can be cultivated mainly in the lower mountain areas. Pears and wild peppers can be grown in the middle and high mountains. The northern areas of the region are suitable for the growth of apples, grapes, strawberries, walnuts, cherries, and wild peppers. The developing industrial base in the south will provide the main market for fruits from the southern part of the region and Chengdu and the northern tourist centres will provide markets for fruits cultivated in the north.

Herbal medicines. There are a variety of **herbal medicines** in this region and species differ from the south to the north. In the south, villous amomum (*Amomum villosum*), galanga resurrection (*Kaempferia galanga*), common andrographis (*Andrographis paniculata*), and wrinkled gianthyssop (*Agastache rugosus*) are the main plants found and, in the northern area, tinata pinellia (*Pinellia terata*) and rough gentian (*Gentiana scabra*) are the principal medicinal herbs (Table 27.5).

Flowers. The **flowers** that are most marketable are *bailan* (*Michelis alba*) and jasmine (*Jasminum sambac*) in the south and roses (*Rosa* spp.) in the north.

Edible fungus. Species such as black fungus (*Auricularia auricula*), white fungus (*Tremella fuciformis*), and fragrant mushroom (*Lentinus edodes*) can be cultivated as cottage industry products on family farms. The farmers have already had a lot of experience in cultivating fuling (*Poria cocos*). It can be established on a planned basis.

Animal and Insect Products

Sericulture can be further developed in the south of the region and apiculture in the north. The musk deer and bear can be bred in the mountain districts and the domestication and breeding of ermine (*Mustela erminea*) and coypu (*Myocastor coypus*) placed on a firm footing and encouraged throughout the area.

Table 27.5. Yields from major medicinal plants in Western Sichuan, 1950–1985

Year	FC (ton)	RO (ton)	NC (ton)	CS (ton)	Musk (kg)
1950	36.53	0.26	0.0	6.67	67.90
1960	66.68	11.69	601.1	6.47	755.50
1970	84.79	12.56	211.8	22.07	689.70
1975	59.71	18.21	562.2	12.60	481.15
1980	60.67	4.48	41.4	2.85	792.45
1983	68.95	3.68	121.4	4.41	164.00
1985	149.87	10.31	650.5	26.60	266.00

Key: FC = *Fritillaria cirrhosa*
RO = *Rheum officinale*
NC = *Nardostachys chinensis*
CS = *Cordyceps sinensis*

Source: Ao Chepu et al. 1988

Cottage Industries Based on Processing

Currently, processing industries are mainly concentrated on the basic processing of herbal medicines, grain, fodder, edible fungi, wild vegetables, and some simple household necessities that have a wider national market.

Gathering of Wild Plants

Both edible and medicinal fungi have remained relatively unexploited. In the 1960s, there was a market for lingzhi mushroom (*Ganoderma lucidum*). Its annual yield is 20 tons. The 1980s saw an increase in the market for songrong mushroom, and the annual yield of salted fungus from this species in 1988 was more than 1,100 tons. The total output value is 100 million yuan.

In recent years, vegetable fern (*Pteridium aquilium*, var. *latiusculum*, a kind of wild vegetable) has been introduced as a new product. The approximate preliminary statistics for 10 counties, including Dechang, Maoxian, Wenchuan, Ludin, Mianning, and Xichang, show that the annual yield will be over 500 tons in 1990, selling at 4,000 to 5,000 yuan per ton. This product alone has increased the total output value of the area by 200 to 250 million yuan.

Herbal medicines have been the basis for the principal subsidiary occupation of villagers in this region and for a long time the sale of such products has constituted the principal source of income.

Finally, there are many varieties of wild fruits, which can be used in agro-based cottage industries, such as roxburgh roses (*Rosa roxburghii*), seabuckthorn (*Hippophae rhamnoides*), Kansu crabapple (*Malus Kansuensis*), and Tibetan crabapple (*Malus transitoria*).

Project Implementation

Local Government Support

Agro-based cottage industries offer many important prospects to assist underdeveloped areas in overcoming poverty and backwardness. The local governments in such areas need to offer every possible support to these industries by popularizing them, making funds available for their development, and introducing the technological skills needed. An example of measures that can be undertaken by local governments can be shown by using the case of the exploitation of vegetable fern in the piedmont of Erlang Mountain in Tianquan County. To develop this industry successfully a number of measures were taken by the Government of Tianquan County, and these have been described below.

- A professional body of 100 members was formed by the County's Native Product Company to be responsible for the production of vegetable fern.
- A production plan, with duties assigned to every team member, was decided upon before the plants were gathered by the villagers. The professional body divided the work up among the families and individuals involved and delegated the responsibility for that work.
- Information about the project was disseminated to every household.
- A training course in the techniques of gathering and processing was held, including demonstrations in the field. The Native Product Company, which had delegated the responsibilities for production, was able to buy 30 tons of vegetable fern within a very short period of time.

  In Miyi County, the local government reduced taxes on the sale of this product or even remitted them. This encouraged the industry, improved its marketability, and increased the income of the villagers.

Production and Processing Arrangements

The *Yi* ethnic group, which inhabits Lianhe Village in Mianning County, had over 10,000 wild mulberry bushes in their village. These were not used to breed silkworms, but rather the leaves were gathered to feed the pigs. The local government encouraged silkworm breeding by:

- paying compensation for cocoon yields of less than 25 kg on one sheet, and
- engaging a technical advisor to instruct the *Yi* in silkworm breeding.

As a result of these measures, 60 households bred more than 40 sheets of silkworms. The total yield of the cocoons was as high as 1,400 kg, with an approximate value of 13,000 yuan, bringing in an average earning of 200 yuan per household.

The local government disseminated facts about silkworm breeding and popularized the occupation. The next measure taken was to plant more mulberry bushes on barren and waste land. In the autumn of 1989, 150,000 bushes were planted under the guidance of the Agricultural Bureau. It is planned to plant 200,000 bushes in 1990, and that, by 1992, every family will have two sheets of silkworms. This has brought about improvements in the economy of the County. Silk filature mills have been built in both Mianning and Ningnan counties. The annual yield from the mill in Ningnan alone is about 200,000 kg and the mill employs 600 people. The local government's support has encouraged the establishment of linkages in order to achieve maximum economic benefits.

Application of Advanced Science and Technology

Advanced science and technology is essential in the optimum development of agro-based cottage industries. Because such technology was available, Ganzi Autonomous Prefecture was able to process goat wool, which fetches 30 yuan for every 100 g. It is expensive, because it is difficult to process this wool. This activity has developed to a stage at which every family has 10 goats. Each goat provides 100 g of wool which brings in 300 yuan per year per family. Likewise, improvements in salsifying techniques have led to an expansion in the songrong mushroom processing industry. More importance has to be given to science and technology in the promotion of agro-based cottage industries.

Support Services

The development of agro-based cottage industries requires a concurrent development of services. These should include disease prevention and control, seed selection, storage, desiccation, packing, and forwarding services. This way the quality of products can be guaranteed and cottage industries firmly established.

Market Research and Feedback

To develop agro-based cottage industries, market feedback is necessary. Unless the products keep up with supply and demand, agro-based cottage industries will not establish a firm footing. To ensure that this happens, market research to determine supply and demand is essential. Products should be modified and changed continuously according to market fluctuations.

REFERENCES

Ao Chepu et al. *The Historical Statistics of National Economy and Social Development in the Area of Minority Nationalities of Sichuan Province*. Chengdu: Sichuan People's Publishing House, 1988.

Jodha, N.S. 'Mountain Perspective and Sustainability: A Framework for Development Strategies.' Chapter 2, Vol. 1, this publication.

28

THE SMALL FARMERS' DEVELOPMENT PROGRAMME IN NEPAL: INSTITUTIONAL INITIATIVE IN POVERTY ALLEVIATION

K.S. Sharma

Kul Shekhar Sharma is a member of the Board of Directors of Integrated Development Systems, Kathmandu, Nepal.

CONTENTS

INTRODUCTION

The Small Farmers' Development Programme (SFDP) is the first development programme in Nepal directed towards a specific segment of the population rather than towards a specific sector of the economy such as agriculture, irrigation, or forestry. It is targeted at the rural poor, i.e., the small farmers, the tenants, the landless, and the artisans. Small and marginal farmers were ignored by general agricultural programmes such as extension, research, and credit in their bid to boost agricultural production, rather than to alleviate poverty—a much more difficult and complex task.

Generally, extension and credit agencies were inclined, quite understandably, to overlook marginal farmers because of the extremely small size of their landholdings and their very low contribution to total agricultural production, even though they heavily outnumbered the better-off farmers. Instead, these agencies tended to concentrate their efforts on farmers with larger holdings where substantial gains in production could be made. General agricultural programmes were, as such, not geared to the needs and limitations of poor and marginal farmers. Extension messages often had little relevance to the conditions of the small farmer, while research programmes had not addressed the need to maximize income from very small holdings through the development of integrated farming packages suited to them. In the field of credit, there was a built-in bias among credit institutions favouring larger loans to bigger farmers, as such loans were easier and less costly to administer. On the other hand, smaller borrowers found institutional credit uneconomic because of the high transaction costs involved. A programme aimed at the poorest farmers, and tailored to their needs and problems, was therefore clearly needed. The SFDP seeks to meet this need.

Even though the SFDP is implemented by the Agricultural Development Bank of Nepal, it is much more than a credit programme. It is more of an integrated or community development programme, credit being an entry point to reach the target group. Conceived and initiated as an action-cum-research project in two areas, one in the *Terai* (plains) in 1975 and another in the hills in 1976, the SFDP has now grown into an effective national programme reaching 110,000 member-farmers in 456 project areas of all the 75 districts of the country (see Table 28.1). The programme has been visibly successful in many project areas, although several weaknesses have surfaced in the course of its implementation and expansion. An effort will be made here to describe the programme in its essentials and to analyse and assess it, both in concept as well as in implementation, and with special reference to the situation prevailing in the mountain areas of Nepal.

It must be remembered here that the SFDP is not a specific response to the mountain specificities of Nepal. It is a country-wide programme aimed at alleviating poverty which is pervasive throughout the country, but is concentrated more heavily in the mountains. However, the programme seeks to address the needs arising from many of the mountain characteristics such as marginality, inaccessibility, and fragility. It should be emphasized here that marginality as it applies to people is not exclusively applicable to the mountain areas of Nepal.

Table 28.1. Growth in SPOs, groups, members, and loan disbursement under SFDP

Year	No. of SPOs (cumulative)	No. of groups (cumulative)	Groups per SPO (average)	No. of members (cumulative)	Members per group (average)	Annual Loan Disbursement (in thousands of rupees)
1975/76	2	28	14.00	443	15.82	453
1976/77	15	128	8.53	1,443	11.27	1,120
1977/78	17	296	17.41	3,205	10.83	2,515
1978/79	24	461	19.21	5,121	11.11	3,357
1979/80	29	634	21.86	6,623	10.45	3,865
1980/81	30	786	26.20	7,974	10.15	4,953
1981/82	54	1,276	23.63	12,831	10.06	12,467
1982/83	84	2,239	26.65	21,319	9.52	25,580
1983/84	121	3,196	26.41	29,552	9.25	42,469
1984/85	162	4,554	28.11	41,603	9.14	70,363
1985/86	250	6,061	24.24	52,864	8.72	96,718
1986/87	298	7,463	25.04	63,112	8.46	99,880
1987/88	346	9,188	26.55	78,518	8.55	149,866
1988/89[a]	416	11,596	27.88	95,968	8.28	202,819
1989/90[b]	456	13,898	30.48	110,570	7.96	203,063

[a] 1988/89 figures are provisional.
[b] Figures for 1989/90 pertain to first nine months of the fiscal year.

Source: Annual Report of the SFDP for 1987/88, ADBN.

SFDP OBJECTIVES AND APPROACH

The main objective of the SFDP is to improve the social and economic well-being of the rural poor by introducing and developing institutions that can enhance their capacity to participate more fully in economic and social programmes designed to benefit them. The SFDP has introduced the concept of group formation with a view to creating a structure and mechanism for demanding, receiving, and using, more effectively, the development services and inputs offered by government agencies, and to enabling them to move towards self-reliance. More specifically, the SFDP seeks to:

(1) build up the institutional base for organizing the disadvantaged and the rural poor into small farmers' groups around common economic activities and/or production resources;
(2) provide supervised credit in order to enable these groups to undertake a diversified range of income-generating activities that are tailored to suit their needs and capabilities;
(3) adapt the local services' delivery mechanisms to the needs of the rural poor;
(4) provide training and technical assistance to ensure effective use of the facilities provided by the project; and
(5) develop self-reliance among these groups so as to enable them to plan and carry out their own projects (IDS 1989a).

The underlying idea is that the small farmers, tenants, and the landless, by organizing themselves into small cohesive groups, will be better equipped to receive credit and other services offered by public and private agencies, and that their capacity to use credit and improve productivity will be enhanced if they have greater access to various social services and to types of training that upgrade their skills. The programme also seeks to enhance their involvement in various activities such as community development, resource conservation, community forestry, population education and family planning, sanitation, and women's development. The SFDP also promotes the creation and development of community assets by constructing small irrigation facilities or school buildings through group and inter-group action. It also seeks to enhance the economic welfare of the poor by discouraging wasteful expenditure on social ceremonies, as well as drinking and gambling. In this way, credit, production support, community development, and social reforms are all integrated into the SFDP. Group action is the main vehicle used for achieving the objectives of the programme.

SITUATION IN THE MOUNTAINS

It may be relevant at the outset to describe briefly some of the important characteristics of the mountain areas of Nepal, for a better understanding of the SFDP, as it operates in these areas.

Geographically, the country can be broadly divided into three belts: (1) the Southern belt or *Terai*, covering 23 per cent of the physical area, (2) the middle hills covering 43 per cent of the area, and (3) the high mountains (above 3,000 m in altitude), covering the remaining area of the country. In this paper, both the Middle Hills and the High Mountains will be referred to as mountains. Some of the characteristics of the mountain areas in Nepal are outlined in the following passage.

First, most of the farmers in the mountains are marginal. The highest concentration of absolute poverty, estimated at 48.1 per cent, exists in the rural hills of Nepal as against only 29.4 per cent in the rural *Terai*. The average size of operational holdings of more than the lowest 50 per cent of the farmers in mountain areas is less than one-fifth of one hectare, often with less fertile soil and lesser access to irrigation. As the small pieces of land owned by such farmers cannot provide adequate support to their families, they have to take recourse to seasonal migration (estimated at 30% to 40% of the population) to the *Terai* or to India, in order to provide for their basic needs. Second, access to most of the areas in the mountains is extremely difficult. Difficulties are caused not only by widely scattered settlements throughout the mountains but also by difficult terrain, and by a dearth of sufficient and dependable trails, tracks, and bridges. It is estimated that two-thirds of the mountain population live more than one day's walk from a motorable road. Third, the mountains of Nepal are by nature very fragile and erosion-prone because of their topography and their recent (geological) formation. Encroachment by man in search of food, fodder, and fuelwood has made matters worse. Fourth, the mountain areas are interspersed with many valleys with their own micro-climates, often diverse agricultural patterns and practices. Fifth, the carrying capacity of the hills appears to have been reached and even exceeded in many areas. Significant out-migration from the hills is taking place in search of employment. Productivity is falling as more and more marginal and unsuitable lands are being brought under cultivation to meet the growing needs of a rapidly rising population, with disastrous consequences for the environment. Environmental degradation in the mountains appears to be both a cause and a consequence of low agricultural productivity. Forest lines continue to recede from the settlements causing villagers to spend more time in gathering fuelwood and fodder. Sixth, per capita availability of cereal grains has been declining in most of the mountain areas. Almost 45 per cent of the households are now estimated to consume less than the recommended levels of food.

It is in this setting that the SFDP, essentially a poverty alleviation programme, is being implemented.

INSTITUTIONAL SET-UP OF THE SFDP

Implementation Framework

The SFDP is administered by the Agricultural Development Bank of Nepal through its network of 456 sub-project offices (SPOs) scattered over the entire country. There were 220 SPOs in the mountains and 124 in the *Terai* at the end of 1987/88. The SPOs are headed by bank officials, designated as Group Organizers (GOs), who are mainly responsible for implementing the programme. The GOs are assisted by support staff as needed. Staffing depends upon the nature and volume of activities in the SPO. Most of the SPOs, especially the new ones, are manned only by one GO and a peon. There are Women's Group Organizers working under the GOs in 63 SPOs where women's development activities are also undertaken. The SPOs are supervised by 'control offices' which may be district branches or sub-branches of ADB/N. Zonal offices of the bank also extend general supervision over the SPOs. Overall direction and control is exercised by the SFDP division at the centre.

Procedure for Group Formation

As soon as an SPO is established in an area, the GO organizes a meeting of the villagers, to explain to them the objectives of the SFDP, the procedure for organizing groups, their working methods, and the benefits of joining the group. The GO then organizes a preliminary household survey to identify the small farmers in the area. Households with a per capita income of Rs 1,200 qualify as small farmers (ADB/N 1987). In the beginning, the criterion of the size of holdings was adopted but was later abandoned in favour of the income criteria.

Farmers wishing to form a group have to apply to the GO and sign a memorandum pledging:

- to discuss, plan, and undertake social/community and economic development programmes for the group or its members and to borrow from the SPO if necessary,
- to be liable collectively for that portion of the loan that a member of the group may not be able to repay,
- to exercise collective pressure on the member who defaults on his repayment and not to demand further loans from the SPO until the defaulting member repays his loans, and
- to collect savings from group members regularly (ADB/N 1987).

A detailed survey is then made of the farmers who decide to form or join the group. Farmers are organized into small groups of 5 to 12 members each. In the earlier stages of the programme, the groups were larger, but smaller groups are now considered to be more effective. A group, on an average, consists of eight to nine members and a SPO has 25 to 30 groups. Soon after a group is formed, it elects a leader who is literate and who is considered to be a man of integrity, having the motivation for community service. The group then proceeds to discuss and prepare its social and economic programme based on the skills and capabilities of its members. The GO helps the group to draw up such plans and to keep records of the meetings that must take place at least once a month.

More than 14,000 small farmer-groups, with a total membership of 110,000, had been formed by March 1990. Details regarding the number of SPOs, groups, and members in the mountains and the *Terai* in 1987/88 are given below. The average number of groups and members per SPO is much lower in the *Terai* than in the mountains, reflecting the relative inaccessibility of the mountain areas.

Particulars	Mountains	*Terai*	Nepal
No. of SPOs	4,769	4,476	9,188
No. of Groups	222	124	346
No. of Members	40,921	38,152	79,073
Groups per SPO	21.5	36.1	26.6
Members per SPO	184	307.7	228.5

The SFDP now covers nine per cent of the estimated 1.2 million rural households below the poverty line. Nearly 15 per cent of the groups formed so far are female.

The Group Organizer

The GO is the key figure in the implementation of the SFDP. It is he who motivates, guides, and assists the small farmers in organizing groups, in operating them, and in formulating viable investment proposals. He acts as a liaison between the groups and various development agencies, at the local level, to make these agencies aware of the needs of the small farmers, and facilitates the supply of inputs and services such as fertilizers, credit, extension advice, veterinary services, and any other services and materials needed by group members.

He sanctions loans to the group, and arranges disbursement. He also impresses upon them the need to economize by cutting down wasteful social expenditure and by giving up undesirable habits such as drinking and gambling. He also helps the small farmers on any other matter about which they may call upon him for assistance. In short, the GO is the 'friend, philosopher, and guide' of the small farmers. The success of the programme largely depends upon his ability, motivation, and resourcefulness.

The GOs are selected mostly from among existing officials working in the Bank. They are given three months' intensive training, both in the classroom and in the field, before their assignment. They are also attached to an SPO for three months to enable them to acquire a first-hand knowledge of the working of the SFDP. Topics of training include: the SFDP concept and operations, group dynamics, household survey techniques, credit policies and procedures, preparation of investment plans, analyses of loans and repayment capacities, and farmers' training procedures.

Loan Operations

When a group is formed and has formulated its plans, it applies to the SPO for loans to implement these plans. Since the GO has the authority to sanction practically all the loans needed by individuals and groups, the farmers are spared the long, tedious, and time-consuming formalities that a bank loan normally involves, and they receive the loans promptly.

Loans are provided on a group guarantee basis even though most of the economic activities are undertaken on an individual basis. Loans are also provided for group and inter-group activities. Purposes for which loans may be granted are crop production, irrigation, livestock, agricultural mechanization, horticulture, appropriate technological measures (e.g., *gobar* gas plants and water turbines), cottage industries, agro-industries, agricultural marketing, small businesses, limited household consumption, afforestation and fodder planting, and purchase of land and construction of houses for the landless. The terms of the loans are fixed according to the estimated returns from the investments and the capacity of the farmer to repay. Normally, loans are extended only to the members of those groups that have not defaulted on repayments. The GO may, however, on the recommendation of the group, sanction new loans to defaulting members if he is convinced that a valid reason exists for default and that the new loans will enable them to repay the old as well as the new loans on time (ADB/N 1986). The maximum amount of loans available to a small farmer is Rs 30,000. Interest rates on the loans are the same as those charged to other farmers, but small farmers are given considerable interest rate and capital subsidies on irrigation loans. Collateral is not essential but is taken from those who have property.

Loans disbursed to small farmers for various income-generating activities during 1988/89 amounted to Rs 203 million, with an average amount of Rs 2,000 per member (Table 28.2). During the earlier years of the SFDP, the bulk of the loans went to livestock raising, but its share in the total loans has declined during recent years (Table 28.3). Livestock, crop production, and farm mechanization (which includes bullocks), in that order, claim the largest share of the total loans. Recovery rates were around 60 per cent until 1986/87, but dropped to 47.4 per cent in 1987/88, to rise again to 51.8 per cent in 1988/89 (Table 28.2). These rates, low as they are, are still better than the repayment rates of the non-SFDP loans of the ADB/N. Recovery rates by the purpose for loans are given in Table 28.4. Livestock loans have the largest overdue amounts.

Training of Small Farmers

The training of small farmers is an important component in the success of the SFDP. The training programmes address the need to upgrade the skills of the small farmers with a view to increasing their farm production and incomes. Five regional training centres have been established in the five regions of the country for this purpose. Practical training is given in subjects such as crop production, vegetable farming, livestock management, veterinary care, poultry farming, bee-keeping, sericulture, horticulture, maintenance of agricultural equipment, weaving, handloom operation, sewing, bamboo furniture, and bag-making. Farmers are also trained in the group concepts, group functioning, group savings, and record keeping. This training is expected to help the farmers to use their

Table 28.2. Growth in loan disbursement, loan collection, loans outstanding, loans overdue, and loan recovery rates under SFDP

(in thousands of rupees)

Year	Total members	Loan		Loan collection amount	Loan outstanding		Overdue loans		Recovery rates (%)
		Total amount	Per member		Total amount	Per member	Total amount	Per member	
1975/76	443	453	1.02	9	444	1.00	—	—	—
1976/77	1,443	1,120	0.78	124	1,440	1.00	—	—	—
1977/78	3,205	2,515	0.78	624	3,331	1.04	137	0.04	82.0
1978/79	5,121	3,357	0.66	1,189	5,499	1.07	874	0.17	57.6
1979/80	6,623	3,865	0.58	1,566	7,798	1.18	546	0.08	74.1
1980/81	7,974	4,953	0.62	2,548	10,203	1.28	2,714	0.34	48.4
1981/82	12,831	12,467	0.97	3,460	19,210	1.50	4,053	0.32	46.1
1982/83	21,319	25,580	1.20	6,879	37,911	1.78	4,943	0.23	58.2
1983/84	29,552	42,469	1.44	14,914	65,466	2.21	8,865	0.30	62.7
1984/85	41,603	70,363	1.69	25,284	110,545	2.66	19,745	0.47	56.2
1985/86	52,864	96,718	1.83	40,376	166,518	3.15	31,399	0.59	56.2
1986/87	63,112	99,880	1.58	51,292	222,509	3.53	28,027	0.44	64.7
1987/88	78,518	149,866	1.91	75,268	299,884	3.82	83,628	1.07	47.4
1988/89	95,968	202,819	2.11	104,640	402,249	4.19	97,443	1.02	51.8

Source: Agricultural Development Bank of Nepal.

loans better. Members of the women's groups receive training mostly in sewing, weaving, knitting, handicrafts, vegetable farming, and the raising of small animals. Training needs within an SPO are assessed on the basis of requests received from the GOs concerned and finalized in consultation with the officials concerned of the ADB/N. Trainees are paid daily or monthly allowances during the training period, which can last from three days to three months. The cost of training varies from Rs 100 to Rs 4,000 per trainee.

A large number of farmers have been trained in the Regional Training Centres (see table).

Year	No. of training courses organized	No. of farmers trained
1984/85	109	1,221
1985/86	189	4,208
1986/87	275	7,615
1987/88	179	3,425
Total	752	16,469

Group Savings Scheme

One important instrument of self-reliance, initiated under the SFDP, is the institution of group savings. Members of a group make monthly contributions to a group savings fund on an agreed basis according to their capacity to save. Group savings can be used in financing individual and group projects, but most often they are used for lending to individual members to enable them to tide over exigencies arising from illness or death, loss

Table 28.3. Loan investment in SFDP by purpose and year

(in thousands of rupees)

Year	Crops	Live-stock	Farm mech.	Irri-gation	Agro & cottage industries	Horti-culture	Others	Total
1975/76	19	261	163	9	0	1	0	453
	(4.19)	(57.62)	(35.98)	(1.99)	0.00	(0.22)	0.00	(100.00)
1976/77	86	561	358	33	21	0	61	1,120
	(7.68)	(50.09)	(31.96)	(2.95)	(1.88)	0.00	(5.45)	(100.00)
1977/78	385	1,392	365	118	117	62	76	2,515
	(15.31)	(55.35)	(14.51)	(4.69)	(4.65)	(2.47)	(100.00)	(100.00)
1978/79	369	1,579	549	331	227	47	257	2,257
	(10.99)	(47.04)	(16.35)	(9.86)	(6.76)	(1.40)	(7.66)	(100.00)
1979/80	458	1,728	752	344	157	143	243	3,865
	(11.85)	(44.71)	(19.46)	(8.90)	(4.06)	(3.70)	(6.29)	(100.00)
1980/81	674	2,365	,068	111	274	226	235	4,953
	(13.61)	(47.75)	(21.56)	(2.24)	(5.53)	(4.56)	(4.74)	(100.00)
1981/82	198	5,995	2,940	0	873	273	399	12,467
	(1.59)	(48.09)	(23.58)	0.00	(7.00)	(2.19)	(3.20)	(100.00)
1982/83	4,595	12,391	3,677	1,556	2,176	455	730	25,580
	(20.64)	(48.44)	(14.37)	(6.08)	(8.51)	(1.78)	(2.85)	(100.00)
1983/84	8,764	18,587	6,493	2,173	3,682	841	1,929	42,469
	(20.64)	(43.77)	(15.29)	(5.12)	(8.67)	(1.98)	(4.54)	(100.00)
1984/85	14,489	30,323	9,291	3,824	6,725	1,829	3,882	70,363
	(20.59)	(43.10)	(13.20)	(5.43)	(9.59)	(2.60)	(5.52)	(100.00)
1985/86	20,071	69,733	13,621	5,961	7,242	2,524	7,566	96,718
	(20.75)	(72.10)	(14.08)	(6.61)	(7.49)	(2.61)	(7.82)	(100.00)
1986/87	23,548	40,733	1,227	5,602	6,786	3,289	7,572	99,880
	(23.58)	(40.78)	(1.23)	(5.61)	(6.79)	(3.29)	(7.58)	(100.00)
1987/88	39,980	57,398	17,379	7,777	10,587	5,412	11,664	150197
	(26.62)	(38.22)	(11.57)	(5.18)	(7.05)	(3.60)	(7.77)	(100.00)

Note: Figures in parentheses indicate the row percentages.

Source: SFDP Division, Agricultural Development Bank of Nepal.

of assets, natural calamities, social expenditures, and other unforeseen expenditure. More than Rs 10 million have so far been collected as group savings under the SFDP. Group savings have exceeded Rs 100,000 in a number of SPOs. This has contributed significantly to the development of a sense of self-reliance and self-confidence among the members.

Social Services

Social development activities are important components of the SFDP. These activities, which enhance the economic and social well-being of the farmers, include the promotion of adult and female literacy, health care for women and children, population education and

Table 28.4. Purposewise loan disbursements and recovery rates in SFDP (1988/89)

(in thousands of rupees)

Purpose	Investment	Collection	Overdue	Recovery Rate %
Cereal crops	33,214	21,595	12,972	62.5
Cash crops	24,845	14,384	10,342	58.2
Marketing	12,820	8,320	5,151	38.2
Farm mechanization	23,042	11,217	14,433	43.7
Irrigation	13,737	5,613	4,811	53.8
Bio-gas	831	210	389	35.6
Land development	4,343	1,071	1,039	50.8
Agro-industries	11,835	6,675	7,293	47.8
Godown and cold storage	134	121	47	72.0
Livestock	62,767	34,520	40,303	46.1
Horticulture	7,543	899	635	58.6
Tea and coffee	66	15	28	34.8

Note: The figures given are provisional.

Source: Asian Development Bank, Progress Report of SFDP for 1988/89.

family planning, supply of drinking water, promotion of improved stoves, sanitation, and construction of school buildings. Women's development activities are also undertaken in at least 63 SPOs where women GOs have been appointed. These activities include various income-generating activities carried out exclusively by women.

Coordination with Other Agencies

Being a multidisciplinary integrated development programme, the SFDP has to depend for its success upon cooperation from relevant government line agencies. Important among these agencies are:

- the Ministry of Local Development—social service;
- the Department of Agriculture—agricultural extension services;
- the Department of Livestock Development and Animal Health—veterinary services, breeding, etc;
- the Department of Cottage and Village Industries—training and technical assistance in the cottage industries component;
- the Agricultural Inputs Corporation—supply of chemical fertilizers and improved seeds;
- the Department of Forests—supply of seedlings, technical advice regarding forestry; and
- the Dairy Development Corporation—marketing of dairy products.

Cooperation with these agencies is achieved through coordination committees at the central and district levels, formed under the Chairmanship of the Governor of Nepal Rastra Bank and the Chief District Officer, respectively.

Financing Agencies

The planning and implementation of the SFDP, during its initial experimental stage, was undertaken with the financial assistance of the Food and Agriculture Organization/United Nations Development Programme in 1975. The International Fund for Agricultural Development (IFAD) came forward with the first major financing of the programme ($12.5 million concessional loan and $1 million grant) in 1981. IFAD also financed a second SFDP loan ($26.5 million) in 1985. A total of 227 SPOs have been financed by the IFAD under SFDP I and II. The remaining SPOs have been financed by various other donor agencies under integrated rural development projects—Koshi Hill Area Rural Development Project (British), High Area Development Project (ADB), Rapti (USAID), and Karnali/Bheri Integrated Rural Development (Canadian International Development Agency). In addition, UNICEF, UNFPA, and the German Technical Assistance Programme are financing various social service and training components. IFAD is financing another project for the development of household forestry and forage and livestock in the hills, through the SFDP framework.

Another major project, SFDP III, is being negotiated with the Asian Development Bank.

The Viability of SFDP

The SFDP, being more or less a community development programme, has high overhead costs and this makes it costlier than a normal credit programme. Administrative costs of the SFDP in 1987/88 were calculated to be quite high (11.8% by the Feasibility Study of SFDP III). This explains why the SFDP accounted for 23.65 per cent of the total administrative costs of the ADB/N, even though SFDP loans were only 13.8 per cent of the total outstanding loans of the ADB/N (IDS 1989a). An analysis of the profit and loss of the SFDP, as calculated by the Feasibility Study, reveals that 'the viability of the SFDP operations is dependent upon the availability of funds at very low rates of interest, and not at the present rates of mobilising domestic funds by the banks'.

It is often argued that the overhead costs of the SFDP can be reduced by increasing the number of groups and members and the volume of loans per SPO. This will, however, involve a trade-off between reducing costs and reducing effectiveness of the programme. Financial viability should not be a decisive factor in judging the effectiveness of SFDP because of its contribution to social development and poverty alleviation, a declared goal of the government.

GENERAL IMPACT OF THE SFDP

The SFDP is considered to be one of the very few successful and promising programmes undertaken for poverty alleviation in the country, in spite of the various problems that have beset it during the last few years. It was the first time that the disadvantaged and the poorest, who had no collateral to offer, gained access to institutional credit. 'It has resulted in significant improvements in the social and economic welfare of the rural poor, and holds greater promise for the future' (IDS 1989b). The programme has demonstrated that it is possible to reach the poor, to organize them, and to motivate them for higher performance. A special follow-up mission, fielded by IFAD in 1989, felt that 'despite the problems affecting the project there is no doubt that it has been, and continues to be successful. It is generally regarded as the only project addressing the needs of the rural poor'. Encouraged by the relatively successful operation of the SFDP, many donor agencies have channeled their activities through small farmer groups.

The most recent evaluation of the SFDP, based on field surveys, was carried out by APROSC in 1987. Among earlier evaluations are the one undertaken by APROSC in 1978 and another carried out by Nepal Rastra Bank in 1980. All three evaluations point to the positive impact of the SFDP. Other evaluations at the micro-level have come out with similar findings.

The 1987 APROSC evaluation, the most comprehensive evaluation so far, is based on the field surveys of six SPOs in the mountains and six in the *Terai*, covering a total sample of 576 households. Notwithstanding some methodological deficiencies of the study, its findings are not considered to be wide of the mark. Some of the findings are listed below.

- The percentage of farmers using improved seeds has increased from 37 per cent to 62 per cent in the hills.
- The amount of fertilizers used went up by 49 per cent.
- Yields of major crops have increased appreciably.
- The number of fodder trees per livestock unit increased.
- The number of milch animals owned by small farmers and the volume of milk produced also increased.
- Savings per member increased substantially.

All this has resulted in significant increases in the total family income of the small farmers covered by the SFDP. The average farm income of the beneficiary households in the hills increased by Rs 1,637 per household, or by 4.7 per cent per year, while off-farm income, constituting 35 to 45 per cent of the total household income, rose by 11.9 per cent per year in the mountains and 16.14 per cent in the *Terai*. Income distribution within the households is also considered to have altered in favour of women through their income-generating activities under the programme (IFAD 1989). A study on employment generation came out with the finding that underemployment in the surveyed areas declined from 48.3 per cent to 24.6 per cent after project implementation (APROSC 1989).

Observations of the IFAD mission for completion evaluation (1989) and the IDS feasibility study of SFDP III (1989) saw real improvements in productivity, in the living standards, and in a heightened social awareness among small farmers. They had become conscious of the new opportunities offered to them by the SFDP and were showing genuine interest in seizing these opportunities. They had also begun using improved inputs, improved cultural practices, and integrated farming systems (IDS 1989b). Some of the small farmers with less than one hectare of rainfed land or less than half a hectare of irrigated land were found to be making sufficient income to meet their basic needs by combining crop production with livestock, poultry, and horticulture. On the social front, significant progress had occurred in the percentage of children enrolled in schools. Drinking water supplies had improved. More women, in areas covered by the programme, were using family planning methods. Over 50 per cent of the women surveyed thought that their social recognition and prestige had been enhanced because of the programme

(CWD 1986). In many communities, member farmers had given up gambling and drinking and had agreed to curtail expenditure on social ceremonies and festivals (ADB 1989).

The most notable achievement of the SFDP, however, is the creation of an institutional base for organizing the rural poor into groups with the purpose of demanding, receiving, and using credit and other services offered by government development agencies more effectively. It has also developed a sizeable cadre of experienced, capable, and innovative field workers prepared to work and live with the farmers and expose them to new ideas in various fields.

The main factors leading to the success of the SFDP are considered to be:

- the organization of small farmers into groups with a view to developing feelings of group solidarity among them,
- the level of personal attention given by the GOs and other SPO staff to motivate and help the small farmers, and
- the delivery of credit to the doorstep of the small farmer.

Notwithstanding all its successes, the programme has, during its rapid expansion, developed many weaknesses, some of them quite serious. What is needed now is an in-depth examination of the programme to find out the causes underlying these weaknesses and to rectify them. Some problems are noticeable in the conceptual aspects of the SFDP also. These too have to be addressed without much delay.

SOME WEAKNESSES IN THE PROGRAMME

As a credit programme, the most glaring shortcoming observed in the SFDP is the very low repayment rate of 51.2 per cent. This directly affects its viability and is the result of several other weaknesses in implementation. Repayment rates show a declining trend with the ageing of the SPOs. Repayment performances of the older SPOs are the lowest, according to an analysis carried out by the ADB/N. Average repayment rates of SPOs that have been in existence for less than three years are 75 per cent, while those from SPOs that have been in existence for 10 years or more are only 39 per cent. Overdue loans are also concentrated in a few SPOs. In the mountains, 6 out of 222 SPOs account for more than 26 per cent of the overdue amount, while in the *Terai*, 4 out of 124 SPOs account for 28 per cent of the overdue amounts. A variety of reasons are considered to have contributed to the low repayment rates. Many of these reinforce each other. Some of them are described below.

Some of the surveys suggest that the eligibility criteria laid down for the selection of small farmers may not always have been met. Cohesion and identity of interests are found lacking in many of the groups. Besides ethnic and occupational diversities, many of the groups have wide disparities in the incomes of their members. All these have resulted in a weakening of the group responsibility for repayments. The concept of group responsibility has also not been enforced uniformly by all the GOs.

The SFDP is a programme requiring intensive supervision and monitoring. It necessitates not only supervision of the groups and its members by the GO but also supervision of GOs by control offices. Adequate supervision appears to be lacking at both levels. The number of groups within a large number of SPOs has expanded beyond the limits of effective supervision. ADBN's policy of increasing the number of groups per SPO, it appears, is aimed at reducing the overhead costs and enhancing the cost-effectiveness of the programme. However, it involves an important trade-off, as recovery rates appear to decline with an increase in the number of groups in an SPO, as seen from the following table.

Regions	Average number of groups in SPOs having		
	70%	80%	90%
Mountains	19 (88)	16 (66)	13 (41)
Terai	26 (38)	19 (31)	16 (18)
Total	21 (126)	17 (97)	14 (59)

Note: Figures in parentheses indicate the number of SPOs.

It would be desirable to take this consideration into account in determining the limits of effective supervision by an SPO. Advantages gained from increasing the number of groups and the amount of loans per SPO may be more than offset by higher delinquency in repayments.

The training of GOs has lagged behind the fast rate of expansion of the SPOs. A number of GOs who had not been given pre-assignment training felt that they had not been receiving adequate guidance from above and that they had been left to their own devices in search of ways to achieve the objectives of the programme.

The training of beneficiary farmers has considerable scope for improvement, and a suitable system of assessing training needs has yet to be developed. In the absence of such a system, it is not certain whether existing training programmes are adequate or geared to the farmer's need.

A lack of government commitment has resulted in various line agencies not focussing their efforts on the small and marginal farmers. Among other things unavailability of services and inputs during crucial times may have contributed to the increasing amount of overdue loans. Coordination mechanisms set up under the programme have not worked effectively. What little coordination exists at the local level is largely a result of the personal efforts of GOs and other conscientious officials of the line agencies, and not a consequence of government policy.

SFDP FROM THE MOUNTAIN PERSPECTIVE

Since the SFDP is a programme aimed mainly at marginal farmers and other rural poor found both in the mountains and the *Terai*, it addresses one of the mountain specificities, namely, marginality. Other characteristics of the mountain areas also appear to have been taken into account in the implementation of the SFDP. There are less SPOs in the mountains than in the *Terai* mainly because of the difficulties of supervising a larger number of groups owing to relative inaccessibility within the mountain regions. The average number of groups in a mountain SPO is 21.5, in comparison to a much higher number of 36.1 in the *Terai*. The policy of approving loans without collateral is also appropriate for the mountains where a large number of the poor have no collateral to offer.

Certain programmes actively promoted by the SFDP are also particularly suited to the mountain areas. One example, is the promotion of sprinkler irrigation which, besides economizing on the use of water, does not tie up scarce land resources in bunds and channels that are needed in regular surface irrigation. Also, sprinkler irrigation carries no risk of soil erosion as do other gravity schemes in the mountains. So far, 2,000 sprinklers, irrigating approximately 600 ha of land, have been financed. Small farmers in the mountains find it viable to use sprinklers because of the high capital subsidy given to irrigation. Similarly, 600 mechanical water turbines have been installed to harness locally available energy sources. These turbines supply energy to a large number of agro-processing units scattered throughout the mountains. Nearly 70 turbines have also been installed for generating electricity. ADB/N is planning to install more turbines in mountain areas where sites are suitable not only to provide power for agro-processing units but also for cottage industries and even for domestic cooking.

There is a need to develop a sound land use policy in the mountains, not only because of the limited availability of agricultural land but also because of the need to protect the environment from further deterioration. Such a policy will necessitate, first, massive reforestation efforts and, second, the optimal use of available agricultural land. Small farmers' groups can be used to achieve both of these. A project for the development of forage, forestry, and livestock, enlisting the efforts of small farmers for reforestation, and, at the same time, providing them with employment and income opportunities, may soon be initiated with assistance from IFAD. Under the project, degraded forest lands will be given to landless and marginal farmers on a long-term lease of 50 years. They will be provided with loans for growing fodder, fuel, and fruit trees on such lands. When the fodder grass and trees begin yielding harvests, the farmers will also be given loans for livestock raising. The project will thus serve the twin purpose of protecting the environment and providing income to farmers through the sale of livestock and its products.

The second need, that of making optimal use of agricultural land, will necessitate the development of agricultural technology packages suited not only to the mountain areas but also to the economic condition of small farmers living there. Such a technology will have to be based on: (1) the least possible use of external inputs (not only because their timely supply cannot be guaranteed owing to the factor of inaccessibility but also because marginal farmers may not be able to afford them) and (2) the whole farming systems approach suited to different types of land in the mountains. Sustained success of the SFDP over the longer term is contingent upon the development and propagation of such technology.

There is also a need to promote non-farm employment as a means of reaching the poor in the mountains. Non-farm incomes are estimated to constitute, at present, approximately 33 per cent of the total income of the mountain regions of Nepal. This percentage is expected to be much higher for small farmers in view of their miniscule landholdings. Off-farm activities will have to be a major source of income for them. Traditionally, cottage industry activities such as utensil making, tool making, shoe making, and tailoring have been carried out mostly by people belonging to lower castes for whom poverty seems to be most concentrated. They have traditional skills in these areas, but their trades have languished for want of adequate financial and market support. There is substantial scope for reviving such crafts and turning such households into independent producers of different commodities that are in demand. Financial, technical, and training support for such occupational groups in the hills can contribute significantly in the alleviation of poverty and the development of self-reliance in these areas.

It has already been mentioned earlier that livestock still claims the largest share of SFDP loans. This component, however, has one of the lowest repayment rates. The basic flaw is that while large amounts of loans were given for livestock, the need for development of fodder to support the livestock was overlooked. The availability of animal feed in Nepal at present is much less than its demand, considering the sources of its supply, namely, agricultural land, pastures, forest land, shrubland, and plantation areas. It would be desirable in the future to ascertain the availability of fodder before livestock loans are sanctioned. In the hills, loans for small animals such as goats and sheep, and poultry, for small and stall-reared animals such as pigs, rabbits, and young buffalo heifers and for fodder development could be supported with advantage.

THE NEED TO STRENGTHEN THE SFDP

The SFDP has now completed 15 years of operation. It has made significant strides during this period. Its potential as a poverty alleviation programme is no longer in doubt, even though some serious shortcomings in the programme have come to the surface. It is time that fundamental issues relating to the programme be reviewed carefully and that appropriate corrections be made with a view to putting it firmly on the road to achieving its real objectives. Some of the problems have already been mentioned earlier. Some conceptual issues will now be discussed briefly.

The nature of motivational activities carried out under the SFDP, limits the pace of its expansion within an SPO. The process of inducing changes in the outlook and behaviour of the poorest and the most backward members of the community is bound to be slow. Persistence in efforts and patience in waiting for results are needed. A target-driven approach to the implementation of the SFDP appears to have created a pressure on the GOs to regard lending as an end in itself, rather than as a means, and to lose sight of the real objectives of the programme. It is important that the GOs should basically play the role of catalysts, rather than that of target-achieving implementors.

For these reasons, it may be appropriate to consider SFDP more as a process than a programme or a project, in which achievements of targets, in terms of numbers, is the most important objective. Instead, it should concentrate on its real goals of self-sustained and self-reliant development of the poor. Efforts for achieving this goal have, however, not been adequate. More than 40 SPOs have existed for more than a decade but many of them, instead of achieving self-reliance, have moved away from it, accumulating heavy amounts of overdue loans. Urgent steps are needed to give the programme a new conceptual orientation to help it achieve the goal of self-reliance among small farmers. The dependence of small farmers on SPOs will have to be gradually reduced and phased out. One way of doing this would be to replace SPOs, within a period of time, by small-farmers' associations which could include several small-farmer groups in a contiguous geographical area. Such associations should be responsible for the disbursement and collection of loans, supply of technical services through lead farmers, supply of inputs, marketing of outputs, and mobilization of savings. The role of the bank will then be confined to monitoring the activities of these associations, helping and advising them in their work, and providing them with credit funds. Such funds should be given at rates of interest which should allow these associations sufficient margin to cover the cost of lending to their members, and thus make them viable units. The replaced SPOs could then move on to other adjoining areas, and thus accelerate the expansion of the programme.

On the planning and implementation side, the SFDP should adopt the system of rolling work plans, under which a plan for the next three years should be prepared every year, based on actual performance during the preceding year. Such rolling plans should enable ADB/N to plan and adjust its activities on a more realistic level and not on the basis of predetermined and rapid targets extending over a five-year period.

Serious and urgent efforts are needed to correct the weaknesses that have crept into the SFDP and to put it on the desired course. Among all existing government programmes, the SFDP remains a programme that holds the highest promise for poverty alleviation.

LESSONS FROM THE SFDP EXPERIENCE

This brings us to the lessons that can be drawn from the SFDP experience, not only for the development of the mountain region but also for the upliftment of the poor who live in these regions. Some of the lessons appear to be as follows.

Groups have proved to be effective instruments in engendering a sense of responsibility among group members for making collective efforts for common well-being. The question is what kind of group would be most suitable for promoting internal efforts for change rather than a heavy dependence on extraneous factors for change. The villages have long been functioning as cohesive social and economic units in the past, and may well continue to do so in the future. And this factor should not be ignored in deciding about the kind of groups necessary if they are to become effective as instruments of change. Considered from this perspective, the formation of groups representing the whole village, rather than the people belonging to a particular economic level in the village, may be more appropriate for inducing changes in the village. The factor uniting the different members of such a group would be a sense of belonging to the village, which has probably been functioning as an organic unit for hundreds of years. Such groups will be much larger than the present average group of approximately eight members, and may generally consist of 30 to 60 members or even more. Large farmers who do not blend with the economic and social milieu of the village can be excluded from the groups. Within such large groups, efforts should be concentrated on the poorest members, although other members should not be ignored. This approach has been tried in one of the grassroots development programmes in Nepal and found to be effective.

The SFDP experience also underscores an urgent need to develop technologies that are suited to the needs and limitations of the marginal farmers in the mountains and that are acceptable to them. Such technologies should be based, first, on the least possible dependence on external inputs; second, on maximizing the productivity of land, which is the most scarce resource in the mountains; and third, on the need for conserving resources and protecting the environment. There is also a need to train small farmers in such technologies on a massive scale.

In the non-farm sector, there is a need to upgrade the traditional skills of the occupational groups and craftsmen who have been practising their trade for generations, with a view to making their products more competitive. Such efforts should also be supported by financial and marketing assistance.

The SFDP has also demonstrated the need to make the process of development in the villages self-sustaining and not overly dependent on assistance from extraneous sources. Initial interventions by catalysts, such as the GOs, are necessary. The main role of such catalysts should, however, be to create institutions that reduce the need for external assistance and ultimately eliminate the need for their own services. Such institutions could be in the form of local organizations and associations which, while striving for the welfare of their members, could also serve as conduits for the supply of government services to individual members of the village, not only in the fields of agriculture and cottage industries but also in social services such as adult literacy, primary health care, family planning, and drinking water. All government agencies will then need to deal with these organizations and not with individual members in the villages.

Finally, no 'cut-and-dried' strategy for the upliftment of the poor in mountain areas has yet been developed. A search for such a strategy will necessarily have to follow a process of trial and error. Presently, efforts are being made in this direction by several government and non-government agencies in their own way. The results of all such efforts should be carefully studied, evaluated, and disseminated in a quest for appropriate strategies.

REFERENCES

Agricultural Development Bank of Nepal (ADB/N). *Procedure for Loan Investment and Collection* Kathmandu: His Majesty's Government Press, 1986.

Agricultural Development Bank of Nepal (ADB/N). *Policies and Procedures Relating to SFDP*. Kathmandu: His Majesty's Government Press, 1987.

APROSC. Evaluation of IFAD-Assisted Small Farmers' Development Programme (Phase I), December 1987.

APROSC. Generation of Employment by the Small Farmers' Development Project, February 1989.

Asian Development Bank (ADB). Appraisal of the Third Small Farmers' Development Project in Nepal, December 1989.

Centre for Women and Development. An Evaluation of the Women's Development Project Under the Small Farmers' Development Project, October 1986.

Integrated Development Systems (IDS). Feasibility Study of Third Small Farmers' Development Project, Nepal, August 1989a.

Integrated Development Systems (IDS). Assessment of Poverty Alleviation Programmes in Nepal, 1989b.

International Fund for Agricultural Development (IFAD). Completion Evaluation Report of SFDP, April 1989.

29

THE AGA KHAN RURAL SUPPORT PROGRAMME: AN APPROACH TO VILLAGE MANAGEMENT SYSTEMS IN NORTHERN PAKISTAN

T. Husain

Tariq Husain is currently Managing Director of Development Research and Management Services, Islamabad, Pakistan

CONTENTS

INTRODUCTION

The overall objective of this paper is to explore those issues in community organization and resource management that represent the common concerns of the International Centre for Integrated Mountain Development (ICIMOD) and the Aga Khan Rural Support Programme (AKRSP). These concerns are related to the design and implementation of rural development programmes. They are motivated, in the case of AKRSP, by a desire to develop a model for high-mountain development in Northern Pakistan that can enable the people of the region to improve their incomes in a sustainable and equitable manner. ICIMOD's motivation is to acquire lessons for the future by comparing the rich and diverse experiences in rural development in the Hindu Kush-Himalayan region.

ICIMOD and AKRSP share the following broad perspectives:

- that sustainable productivity and sound environmental management are long-term goals of rural development;
- that particular attention needs to be paid to organizational structures at the project and community levels, in addition to the attention that is normally paid to technical and financial constraints; and
- that organizational resources are needed to facilitate implementation and enhance local participation in rural development programmes.

These perspectives have been the basis for collaborative work between AKRSP and ICIMOD since 1985. A pilot study on the interrelationship of community management of rural resources, with accelerated development efforts, conducted in the project area of AKRSP, has been one of the major collaborative activities of the two organizations (Husain et al. 1990). This paper draws on the results and understanding generated by the above study. The paper first introduces AKRSP and its approach as well as its mandate area, namely, Gilgit District, with focus on its natural resource base and the institutional framework affecting resource management. The second part of the paper presents village-level evidence on the management of natural resources as revealed by the case studies carried out in nine villages. The third section presents conclusions and general inferences from the case studies to reflect on the AKRSP model of institutional innovations.

INTRODUCTION TO GILGIT DISTRICT AND AKRSP

Introduction to Gilgit District

The project area of AKRSP consists of the three northernmost districts of Pakistan—Gilgit, Chitral, and Baltistan—situated between longitudes 71°2′E and 75°4′E and latitudes 35°3′N and 35°6′N; the region borders on India, China, and Afghanistan. The area covers 69,200 sq. km and has an estimated population of 830,000, scattered over 1,030 villages (AKRSP 1987b). The region is extremely mountainous, since it is at the intersection of four of the world's highest mountain ranges—the Himalaya, the Karakorams, the Pamirs, and the Hindu Kush. Within this region, Gilgit District is the most privileged, in terms of accessibility and overall development. Its area is 28,500 sq. km with an estimated population of 286,000 living in more than 300 villages. There are some 30,000 farm families in Gilgit District and an urban population of about 40,000 (AKRSP 1987b).

The physiography of the region is rugged and hilly, with steep, heavily dissected slopes and water courses along the slope faces and valley bottoms. The terrain is naturally unstable and rockfalls and landslides are common occurrences. The soil is mixed with stones and boulders, and the depth, aspect, and location of seepage areas have more influence on production potential than the variation in the parent material. These soils are low in clay content, and, due to extreme dryness, are very low in organic content. They are, however, suitable for a large number of annual and perennial crops.

The region lies just outside the monsoon area in a partial rain shadow. It receives about 100–900 mm of rain annually, mainly as snow in the winter months. Agricultural production is sustained by irrigation with the glacial melt. The region can be best described as having an arid, continental-type climate. The growing period at 1,500 masl is estimated to be 325 days, and at 3,000 masl it is 195 days (AKRSP 1987b). The nine villages selected for this study fall within this range.

The district is connected to the rest of Pakistan, and to China, via the all-weather Karakoram Highway (KKH), which was formally inaugurated in 1978. There are scheduled daily flights between Gilgit and Islamabad, 600 km away by road.

Agriculture is by far the largest economic activity in Gilgit District and is the principal means of livelihood of 85–90 per cent of the population (detailed descriptions of the region's agriculture are given by Staley 1982, Saunders 1983, and Whiteman 1985.) The region supports a range of farming systems, but nearly all contain common elements: cereals, grain, legumes, fodder crops, small livestock, fruit and nut trees, vegetable crops, and fuel trees. Broadly, all the farming systems in the region can be described as arable crop systems with trees and livestock. Indeed, farming in the region shows many of the symptoms of transition from a subsistence economy to a semi-commercial economy. The following picture of Gilgit's traditional agriculture by an agronomist (Whiteman 1986) describes the farm household and its resources succinctly:

> A typical village will contain about 60 households with an average family size of eight people and an irrigated area of 0.75—1.0 hectares in double-cropped areas (up to about 2,000 m altitude) and 1.5 to 2.0 hectares in the single-cropped area. Wheat is the dominant crop; maize became popular some 55 years ago and has largely replaced the earlier-maturing Panicum and Setaria millets and buckwheat that now persist only at the upper end of the double-cropping zone. Up to a quarter of the land may be under fodder crops, mostly lucerne for hay for winter use and shaftal clover for green cutting in spring. Pulses are rarely grown, and a small area is allocated for vegetables and potatoes. The area is deficient in grain and up to a third of the wheat that is consumed is from a subsidised quota. Yet in spite of the shortage of land for cereals, a range of multipurpose trees for fodder, fuel, timber, and fruit are grown along field boundaries around the house and on any steep but irrigable land. Poplar, willow, mulberry, apricot, and Russian olive (Eleagnus) are the most common, with walnut, peach, grape, apple, almond, pear, fig, and pomegranate widespread. There will be a pair of oxen, one or two cows, a calf, perhaps 20 goats, 10 sheep, 15 hens, and a donkey.
>
> About 1,500 metres higher there is a sparse communally grazed alpine pasture about two days' walk away where the livestock are taken for a four-month period in summer. Farther up the mountain, in small side valleys, are stunted, gnarled remains of open juniper forest with a little birch heavily overcut and

> grazed that provides the firewood for cooking. Between the village and the mountain pasture is often a small meadow or barley field wherever the valley becomes wide enough.
>
> This little scenario depicts the total resources available to meet all family needs for house construction material, food, furnishings, woollen garments, dairy products, livestock fodder, and farm implements as well as cash for small sundries (paraffin, tea, matches, salt), though these are more often bartered for. Nowadays most households have a male member working part-time or full-time outside the area as a source of cash. Despite the material poverty and frugal life, there is a robust quality apparent in a life lived in equilibrium with an adapted farming system from a consistent resource base.

Changes taking place in rural markets and agricultural technology are affecting the above situation in significant ways. With improved communications, a majority of the district's farmers are now using tractors, threshers, and new varieties of wheat; an even larger proportion use chemical fertilizer. Timber is imported in large quantities from the neighbouring district of Diamer; wheat, rice, dairy products, vegetables, cooking oil, livestock, kerosene, liquefied petroleum gas, cement and construction material, and a number of other items of daily use are now supplied from the plains of Pakistan. Able-bodied men migrate in large numbers within the region, following agricultural, construction, and tourism activities, or they go down-country in winter to work for cash. Increasingly, women are becoming involved in running the farm household in association with old men and children. Small hydroelectric units provide night-time electricity for lighting. New roads connect remote valleys to the KKH. Education is becoming more widespread. The value of time is rising, and labour-intensive activities are increasingly being performed in less labour-intensive ways or else given up. Large amounts of credit are being made available for agricultural development, construction, and commerce. In short, the allocation of resources in Gilgit is being subjected to rapid and pervasive change. After centuries of isolation and low-income equilibrium, the region's rural economy is characterized by constant change and the opportunity to reallocate resources from low pay-off options to high pay-off ones.

AKRSP: Organization, Objectives, and Approach

The AKRSP was established by the Aga Khan Foundation in 1982 as a private company limited by guarantee. It is a non-profit, non-sectarian Pakistani organization, with its own Board of Directors for policy-making and direction and a management group in the project area (headed by the General Manager) for day-to-day operations.

Although seed money for the AKRSP was (and is) provided by the Aga Khan Foundation, the company has received generous financial assistance over the years from several donor groups and development agencies. Finally, AKRSP has, in a few significant instances, pooled its resources with development funds put at the disposal of elected representatives in Gilgit by the government.

AKRSP started operating in Gilgit in December 1982 and subsequently extended its operations to the districts of Chitral and Baltistan. Its project area now includes followers of three major Islamic traditions in roughly equal proportions. All its staff are Pakistanis, and all the field and support staff are recruited from the project area. AKRSP's activities now extend to about 800 village organizations (VOs, nearly half of them in Gilgit District) and include programmes for social organization, women–in development, physical infrastructure—particularly for irrigation and communications—agricultural and livestock research, extension input supply, appropriate technology, commercial and industrial development savings and loans, resource management—particularly forestry and pasture development—and training in a wide range of practical and managerial skills. In addition, AKRSP is working with government and private agencies to provide VOs with access to basic health coverage, education, and improved living conditions. Wherever feasible, AKRSP provides services, through existing private or government entities, and works to create effective links between these and the VOs rather than duplicating the work of existing organizations.

AKRSP's Second Phase Strategy Paper describes the programme's objectives in the following words (AKRSP 1987a).

> The broad objective of AKRSP is to increase the capacity of local people to identify and utilise opportunities and to solve their own problems so that they can plan and implement development programmes leading to increased incomes and employment (without significantly increasing inequalities); to improved health, nutrition, education and living conditions; and to improvements in the sustainability and productivity of the environment. Thus AKRSP is designed to promote development in an equitable and sustainable manner. It is also conceived, from the outset, as a self-liquidating organisation, able to work itself out of a job in any location within approximately ten years. The aim is to leave in-place local institutions capable of facilitating further progress into the future.

The basic planning tool for AKRSP is a series of diagnostic dialogues carried out with villagers (AKRSP 1983). The General Manager initiates the first dialogue, explaining the objectives and methods of AKRSP and inviting the villagers to identify a project that could be undertaken and maintained by the villagers for the benefit of the village as a whole.

The second dialogue determines the feasibility of the project under the technical supervision of a competent senior manager. Field operations are managed by the Social Organization Unit and the products of the second dialogue are blueprints and cost estimates for the project.

The third dialogue starts with a discussion of the finalized scheme. The terms of partnership between AKRSP and the villagers are also discussed and AKRSP describes the form and extent of assistance it can provide, and villagers explain how they will plan and implement the scheme, develop skills, meet regularly as a disciplined organization, and establish group savings. If successful, the third dialogue results in a village-level project for the VO.

The key concept in AKRSP's approach is that of the VO—this is a broad-based coalition of all those village residents whose common economic interest is best served by forming a multipurpose development organization. The VO is the executing agency for all village-level projects sponsored by AKRSP and its collaborators. This institution is established, in the first instance, by the promise of a grant (an average Rs 150,000) for a village- level Productive Physical Infrastructure (PPI) Project. Since farmers attach great importance to improvements in their common physical assets, the investment by AKRSP initiates a process of disciplined organisation and collective management in the village. In turn, the formation of the VO enables the village to complete the PPI project more quickly and cheaply than would be possible otherwise. There is, thus, a symbiotic relationship between the village organization and the grant-funded PPI; each enhances the effectiveness of the other and results in income generation for the villagers. The new social organization (the VO) is aided by the catalytic effect of the new economic infrastructure (the PPI) that the VO is implementing. Together, the VO and the PPI become vehicles and stimulants for local income and employment generation.

During the First Phase (1983–1986), the principal focus of AKRSP was the establishment of village-level institutions for managing development and the funding of essential local, infrastructure projects, one per VO, chosen by the VOs. During the First Phase, both AKRSP and the villagers invested in various types of productive common property on a very large scale. This experience demonstrated the potential for community management of financial resources and physical assets such as irrigation channels, link roads, and reservoirs. Besides contributing to widespread increases in income, the collective management of these resources has helped shape the VOs as institutions for village development.

Village Organization and Resource Management

To build upon the experiences of the First Phase, AKRSP's Second Phase strategy lays down the objective of improving the integrated management of resources at three levels—farm, village and valley/watershed. This would include work on farming systems, integrated livestock-cropping-pasture systems at the village level, and contributions to valley planning and watershed management (AKRSP 1987a). The pursuit of this objective is expected to lead to:

- improvements in the productivity and sustainability of natural resources, i.e., greater sustainability of natural resource use together with increases in farm incomes, and
- a greater capacity among the villagers for managing their common resources.

AKRSP's existing programmes have begun to address issues of:

- land use and the development of new land;
- irrigation development and water management;
- forest management and forestry development; and
- livestock and pasture development.

In implementing these programmes, the AKRSP has benefited from collaboration with relevant government agencies, as well as the International Wheat and Maize Improvement Centre (CIMMYT), the International Union for the Conservation of Nature and Natural Resources (IUCN), ICIMOD, the International Institute for Environment and Development, and the International Irrigation Management Institute (IIMI). Collaboration with these agencies is a response to the realization that the development of village management capacity in the future requires 'a growing sophistication in the identification and analysis of opportunities and problems, and in the development of entrepreneurial response and of internal mechanisms of management and control' (AKRSP 1987a).

AKRSP's experience with resource development programmes in the first phase led to the recognition of the amount of women's involvement in, and dependence on, the region's natural resources. To a varying but increasing degree, rural women are involved in or affected by the management of land, water, forests, pastures, and livestock. In areas where men have been attracted to off-land employment opportunities, it is particularly important to improve the efficiency of the time and management inputs provided by women, in order to equitably improve the productivity and sustainability of natural resources. To this end, AKRSP's work included the sponsorship of the Workshop on Women and Resource Management in Gilgit, in November 1987 (Hussein and Karmali 1987, Magrath 1987).

World Bank Evaluation of AKRSP's First Phase

While it is too early to assess the new directions of AKRSP's second phase, the first phase was evaluated by the Operations Evaluation Department of the World Bank, in September 1986. The evaluation report has since been published (World Bank 1987).

The World Bank report concluded that AKRSP's achievements 'are largely attributable to the effectiveness of the institution-building efforts at the village level.' It observed that the management principles that are critical to this effectiveness include:

(1) The principle of the primacy of the VO. The VO is the focal point for all AKRSP activities but its sovereignty is sacrosanct, although AKRSP is firm in keeping to the agreed conditions of the partnership. The VO and AKRSP are seen as contractual partners insofar as activities of the VO are supported but never undercut.
(2) The principle of continued attention to innovations. Villagers and staff of AKRSP alike are encouraged to innovate, using a trial and error approach that is carefully monitored. The effect is to create a 'learning environment' of active improvisation and innovation.

The World Bank report also points out that the 'pursuit of these principles is aided by the flexibility of AKRSP as a small, independent non-government organisation, relatively free of fixed procedures, hierarchical clearance, or internal constraints on actions. This flexibility facilitates the ''working'' method of experimentation, adaptation, and trial and error innovation that is the hallmark of the program.' The following characteristics of the project area appear to have worked to AKRSP's advantage:

(1) institution-building could proceed with little or no competition, in something of a political and administrative vacuum;
(2) a tradition of cooperation in the villages that is consistent with the VO approach; and
(3) the high proportion of Ismaili villages in Gilgit District, favourably disposed to an Aga Khan-supported programme, gave an initial impetus which was invaluable, though only about one-third of the population of the project area is Ismaili.

The World Bank commended the institutional model of AKRSP that combines village organization and PPIs at the village level. At the same time, it found that the 'production model' was less well studied and conceptualized than the 'institutional model' and several changes were recommended in this area. For example:

(1) environmental and resource constraints are a major issue, and, while much is being done, further attention to this issue is needed;
(2) institutional development within and beyond the VO, especially relating to land and water use, warrants support.

The World Bank report summarized its understanding of AKRSP principles in the following list:

(1) Small farmers in isolated communities require a village organization to overcome the disadvantages of everything being on a small scale.
(2) VOs can be used successfully to promote formal savings and credit by individuals and the group, provided that control of the savings and credit remains with the group.
(3) VOs can be employed to promote genuine participation in planning and implementation of rural development.
(4) Villagers can be effectively organized initially around economic, rather than social, sector activities.
(5) A PPI Project is an effective entry point and catalyst for the organisation of villagers.
(6) In order to implement a PPI project efficiently and without exploitation, when village labour is employed it should be paid.
(7) Regular savings, however small, are an essential part of the discipline of collective management and finance of development.
(8) Members of the VO can acquire the necessary organisational and technical skills, for which other villagers are prepared to pay, to serve themselves and their community.
(9) The VO following these principles can take continuing responsibility for sustainable development of the resources at its disposal.

A direct operational implication of these principles is that the VO is the missing link between conservation and development, between income generation from a resource and its sustainable use over time. This can be considered an extension of AKRSP's first phase approach to its second phase concerns with sustainable resource management, particularly the management of change through institutional and technological innovations.

Institutions, Laws, and Natural Resources in Gilgit

The Context of Institutional Change

Like many Third World communities, Gilgit is subject to the forces of social fragmentation, disintegration of values and institutions, and the alienation of social and economic life from the values, institutions, and resources of rural communities. These forces represent both a constraint on and an opportunity for institution building.

In Gilgit, land and irrigation development, as well as control over forests and pastures, were traditionally spearheaded by feudal chiefs such as Mirs and Rajahs. They could use the authority of the State to induce or constrain their subjects (through forced labour and transfers, exile, and punishment) to construct new channels, rehabilitate old ones, develop new land, restrict the exploitation of forests, and enforce rules for summer and winter grazing. There was a system, therefore, for maintaining and increasing society's vital physical infrastructure and the natural resource base.

A general decline in feudal authority commenced with the arrival of the British administration in 1892. This decline appears to have become more pronounced in the last 35–40 years. The feudal states were formally abolished in 1974. The effect of the decline in feudal authority is evident in the slow pace of irrigation and land development and a diminishing natural resource base. For example, despite growing populations, no land settlement schemes were undertaken that matched the size of projects sponsored by the Mirs.

Whereas the Mirs had helped establish new villages, AKRSP's irrigation development programme has opened up additional land to existing villages in magnitudes that are at least as significant as the achievements of the *Mirs*. There is a significant difference, however, between how villagers perceive irrigation development and how they perceive forest and pasture management. This perception has to do with the perception of ownership. As feudal chiefs were replaced by government administrators, the forests and pastures of the feudal states became the *de jure* property of the Government of Pakistan, acting through the Forest Department of the Northern Areas. Irrigation channels, however, and lands contiguous to villages remained outside the government's domain. Thus, when AKRSP arrived on the scene in Gilgit, it found the villagers keen to improve their irrigation infrastructure, but it has had difficulty organizing villagers to improve the management of their common natural resources. The situation now is that the Forest Department has *de jure* jurisdiction over much of the forest and pastures but the actual position resembles open access. Overexploitation is observed and there is little or no investment in sustainable management.

AKRSP Experiences with Common Property Management

In the last five years, both AKRSP and the villagers have been challenged to devise new rules and conventions for the management of village resources, sometimes in an ambiguous legal and institutional situation. This has happened particularly when the traditional status of a resource has undergone change or when new assets have been introduced.

An early example of great interest was that of land development. AKRSP-sponsored irrigation channels assisted villagers in converting low-productivity, winter grazing land into potentially high-productivity, multiple-use farm land. The grazing land, by tradition, belonged to the entire village, so all households descended from those who established the village had equal rights to it. But how were the villagers to implement AKRSP s principle of private ownership and collective management on this now irrigated new land? The villagers responded with a full range of options on various combinations of ownership and collective management.

At one extreme, some villages simply divided up the new land by handing the plots over to individual households which then developed the land through their own resources. However, even these villages generally approached AKRSP for land development loans through their VOs. At the other extreme, Khaiber Village in Upper Hunza has a VO that is the regional leader in terms of land and labour specialization. The new land there is being developed as a single farm, and portions of it will be transferred to individuals for farming after it has been fully developed. The VO will continue to own the fruit orchard and the fruit-cum-forest nursery on the new land. Women have been trained to manage the nursery. All irrigation on the new land is undertaken by three specialists. There are various other village specialists, as well, and all are remunerated by the VO.

In between these two types of management system, there are wide variations in what the villagers have adopted. By and large, new land is divided up (usually equally, according to traditional rights) among individual households, but specific inputs may be managed collectively. These inputs include loans for land development, transport and implements for land development, fertilizer, seed and saplings, the services of village specialists, and, quite often, labour pooled among neighbours. In terms of collective management issues,

AKRSP's first phase was dominated by the land development process. The major lesson for AKRSP was that it should not insist on the VO treating its new land as a single farm. It should, instead, encourage the rapid and equitable development of land through collective management of critical inputs.

A multiplicity of issues arose in the second phase as the VOs began to tackle non-traditional assets and the supra-village dimensions of collective management. For both, the VOs had to define new rules and conventions. Not surprisingly, they did so, usually with reference to traditional patterns of management. AKRSP has catalogued and discussed these experiences in its Fifth Annual Review (AKRSP 1987c), and the experience with forest and pasture management is too limited to offer operational generalizations at this stage. The case studies discussed may instigate the articulation of a few operational guidelines for AKRSP.

The Legal Situation of Forests and Rangelands

The legal situation of forests and rangelands is governed by the Land Revenue Act, 1967 (XVII of 1967), Section 50, the Forest Act of 1927, and the Northern Areas Wildlife Preservation Act of 1975.

According to the Land Revenue Act, the presumed ownership of forests, quarries, and wasteland rests with the government, unless there is a written record of rights to the contrary completed by or before November 1871. A record of rights was drawn up wherever land settlement took place. In Gilgit District, land settlement took place in only one of the five subdivisions. Hence, in four subdivisions, there is no question of records of rights. After the *Mirs* and *Rajahs* were deposed in 1974, all land without a record of rights was resumed by the Government of Pakistan, Northern Areas' Administration, under the Land Revenue Act. The Administration's Forest Department maintains, therefore, that the region's communities have no claim whatsoever over forests, except as provided by the Department under the Forest Act of 1927. The Department further maintains that the villages have an option only over *shamlaat* forests, i.e., those on land accessible to the village irrigation channels. Finally, the Department maintains that the forests of six of the nine villages selected for this study belong to the State and the remaining three villages are said to have an insignificant number of forests.

The concessions provided to local communities under the Forest Act are listed (CDC 1987). These concessions differ according to the legal status of the forest. The ownership and management of natural forests are of three types:

- private: usually commercially exploited;
- State: State control of local and commercial use; and
- reserved: ownership and management by the State.

The matter of community use rights arises for State forests (category 2 above). Briefly:

- there are no rules for grazing, but it is prohibited in specified areas of National Parks under Section 7 of the Northern Areas Wildlife Preservation Act;
- villagers within five miles of the forest, or with traditional rights over it, can apply for the use of standing timber for domestic purposes upon payment of a concessionary fee;

- such villagers also have free use of any dead, dying, or diseased timber for fuelwood; those living more than five miles away need a transport permit, which is free;
- timber for commercial use may be extracted upon payment of a standard fee; and
- fuelwood obtained by contractors for commercial purposes requires a charge of Rs 5 per 100 kg and a transport permit (fuelwood sells in many parts of the district for one rupee per kg).

There have been recent incidents that have eroded the Forest Department's unqualified control over the use and management of State forests. The most contentious case is that of the Chalt-Chaprote Forest and this is discussed later in this paper. In this case, the Deputy Commissioner of Gilgit, acting on an application by the community, authorized the community to exercise control over the neighbouring forest. Such control was previously completely vested in the Forest Department. Legal support for the orders passed by the Deputy Commissioner may conceivably be found in the Forest Act, but this has not been confirmed by the present author.

Local communities also contend that the procedures specified by the Forest Act before resumption or reservation of forest lands have not been followed by the Forest Department. In particular, it is alleged that villagers were not given the opportunity to establish claims over resumed land, nor was there a land settlement made by any government.

In general, the ambiguous legal situation in Gilgit will continue to plague attempts at improved resource management. The options currently available to the administration are:

- continue with the status quo which will result in a continuing and rapid depletion of forest cover and degradation of pastures;
- seek to enforce the authority of the Forest Department which will lead to confrontation in a sensitive part of the country; or
- offer to work with AKRSP and the VOs, which will be effective if the VOs can devise rules for internalizing the costs and benefits of resource use.

Given the constraints on the Forest Department, there is a recognition among sections of the government that the last option potentially represents the most effective strategy. If this view can be articulated as official policy, then AKRSP and the VOs will need to respond to the challenge of developing institutions that can demonstrably sustain and improve the natural resources of the district.

THE MANAGEMENT OF NATURAL RESOURCES: VILLAGE LEVEL EVIDENCE

Background to the Village Studies

The Selected Villages

Nine villages in Gilgit District were chosen for an in-depth study of institutional arrangements for resource management. These nine villages are: Broshal, in the Nagar Tehsil of Nagar Subdivision; Khaiber, in the Gojal Tehsil of Hunza Subdivision; Passu, in the Gojal Tehsil of Hunza Subdivision; Roshanabad-Sherabad, in the Aliabad Tehsil of Hunza Subdivision; Rahbat, in the Sikanderabad Tehsil of Nagar Subdivision; Rahimabad I, in Gilgit Subdivision (which has only one Tehsil); Oshikhandass, in Gilgit Subdivision;

Sherqilla, in the Punyal Tehsil of Punyal-Ishkoman Subdivision; and Thingdass, in the Punyal Tehsil of Punyal-Ishkoman Subdivision.

The important features of the nine villages, presented in Table 29.1, are: (1) access—on or off the KKH; (2) agro-ecological zone—one-crop, two-crop, or two-crop transitional; (3) number of AKRSP-sponsored VOs in the village; (4) scale of village—large, medium, or small—and the number of AKRSP-sponsored VOs operating in the village; and (5) whether or not off-land employment opportunities are substantial.

Table 29.1. Some basic characteristics of the nine selected villages

Village name	On KKH?	Agro-eco zone	No. of VOs	Village size	Strong off-land opportunities[3]
Broshal	No	1-crop	1	Medium[2]	No
Khaiber	Yes	1-crop	1	Small	Yes
Passu	Yes	1-crop	1	Small	Yes
Roshanabad-Sherabad	Yes	2-crop transit.[1]	1 ?	Small	Yes
Rahbat	No	2-crop	2	Large	No
Rahimabad	Yes	2-crop	2	Medium	Yes
Oshikhandass	No	2-crop	3	Large	Yes
Shergilla	No	2-crop transit.	3 ?	Large	No
Thingdass	No	2-crop transit.	1 ?	Small	No

[1] Double-cropping extends up to about 1,850 masl, but villages at that altitude cannot expect the second crop (maize) to mature with certainty; these borderline villages are referred to as 2-crop transitional.

[2] A medium-sized village has 100–150 households.

[3] Strong off-land employment opportunities are evaluated subjectively by the author in terms of both seasonal and permanent jobs.

Source: Author

For the study, the organizational structure of each village was examined with respect to a number of natural resources and other common property. The purpose was to analyse the performance of the village vis-a-vis a list of indicators of collective management. Both traditional and non-traditional forms of common property were examined so as to identify the institutional innovations introduced by a village. In particular, the analysis focussed on:

- any outstanding strengths and weaknesses of the VO;
- significant elements of the process of constructing and maintaining the irrigation channel and the subsequent process of land development;
- the organization of a cadre of village specialists who perform specialized tasks for remuneration by the VO;
- innovations in the management of forests and pastures; and
- brief notes on the VO's performance with respect to non-traditional common property such as community-owned tractors and VO-owned hybrid cattle (the latter is the Heifer Project).

Table 29.2 summarizes the presence or absence of selected indicators of collective management in the nine villages. In addition, the case studies provide basic locational and agro-ecological data on each village, supplemented by some statistics on the resource base.

Table 29.2. Indicators of collective management in the nine selected villages

Village	Access to *nullah*[2]	PPI	Land Loan?	Develop. Tract?	Common project?	Heifer
Broshal	Shared	Irr.	Chnl	Yes	Yes	No
Khaiber[1]	Exclusive	Irr.	Chnl	Yes	Yes[3]	Yes
Passu[1]	Exclusive	Irr.	Chnl	Yes	No	No
Roshanabad-Sherabad	Shared w/5 VOs	Irr.	Chnl	No	Yes	No
Rahbat	Shared w/8 VOs	Irr.	Chnl	No	Yes	No
Rahimabad I[1]	Shared w/4 VOs	Link rd./ Irr.	Chnl	Yes	Yes?	Yes
Oshikhan-dass[1]	No *nullah*	Sed.	Tank	No	Yes	No
Sherqilla[1]	Shared w/2 VOs	Irr.	Chnl	Yes	Yes	No
Thingdass[1]	Shared	Irr.	Chnl	Yes	Yes	No

[1] The village also has at least one cooperative society other than the AKRSP VO; Oshikhandass and Sherqilla have 3–4 cooperatives each.

[2] *Nullah* is the local term for the valley/watershed in which the forests and pastures are located;

[3] A (?) against the 'Yes' for community tractor indicates an unconfirmed statement that the tractor is owned by a village cooperative society.

Source: Author

The Traditional Management System for Village Resources

The majority of villages in Gilgit District are located on alluvial fans or river terraces, dominated by a backdrop of steep mountains with narrow openings into *nullahs* that lead to alpine pastures, glaciers, and snowfields. The *nullahs* contain mountain streams that feed the gravity channels which irrigate the fans and terraces. From cultivated fields, water drains freely (when it is abundant) into rivers that merge into the River Gilgit or the River Hunza, which, after their confluence near Gilgit Town, flow into the River Indus within the boundaries of Gilgit District.

The *nullah* contains one or more alpine pastures and, occasionally, flatter meadows and land sown with barley or potatoes (Kreutzmann 1985). The highest among these pastures are at 4,600 masl, and they are used only for grazing yaks, although they may sustain protected wildlife (including the snow leopard). The migration of livestock to the pastures starts in April and May. Usually men and children accompany the animals to the pastures but in *Wakhi*-speaking areas (including Gojal *Tehsil*) women maintain the dominant role in tending livestock and making dairy products in the pastures. Each stage on the way to the highest pasture has huts for temporary residence, usually next to the watering holes. These resources belong to the villages using the pasture. Barley or potatoes may be cultivated on individual fields. Forest products may be brought down from the *nullah* on donkeys or carried on the back. The return movement from the pastures to the villages takes place in September or October.

The snowfields and glaciers in the *nullah* melt into mountain streams that are tapped for irrigation. The channel head may be several miles from the village, and its maintenance is the collective responsibility of the village. In spring, the entire village turns out to clean the channel before the date for first irrigation. This common effort is part of history and is referred to as *rajaki*. Violators of *rajaki* are required to pay a fine, usually wages for the number of days on which the individual absented himself from *rajaki*. Much of the length of a channel may be lined with trees that are individual property. Routine maintenance during the agricultural year is carried out by one or more *chowkidars* paid through contributions made by the villagers in cash or kind. The *chowkidar* enjoys a high status in the village. In periods of water scarcity (such as at the time of planting in spring) the villagers practice *warabundi*, i.e., a roster of turns by which water is used by each farmer for a specified length of time.

The land beyond the access of the irrigation channel is usually steep and uncultivated, supporting some grass and *hyppophae*. It is usually grazed in winter by free-grazing livestock. This winter 'pasture' is common land. Winter grazing also takes place on other uncultivated land, if any is available, by tradition, in the proximity of the settled village. Significant parts of such land have been converted to higher pay-off uses once irrigation has become available, because such land has represented the natural avenue for expansion in cultivated areas over the years.

Steep slopes often dominate the landscape below the irrigation channels and above the settled villages. With careful irrigation, this land can support lucerne and trees that are planted on individually owned plots running vertically down the slope.

The settled village itself is dominated by houses, individual crop fields, and trees on steeper land. Farming fields are often surrounded by trees. There are well-defined rules governing the distance at which a tree can be planted from a neighbour's field. These rules are meant to ensure adequate sunlight and water to field crops. After the maize harvest in autumn (or after harvest in the single-crop areas), all crop fields may be grazed for stubble. Free-grazing coincides with the arrival of livestock from the alpine pastures. (Some villages are now beginning to ban free-grazing, perhaps in response to the benefits from tree planting on village land.) Steeper parts of the settled village are planted or allowed to regenerate as individual woodlots.

The version of traditional systems, as depicted above, is becoming increasingly differentiated as different villages respond in different ways to the forces of change. Some of the important aspects of this differentiation are brought out in the case studies below.

General Analysis of Village-level Organisations

The VOs sponsored by AKRSP have several features in common. The membership of the VO is open to all households in the village. The general rule is one male per household, but exceptions to this rule may be found in instances where an occasional household contributes two members to the VO. When women participate actively, it is either through their own organization or by attending the VO meeting. In traditional villages, there is little active participation by women. In many Ismaili villages (particularly the *Wakhi*-speaking ones) men and women meet in a joint assembly and, in other cases, women may be represented in the VO by selected (male or elderly female) individuals.

In large villages, there are multiple VOs organized on the basis of neighbourhoods. Where the neighbourhood coincides with an irrigation channel's access area, each VO will have its own land development plan and loan, otherwise, land will be developed jointly by the VOs concerned. Similarly, when one project has to be implemented by several VOs, each VO is apportioned a share of the work by consensus. The multiplicity of VOs within a village does not, at present, affect the management of forests and pastures common to the village.

The VOs were formed initially to implement and maintain PPI projects, start a group savings programme, and nominate and support a cadre of village specialists trained by AKRSP. They initially met every week. Over time, the VOs have acquired a longer-term perspective on village development and now participate in all the programmes offered by AKRSP and collaborating agencies. They also meet less frequently (two to four times each month) now that the vast majority of VOs have completed their PPI projects.

VOs receive a grant from AKRSP for implementing their PPI projects. Most, but not all, VOs were far-sighted enough to save from this windfall labour income and deposit the savings in the VO's group account. These savings were augmented over time by savings from the sale of produce and non-farm employment. The 376 VOs of Gilgit District had combined savings of nearly Rs 24 million by the end of 1987. These savings are used by AKRSP and its collaborating bank as cash collateral against which the VOs are given loans for various development programmes. Rs 39 million had been disbursed as short-term and medium-term loans by the end of 1987, with a nearly flawless recovery record so far.

Unskilled labour for village projects is contributed by the villagers themselves. If the work is to be done without payment, as under the *rajaki* system, then each individual is expected to contribute equally; defaulters will pay the wage cost of their absence. Presence may be voluntary, as with PPI projects, if labour is being paid wages. The tradition is to reserve village-level tasks for the villagers themselves, although that tradition is now changing as more and more market exchange of labour develops. A village will also give preference to its own residents when hiring skilled labour.

Technical services for the VOs come from AKRSP and collaborating agencies and from the villagers themselves. AKRSP has a field unit called the Social Organization Unit (SOU), consisting of a Social Organizer, an engineer, and an agriculturalist. This unit is mobile and provides AKRSP with its technical and motivational outreach to the villagers. The VO itself supports a cadre of village specialists, in practical and managerial skills, who are trained by AKRSP and remunerated for services and supplies by the VO; supplies may be obtained at cost from the AKRSP.

The mobilization of resources from among VO members is subject to a variety of rules (or, in some cases, no rules). For financial resources, contributions from members may be raised by one of the following mechanisms:

(1) a fixed minimum to be contributed by each member;
(2) an equal contribution by each member;
(3) contribution in proportion to perception of benefits;
(4) contribution on the basis of economic status; or
(5) a contribution left to the decision of the individual.

There are no aggregate data on how many of the VOs follow each type of rule. Before AKRSP started interceding with the VOs, the majority of them appeared to be asking for a minimal fixed amount from their members for group savings. AKRSP's suggestion has been to adopt options (3) or (4) and many VOs have responded positively to this suggestion.

In the case of the utilization of loans given by AKRSP, there is a difference between short-term production loans and medium-term development loans. Short-term loans (for fertilizers, plants, marketing, etc.) are given out by the VO according to a household's demand for inputs or contributions to the produce that is marketed. Medium-term land development loans are divided equally among VO members, the rationale being the suggestion by AKRSP that a minimal amount must be available to each member to preserve equitability in the use of a rationed input. Better-off individuals may supplement the loan with their own cash resources.

In the case of village specialists, each user pays a fee that is in proportion to the services used. This straightforward rule applies most commonly to para-veterinarians and plant protection specialists.

Errant VO members are disciplined through a series of graduated measures. An offender who has injured the interest of part or whole of the VO will be asked to render compensation to the injured party. One who breaks a VO's rule for the protection of common property is expected to pay the stipulated fine. A refusal to honour the decision of the VO is met, initially, by an attempt by the elders to convince the offender to obey the decision of the VO. If this and other means fail to bring around the dissenter, then, the traditional penalty of social boycott of the offender's household is imposed. This is considered a severe punishment.

Communications among members of the VO take place formally in the VO meeting. Here, VO matters are discussed, the options offered by outside agencies are examined, and every member has the right to express his opinion. Decisions are reached by consensus or majority vote. VO decisions are communicated to AKRSP by means of a resolution of the VO. The resolution is forwarded to the area's Social Organizer, whose recommendation on it is nearly always respected by the management group. The Social Organizer and his associates on the SOU tour their area almost non-stop and provide the most reliable and effective channel for communications between VOs and the management group of AKRSP. In addition, frequent field visits are undertaken by the management. The VO itself sends its office-bearers and specialists to Gilgit for VO conferences and refresher training in specialist skills. Proceedings of VO conferences (one every month, for about 80 VOs each) are published and sent to each VO through the SOU.

The VOs interact formally or informally with a large number of religious, political, social, economic, and government organizations. It is not possible to sketch out the relationship between the VOs and each of the other organizations active in Gilgit District. In the next two paragraphs, a list of such organisations is presented to illustrate the context in which the VO works.

All villages have regular religious and traditional gatherings in addition to VO meetings. Many villages have a formal religious organization working in the village. The Ismaili villages participate in the programmes of the various Aga Khan service institutions (for health, education, and housing). Many villages, particularly those supported by the Aga Khan Economic Planning Board, have village cooperative societies.

The political structure of the district revolves around the system of Local Bodies and Rural Development (LB&RD). There is a District Council in Gilgit with an Annual Development Plan drawn up on the recommendations of District and Union Council members. Each Union Council covers three to five villages, and there is a Union Council member from practically every village. Elected councillors are provided with technical support by the LB&RD Department of the Northern Areas Administration. Other line agencies operating in the district include; the Agricultural Department, the Animal Husbandry Department, the Forest Department, the Northern Areas Public Works Department, the Health Department, the Education Department, and the Social Welfare Department. In addition, there are commercial institutions, including scheduled banks and specialized institutions for agricultural, industrial, and cooperative capital.

The Situation in Individual Villages

Broshal

Broshal is the highest of the nine villages studied. Its altitude is 2,740 masl (only one crop can be grown each year on a given plot of land), and it is located 130 km from Gilgit and 40 km from the KKH, in the Hoper Valley of Nagar. Broshal lies in one of the more remote parts of Gilgit. Its 105 households belong to the Shia branch of Islam.

The documentation on Broshal and its neighbouring hamlets includes the works of Butz (1987) and Semple (1986) and notes and case studies undertaken by the SOU of the AKRSP.

The following organizations are active in Broshal: the Union Council of the LB&RD System; the Project Committee of LB&RD; the traditional *jirga* (council of elders); the Committee to oversee the *Imam Bargah* (religious place); the AKRSP-sponsored VO, and two committees set up under the aegis of the VO to manage the VO's tractor and enforce the livestock grazing rules of the VO. The Aga Khan Health Services are exploring the terms of partnership under which they can collaborate with the Broshal VO. In addition, there is a government school and dispensary.

Traditionally, as in other villages in the district, Broshal had a council of 7 to 10 elders (the *jirga*) led by the village headman (the *numberdar*). The *numberdar* was appointed by the *Mir* and was also responsible for the collection of taxes from the village. The *jirga* regulated the management of natural resources at the village level, including water distribution and allocation, channel maintenance, movements of livestock to the various pastures and within the village, dates of closure of pastures, etc. This system appears to be in force even today, but the *numberdar* has no official status, and the *jirga* faces competition from other (religious, political, and economic) organizations. For inter-village disputes, the *Mir* was the arbitrator; today, there is increasing recourse to courts and government administration.

AKRSP's intervention in Broshal started with its sponsorship of the Hunono irrigation channel. This channel already existed but was in a state of disrepair and subject to occasional destruction as a result of landslides. The villagers proposed that the channel should be improved, with concrete work where necessary, to increase the reliability of water supplies and reduce the considerable risk to their agricultural production. AKRSP's agreement to this suggestion led to the formation of the Broshal Village Organization in July 1983.

The Broshal VO is led by a strong village activist, the Manager of the VO. An ex-serviceman, the Manager has run the VO with a fair bit of personal authority in support of the AKRSP message of collective management. As a result, the VO has been consistently ahead of other VOs in the valley in accepting AKRSP-sponsored activities, particularly those that require strong collective management. For example, it was reported in Semple's (1986) case study that the bulk (75%) of the VO's savings were raised when the Manager decided to transfer part of the AKRSP grant for the channel to the VO's group savings' account. Voluntary savings were very small and came in response to the VO rule that each member must save one rupee per week. The savings were offered by the VO as cash collateral against a loan provided by AKRSP for the purchase of a tractor, one of the first VO-owned tractors in Gilgit.

The purchase of the tractor led to the first institutional innovation by the VO. The VO set up a tractor committee to manage the day-to-day affairs of the tractor. It appears, however, that the tractor's operations in Gilgit Town (when it is not in use in the village) are in the hands of a relative of the Manager who lives in Gilgit. The committee's existence has been a source of some concern at AKRSP, since AKRSP fears that such committees may take over control of an asset rather than remaining answerable to the general body that elected them. The tractor committee has not, so far, usurped the powers of the VO over tractor affairs. At the same time, the VO has decided that each member will deposit Rs 200 in group savings against future payments for the tractor, whether for maintenance or for loan repayment.

The VO also established a committee to control free grazing in the village. The committee drafted both punitive and preventive edicts. It is possible that the committee is effective in discharging its mandate. It needs to be noted, however, that Broshal experiences some seasonal migration of men. Control over free-grazing reduces the returns from livestock by increasing the labour cost of livestock control. For free-grazing to be controlled, the villagers must realize greater gain from the crops that can be grown on the controlled fields. It is not yet clear whether this trade-off has been resolved in favour of crops and against livestock.

The village has an active para-veterinarian who has earned significant amounts from vaccinating livestock and considerably reducing their mortality rates. This specialist has been remunerated regularly by the VO for his services and supplies.

There are no significant innovations in the management of forests and pastures. The traditional system of the Hoper Valley continues to be in place.

Khaiber

Khaiber village, lying in the single-cropping zone at an altitude of 2,600 masl, and about 180 km from Gilgit Town on the KKH, has perhaps the most remarkable village organization in AKRSP's project area.

Khaiber has 55 households belonging to the *Wakhi* ethnic group and following the Ismailian tradition of Islam. These villagers are highly educated and close-knit. Their VO is led by a superior village activist, the President of the VO.

The documentation available on Khaiber includes four papers prepared for an AKRSP workshop—Abidi (1987), Husain (1987b), Hussein and Karmali (1987), Magrath (1987)—and Caroe (1986), CDC (1987), Meghji et al. (1987), and Semple (1986).

The following organizations are (or have been) active in community-oriented work in Khaiber: the Union Council of the LB&RD System; project committees set up for specific LB&RD projects; the project committee set up to implement the rural water supply project of the Community Basic Services Programme of UNICEF; the Government of Pakistan and the Aga Khan Foundation; a cooperative society; the Aga Khan Health Services; the Aga Khan Education Services; an AKRSP-sponsored village organization; its Project; the *Ismailia* Local Council; and the Ismailian *Tareeqi* Board (for religious affairs). In addition, there is a government school for boys and a school for girls managed under the Aga Khan Education Services. There is also a hydroelectric power station that provides electricity to the neighbouring villages of the Gojal *Tehsil*.

The Khaiber VO is unique because of the extent of collective management practised by it. The AKRSP-sponsored, new irrigation channel has enabled the VO to irrigate and develop a large tract of previously low-productivity, winter grazing land, lying at a distance of 2–3 km from the village. The VO allotted portions of this land for use as cropland, a fruit orchard, and a multi-purpose nursery for fruit and vegetables. All the new land is considered to be the common property of the VO, although the cropland will be assigned to individuals through the traditional system of lottery once it is developed. The VO has hired three *chowkidars* to be responsible for the irrigation of the new land. This is an innovative extension of the traditional practice of hiring a village *chowkidar* to clean and maintain the irrigation channel. Development of the crop land is the responsibility of the VO and individuals are assigned duties by turn to manage this process. The nursery is managed by the women of Khaiber, with the assistance of one male specialist and six women trained by AKRSP. Marketing from the new land is also done collectively by the VO. Women participate regularly in VO meetings and have a say in collective decision-making over common property.

Because of its ability to manage assets collectively, Khaiber was selected to be the recipient of 10 high-yielding hybrid cows which had to be housed in a single unit. This operation is part of a grant from Heifer Project International. The VO sent its nominees for training in basic animal production techniques, it allotted a piece of land (2–3 km from the settled village) for the construction of cattle sheds, and it organized the supply of considerable amounts of fodder that were needed by the new cows. The most recent information available indicates that the Heifer Project cows have the highest milk yields among all eight of the Heifer Project villages sponsored by AKRSP.

The Khaiber Village Organization supports a large cadre of village specialists. One of the earliest specialists was the para-veterinarian. His effectiveness in reducing mortality rates has enabled him to pursue his new speciality as a part-time job. The VO has also invested Rs 550 in an automatic syringe, thereby reducing the time costs of vaccination. Part of this saving has been passed on to VO members through lower charges. Several other specialists in Khaiber pursue their new vocations as part-time jobs, thus testifying to the VO's ability to create new employment opportunities within the village in response to the perception of higher returns for specific farm-based activities.

The issue of changing patterns of profitability has also influenced the village to take steps to stop the centuries-old practice of free-grazing. Villagers are convinced that free-grazing needs to be controlled in order to benefit from the improved marketing opportunities for fruit. Apple trees can now be seen in wheat fields, although previously no tree could last long outside a boundary wall. The village has found it possible to transfer free-grazing animals in autumn to its traditional winter pasture. Thus, an institutional innovation has come about as a result of changing markets and the relatively small cost of institutional change.

The overall trend in the allocation of labour, land, and livestock in Khaiber appears to be one of specialization. This specialization has been carried out with innovations and has reinforced the spirit of collective management. It is possible, taking the example of Khaiber, to see specialization in resource use as an innovative response to changing patterns of profitability and innovations in collective management as vehicles for growing specialization. Numerous jobs have been created in the village as a response to new ways of increasing income from agriculture. This has happened (in contrast to some other villages) despite the availability of off-farm opportunities and a high level of education in the village.

Passu

Passu is very similar to Khaiber in terms of some important features; and yet it represents a development situation that varies substantially from that of Khaiber. Passu, with 67 *Wakhi*-speaking households of the Ismailian tradition, is located at an altitude of 2,440 masl, about 150 km from Gilgit Town on the KKH. It is in the single-cropping zone. Documentation available for Passu includes the four workshop papers cited above for Khaiber, as well as AKRSP (1984), CDC (1987), Conway et al. (1985), Kreutzmann (1985), Saunders (1983), and the World Bank (1987).

The following organizations are (or have been) active in Passu; the Union Council of the LB&RD System; the project committee for rural water supply under the Community Basic Services Programme of UNICEF; the Village Production Group organized by the Integrated Rural Development Programme (IRDP) of the United Nations Development Programme (UNDP/FAO); a multipurpose cooperative; a potato seed growers' association organized by UNDP/FAO to work with a commercial firm (Jaffer Brothers); the VO and its women's group; and the Ismailian Local Council.

According to one hypothesis (World Bank 1987), the distinguishing feature of Passu is that it commands access to the Passu and Batura Glaciers, and the surrounding alpine scenery is popular with growing numbers of tourists, trekkers, and expeditions. According to another point of view, Passu is distinguished by its factional VO and the lack of an acceptable activist within the VO.

Passu's PPI project is a new irrigation channel that takes off from the Batura Glacier and brings water to a large tract of land that was previously used for winter grazing. As a result of the new channel, each household in Passu increased its land holding with an additional 4.5 ha. This channel has succeeded in bringing water to the new land, whereas several attempts before it had failed. It appears that the major reasons for earlier failures were: (1) the lack of proper surveying techniques and (2) the uncertain movement of glaciers. Villagers, using the traditional methods of following the water level, ended up with the channel being too low to have access to any significant area, or else the glacier advanced to a point which made the location of the take-off point too low for necessary access. The AKRSP assisted the VO by putting down a proper alignment. The site survey also used information on the movements of the glaciers collected by Chinese road engineers in the course of their work on the KKH.

The successful completion of the channel led to an expectation on the part of the AKRSP that the Passu VO would take up land development promptly and complete it speedily; this did not happen. The villagers of Passu observed that the process of making a new channel operational for full discharge is a long process that may take five to eight years and their observation has been borne out by the experiences of other mountain communities. The AKRSP believes that the VO's collective management of land is hampered by discord within the VO and the inability to perceive the value of investing in a sustainable source of income from agriculture. It appears that most able-bodied villagers prefer to work as trekking guides in the summer (at about Rs 110 per day), rather than investing labour or cash in land development. Moreover, Passu's land development, unlike Khaiber's, is an individual affair for each beneficiary household, even though the VO has taken a collective land development loan from AKRSP. The use of new land in Passu does, however, resemble that of Khaiber, in that tree (particularly fruit) crops appear to be preferred to annual crops.

Clearly, the residents of Passu are responding to new opportunities by seeking a balance between short-term prospects for cash income and the longer-term payoff to investment in land. In the short term, there is a movement of labour and other resources away from agriculture and livestock. Kreutzmann (1985) observes that many of the huts in the alpine pastures inside the Passu *nullah* now lie vacant, as fewer people make the seasonal trek with their livestock to the pastures—the treks made today are with tourists and for cash income. Thus, the glaciers and their surrounding scenery are being transformed into multiple-use resources, while prior to the KKH they were of importance only to agriculture and livestock.

The financial and entrepreneurial resources of the VO and its members are also subject to the strong dual pressures of competing agricultural and non-agricultural uses. For example, VO savings, normally reserved for investment in agriculture, were used to purchase stocks for an electrical goods store to be operated by the VO for all the neighbouring villages that presently receive power connections. Similarly, there appears to be a reluctance on the part of the VO to nominate villagers for training in specialized functions; most eligible candidates prefer non-farm employment. The case of the neglected VO para-veterinarian indicates, too, that the effect of competing demands on resources is magnified by the factionalism in the VO. On the other hand, the VO has responded with enthusiasm to the highly profitable seed potato production programme introduced by the UNDP/FAO and a commercial firm. In the short run, income from seed potatoes is estimated to equal the income from tourism in Passu. In the long term, both activities are liable to be associated with environmental problems (the potato programme because of sustainability and disease-resistance issues).

The changing patterns of incentives have placed increasing responsibility for farming on women. It is conceivable that specialization in labour over time could make women the farmers of the village, while the men take up more remunerative non-farm jobs. The importance of this transformation is appreciated by the VO and AKRSP, and a conscious attempt is made by both to channel motivational and other inputs to women.

By and large, there is evidence that both social and economic forces are responsible for the substantial difference between Khaiber and Passu. While Passu has much easier access to non-agricultural income, it is also more factional as a village organization. One consequence of the latter is that there is lack of clarity in the VO's medium-term perspective—the balance between traditional resource use and new opportunities has yet to be articulated by the VO and AKRSP. In particular, there is little recognition of the value of specialization in labour for managing the entire range of options available to the village.

Roshanabad-Sherabad

Rosshanabad-Sherabad is a small village of about 20 households, lying on both sides of the KKH in Central Hunza, about 95 km from Gilgit Town. The village lies at an altitude of about 2,000 masl, and maize, the second crop, is used for fodder since it does not ripen as grain. The inhabitants speak Burushaski, the main language of Hunza, and belong to the Ismaili branch of Islam.

Documentation available on the village includes Meghji and Saleem (1987) and Neseem (1986). The following organizations are active in Roshanabad-Sherabad: the Union Council of the LB&RD System; the project committee for rural water supply under the Community Basic Services Programme of UNICEF; and the VO and its women's group (and their smaller nursery and tractor committees). In addition, there is a school nearby and a hydroelectric power station that provides electricity to Central Hunza.

The Roshanabad-Sherabad VO has a strong and well-educated leader and is a close-knit organization. VO membership includes women who participate fully in all VO meetings. Thus, from its inception, the VO has been active in pursuing women's development activities with the same vigour as those for men. In particular, the women have been managing a multi-purpose nursery, defining the procedures for income-sharing from this new common asset; they have also taken up a number of appropriate technology devices, such as nut-crackers (for apricot kernels) and fruit pulpers. Given the same broad pressures for changing gender roles as those that prevail in Passu, Roshanabad-Sherabad seems to have accomplished more in preparing for change by involving women.

The PPI project for this village was an irrigation channel. The VO also took out a loan at an early stage for a tractor. The purchase and operation of the tractor turned out to be a saga of unforeseen circumstances. These events were narrated by VO representatives at a conference of VOs in Gilgit and drew applause from the audience for both humour and relevance. The story illustrates the tremendous institutional innovation and managerial capacity that is required for acquiring and maintaining non-traditional assets and technology.

The Roshanabad-Sherabad VO has, since its inception, tried to develop a complete cadre of specialists for the activities undertaken by the VO. It has, for instance, a marketing team, with individuals nominated for fruit and livestock marketing and others trained in fruit processing and packaging. It also has groups of women working, by turn, on the nursery (this is also observed in Khaiber). Like Khaiber VO, therefore, Roshanabad-Sherabad appears to be moving towards specialization in labour and management.

In a formal sense, the VO is a leader in village planning. It regularly works out (and presents on flip charts) a five-year plan for village development. While the earlier emphasis was on AKRSP-sponsored programmes, the plan now shows education and civic components as well. The plan is fairly basic, in that it lays down targets for products to be marketed, land to be developed, etc. It does not, as yet, show the ways and means for achieving the targets. The planning exercise shows how a basic concept introduced by AKRSP (initially for land development planning) is being extended and redefined by the VO; it points to the possibilities for innovation in planning for village development.

Rahbat and Its Neighbouring Villages

Rahbat is located about 60 km from Gilgit Town and about 5 km from the KKH in the Chalt Valley of Nagar Subdivision, at an altitude of about 1,800 masl. Chalt Valley, with a population of over 4,000 followers of the Shia branch of Islam, includes six villages with nine VOs. While much of the development activity is carried out by individual VOs, issues of natural resource management have entailed cooperation among two or more of the VOs. Thus, it is important to discuss both village-level and supra-village innovations.

The cluster of villages in Chalt has particular significance because of the evolving situation in the Chalt-Chaprote *nullah*. Here, the community of resource users has intervened to take control over the natural forest and pastures of Chalt-Chaprote. This development represents a test case that will challenge the ingenuity of AKRSP, the VOs, and the government in dealing with the issue of community control of natural resources.

Documentation on Rahbat and the neighbouring villages includes CDS (1987), Gohar (n.d.), Gohar et al. (n.d.), Hunzai (1987), and Jan (n.d.).

Rahbat Village has the following development organizations: a project committee set up under the LB&RD System; the Union Council of the LB&RD System; a project committee set up to implement the rural water supply project of the Community Basic Services Programme of UNICEF; the AKRSP-sponsored VO, its affiliated women's group, the forest management committee set up by Rahbat and its neighbouring VOs; and the Aga Khan Education Services.

After visiting the Gilgit area in 1986, a team of workshop participants had recommended that the 'AKRSP could make a valuable contribution by interceding with the Government to return these forests to the status of locally held commons, to be managed by an organisation—complete with enforceable sanctions—established by AKRSP' (Dani et al. 1987). AKRSP's approach has been to act on institutional innovations once they appear to have the interest and confidence of the villagers. Thus, while AKRSP was waiting for villagers to establish a line of approach for new ways of managing natural resources, the villages of Chalt decided to intervene to protect and sustain their natural wealth.

Villagers who were interviewed (CDC 1987) estimated that the Chalt-Chaprote forest is now only one-fourth of what it was about 20 years ago. The rapid depletion of forest and pasture is due to the changes in incentives that started with the construction work on the KKH. This brought about significant increases in the value of forest products; grazing has been particularly damaging to juniper regeneration. There is no doubt that the changes have benefited those in the area who were engaged in the commercial exploitation of the forest but at no little cost to the environment.

In March 1986, the six VOs of the area, acting through 36 representatives, set up a Reform Committee for Forest Conservation. Although there are several activists in the group, perhaps the most influential is a former *numberdar* from Rahbat. The Committee declared an immediate ban on commercial exploitation and domestic requirements were to be met as follows:

- only dead wood would be used for fuelwood, with each household permitted one trip to the forest every week; and
- timber would be made available upon application to the Reform Committee, which would verify the requirements and then apply to the Forest Department for approval.

A gate (or checkpoint) was set up on the road out of the village and was manned 24 hours a day. The *chowkidars* at the gate were remunerated with equal contributions from each household. Offenders were to be fined Rs 25 per *maund* (about 38 kg) of fuelwood and Rs 500 per log of timber. The ban and sanctions are reportedly being enforced effectively.

The ban on commercial exploitation of forest still left unresolved the conflict between livestock grazing and forest and pasture regeneration. In 1987, the VOs proposed a new system of rotation that would reduce the pressures of overgrazing. They also agreed to a suggestion from AKRSP that some new tracks be constructed, to open up hitherto inaccessible parts of the rangeland, and that additional earthen tanks be built to provide water for livestock. The Rahbat VO has set up a five-person pasture development committee.

In response to the initiatives undertaken by the VOs, AKRSP is providing technical and financial assistance for sustainable forest management with community participation. This assistance is outlined in the documents prepared by AKRSP staff and listed above.

It is not yet clear how the fundamental question of authority between the Reform Committee and the Forest Department will be resolved. The Committee's intervention takes over some of the functions of the Forest Department on State-controlled forest. The villagers maintain that they are helping the government enforce forest regulations and that they have the written permission of the former Deputy Commissioner to do so. The Head of the Forest Department maintains that the Committee is a refuge for 'miscreants' bent upon the destruction of forests for their own vested interests. It is believed, however, that the Forest Department is issuing no new permits for commercial exploitation of the Chalt-Chaprote forest.

The villages of Chalt have also undertaken several other supra-village initiatives. Rahbat *Bala* and Rahbat *Paeen* VOs are working together to construct a domestic water supply project as well as a girls' school. The school represents the first instance of cooperation between a non-Ismaili village and the Aga Khan Education Services in the provision of a complete package of educational facilities. Rahbat *Bala* also hires a *chowkidar* jointly with the Chaprote *Paeen* VO for the maintenance of their common irrigation channel.

An institutional innovation at the village level was observed in Chaprote Village. This village had been gifted 10 high-yielding hybrids by the Heifer Project, with the expectation that, as at Khaiber, the cows would be kept in a collectively managed unit. The villagers of Chaprote, however, have distributed the cows to individual households who will share the costs and benefits. The reason given for this system was that it is too costly to pay cash to the attendants who were to look after the cows in the common livestock unit.

The Rahbat VO appears to be a leader, among Shia villages, in involving women in the development programmes available for the region. In addition to the girls' school mentioned above, Rahbat has a multipurpose nursery of the kind present at Khaiber and Roshanabad-Sherabad. This nursery is expected to play a supportive role in plans for sustainable forest management in the Chalt-Chaprote forest.

Most of the VOs of the valley have a full range of village specialists trained by AKRSP. These specialists are likely to include forestry and pasture specialists in the future.

In conclusion, it appears that the villages of Chalt have embarked on a dramatic course of institution building that may have relevance to many other villages in the region. The initiative by the community has placed both AKRSP and the government in a challenging position. Whereas the government needs to articulate a response to an apparent conflict of authority, AKRSP needs to strengthen community institutions with the technical and financial assistance needed to capitalize upon the community's initiative; community intervention needs to be extended into a strategy for sustainable resource management at a high level of productivity.

Rahimabad I

Rahimabad I is located along the KKH, about 30 km from Gilgit Town, at an altitude of about 1,670 masl. It has two VOs—*Bala* (upper) and *Paeen* (lower)—that are organized around separate *jamat khanas* (the religious gathering place for followers of the Ismaili branch of Islam); the combined population is 125 households, mostly from the Ismaili sect but also including a number of Shia families. Both sects are represented among the office-bearers of the VOs. Because of its proximity to Gilgit Town, Rahimabad I is part of a greater Gilgit economic zone supplying produce and manpower to the urban area on a daily basis.

Information on Rahimabad I is available in Hamid (1987), Khan (1985), Meghji (1984), Meghji et al. (1987), and Semple (1986).

The following organizations are active in Rahimabad I in addition to its two VOs: the Union Council of the LB&RD System; the Ismailian Local Council; a cooperative society; and the Aga Khan Education Services. In addition, the village has a government school for boys, a school for girls managed by the Aga Khan Education Services, a government dispensary, and a government veterinary dispensary.

Rahimabad (original name Partab Singh Pura, subsequently Matum Dass) is one of the newer villages of Gilgit and thus there are people in the village (as in Oshikhandass) who can narrate the events leading up to the establishment of the village and the subsequent lengthy process of land development. According to these elders, the settlement of Rahimabad started with the construction of an irrigation channel in 1903. The construction of the channel is said to have been carried out, under the supervision of soldiers sent by Maharajah Partab Singh of Gilgit, as part of an agreement with *Mir* Nazim Khan of Hunza. The *Mir* sent 28 households from Hunza and their descendants inhabit the village today with those of the other original families. During the early stages of land development (1903–1920), the villagers brought fruit and forest trees from Hunza. Thereafter, they established fruit nurseries and obtained other tree cuttings locally.

Since 1903, the irrigation channel has been maintained with the help of a village *chowkidar*. In 1975, the then *chowkidar* applied to the VO for an increase in wages. The village agreed to increase the wages from 2 kg each of wheat and maize grain per household, per year, to 4 kg each of wheat and maize grain per household, per year, plus Rs 200 in cash from the village common fund. The revised wage rate also appears to be in effect today.

Rahimabad I was one of the first villages to form a VO after AKRSP's arrival. Its first PPI project—that for the *Paeen* VO—was a link road, through the length of the village, connecting it at both ends with the KKH. From the very beginning, the issue of compensation for land taken up in road construction dominated discussions between AKRSP and the VO and among members of the VO. Some villagers maintained that AKRSP should follow the policy of the Northern Areas' Public Works Department and pay land compensation at market rates, in addition to the cost of labour and material that is normally included in AKRSP cost estimates. AKRSP maintained that land compensation was an internal matter for the VO to resolve. It took two to three years for the issue to disappear from the agenda of meetings between AKRSP and the VO. The VO decided that no compensation would be paid, since those who bear the loss of land also benefit the most from the road by virtue of their proximity to the road. Many of the affected families appear to support this rationale. Thus, Rahimabad represents an example of a VO internalizing the costs and benefits of public good.

Rahimabad I also provides insight into traditional and new ways of discharging financial obligations in the village. Although the VOs of Rahimabad have taken out and repaid several AKRSP loans, recovery of the first loan of Rs 6,534 for fertilizer was plagued by problems. Recovery of loans was then in the hands of the VO's model farmer, who died suddenly before the loan could be repaid to AKRSP. Some villagers report that he had already collected about half the loan from individual VO members for repayment to AKRSP. The repayment of the loan was taken over by the late model farmer's nephew, as a matter of family honour; the nephew is the current manager of a VO. A meeting of village elders was called to discuss repayment; the elders decided that well-to-do families in the village should make donations towards erasing the loan, since many of the others were too poor to pay. The understanding was that the contributors would be repaid once a second PPI project (a channel for the *Bala* VO) was approved by AKRSP. Thus, eight villagers provided what is essentially bridge funding to the VO in anticipation of an improved cashflow for the VO later on. The second PPI project was approved after lengthy debate between AKRSP and the concerned VO. In the final analysis, some of the debtor VO members have paid off some of the contributors (to bridge finance) by selling produce for cash. It is not known with certainty whether the remaining amount has, in fact, been repaid out of the grant for the second PPI.

The *Bala* VO nominated two young men for training in para-veterinarian and plant protection functions. It was soon discovered that the para-vet was, in fact, redundant, since there is an Animal Husbandry Dispensary in the village staffed by a properly trained employee from the village. Contrary to the expectations of many outsiders, the dispensary appears to be well stocked with necessary drugs and vaccines. The villagers understandably prefer the government dispensary to the VO specialist, since the government provides free services while the VO charges for cost and the specialist's fee.

Rahimabad I is also one of the villages taking part in the Heifer Project. It has been successful, so far, in keeping the 10 cows together at one, collectively managed location. Furthermore, Rahimabad is in the process of developing what little land had been left undeveloped over the years. The pattern of land use on the new land favours tree crops (understandable in view of nearby markets for fruit and wood) and the diversion of labour to urban centres.

In retrospect, Rahimabad I has consistently chosen investment options that reinforce its position on the KKH close to Gilgit. Its first PPI was a link road; it preferred the government veterinarian to the more costly VO para-vet trained by AKRSP; at the same time, it accepted the high cost of upkeep of hybrid cattle in anticipation of later returns from milk marketing; and it has developed land for fruit and forest products that are in great demand locally and nearby. In retrospect, there is little an outsider could have done to improve upon the village's investment decisions in response to changing opportunities. At the same time, Rahimabad and Oshikhandass village represent possibly replicable approaches for agricultural development in other villages that are only now acquiring reliable and cheap access to sizeable markets.

Oshikhandass

Oshikhandass is a large village with 540 households from the Shia and Ismaili sects. It was established in the late 1930s, when 58 families migrated to the location and constructed an irrigation channel under the patronage of the feudal chiefs of the time. The village is situated just southeast of Gilgit Town, about 1 km along a dirt track road from the KKH, and its altitude is 1,400 masl. It is divided into three neighbourhoods (*patees*) that correspond to the ancestral domiciles of the present inhabitants. The neighborhoods are called Jagir *Patee*, Bulchi *Patee*, and Farfoo *Patee*; each *patee* has its own VO, and there are also overlapping women's organizations.

References to Oshikhandas may be found in Caroe (1986), CDC (1987), Conway et al. (1985), and Meghji (1984).

Oshikhandass has the following organizations in addition to its AKRSP-sponsored organizations: the Union Council of the LB&RD System; the Aga Khan Education Services; the Literacy and Mass Education Commission of the Government; three cooperative societies (including one trading in timber); the Ismailia Local Council; the Shia Association; and a *numberdar* system from the days of the *Mirs*.

The original 58 families that settled in Oshikhandass were each given 3 ha of land to develop. (They were not, however, given any share in the rights to the nearby *nullah*, as older villages had prior claim to it.) Additional land was brought under cultivation subsequently. Available information indicates the following rates of change over the last 50 years.

Total cultivated land	2.75% per annum: 275% over 50 years
Population	4.56% per annum: 831% over 50 years
Land/household	1.82% per annum: 58% over 50 years

Subjected to the pressures of in-migration and nearby urbanization, and existing without a natural forest or pasture of its own, Oshikhandass has responded by creating a resource base that is a model for many other villages in a similar.position.

Since the very establishment of the village, its residents undertook a substantial programme of forestry inter-cropping with lucerne, as they had no other natural source of fuelwood, timber, and fodder. They planted trees on the slopes behind the village, as well as within the homestead. The village is, today, a remarkable example of forestry management in the village agro-ecosystem. It is estimated that 80 per cent of the village's cash income now comes from forest products, almost all of it from individual holdings.

While livestock holdings are small, the village is attempting to improve the quality, quantity, and marketing of fruit, vegetables, poultry, and eggs for sale to the Gilgit urban market. Some of this is being accomplished through the VOs of Oshikhandass as well as its women's organization.

The Oshikhandass VO (which later split into three VOs) was one of the first two or three sponsored by AKRSP. It suffered, therefore, from a certain lack of knowledge about the intentions and approach of the management of AKRSP; the villagers simply extrapolated from their knowledge of the other agencies working in the district and paid little attention to the spirit of the AKRSP message. For example, dialogues with AKRSP staff were initially valued more for their recreational content than for discussing development problems and solutions. VO office-bearers were chosen by lottery! Few meetings were held, and the attendance was very thin. The implementation of the PPI project—a sedimentation tank—was ignored by the vast majority of the villagers and work was handed over to a committee; the project suffered from faulty implementation and was finally completed three years after it should have been. As a large and urbanizing village, Oshikhandass has found little in the AKRSP package to interest the majority of its residents.

In turn, development agencies have done little so far to develop a menu of programmes from which villages like Oshikhandass and Passu could choose major initiatives in high-value horticulture, forestry, and agro-based industry (such as wood products, including furniture for the local market). In other words, there is a need to discover linkages between the kind of agricultural production model that AKRSP is trying to articulate, as a follow-up to its institutional model, and a rural-based model of small enterprise.

Sherqilla

Sherqilla, like Oshikhandass, is a large village of about 500 households, with three VOs and a women's organization. Sherqilla lies on a jeep track, about 40 km from Gilgit Town; it takes two hours to complete the journey from Gilgit to Sherqilla. The jeep track is now being widened and improved to take trucks and buses.

Sherqilla is inhabited mostly by followers of the Ismaili sect and a handful of Sunni families. The village was the seat of the former *Rajah* of Punyal, who still lives in Sherqilla. It is located at an altitude of about 1,830 masl and there are years in which the second crop (maize) does not ripen.

The following organizations are active in Sherqilla: the Union Council of the LB&RD System; project committees for LB&RD activities; Aga Khan Education Services; Aga Khan Health Services; three VOs and their women's groups; the Ismailian Local Council; and four cooperative societies. In addition, there is an animal husbandry dispensary managed by the government, a government and Aga Khan school, and a hydroelectric power station to supply electricity to Sherqilla and a neighbouring village.

The PPI project in Sherqilla was an irrigation channel. This channel was constructed by what was then the combined VO of Sherqilla. Since it was not easy for a large village to congregate regularly in one place for VO meetings, the villagers decided to divide into three VOs, based on neighbourhoods (each with its own *jamat khana*). This division took place soon after the completion of the channel. At the time of division, the financial assets of the old VO were also divided by common consent.

Soon after the channel project was completed, the three VOs applied for land development loans. They were the first VOs to receive such loans from AKRSP and helped establish AKRSP policy on land development loans. It was observed that the channel was irrigating unequal land holdings within the settled village. One option was to give out the loan in proportion to the land holdings. The option chosen by AKRSP was to give a fixed amount of Rs 2,000 to every household, on the grounds that this policy represented an equitable sharing of a rationed financial resource (i.e., subsidized credit). Accordingly, every household in Sherqilla received Rs 2,000 in medium-term credit in December 1984. It has been estimated that the actual land development cost has substantially exceeded the amount loaned out by AKRSP; the difference has been provided by individuals through direct or hired labour.

Sherqilla is, in many ways, a microcosm of the evolving situation in Gilgit. One can observe those who have too much land relative to family labour selling undeveloped land to migrants from higher up the valley; new migrants with little or no land creating a local market for grain, pulses, fodder, and dairy products, the landless and other poor working in the village on land development and haulage for wages; those with donkeys specializing in bringing fuelwood down from the forest; female education creating changing expectations among people of all generations; and the prospect of improved road transport generating expectations of bigger marketing efforts and higher cash incomes.

One consequence of change is in perceptions of livestock profitability. Those households whose men are involved in non-farm work are selling off their goats and sheep and retaining cows that can be managed by the women at home. Some households contract out livestock care to professional shepherds (*gujars*), but the cost of that option also seems to be rising. The practice in the past was that the *gujar* family would retain the butter and milk produced from the livestock; the situation now is that *gujars* ask for about 4 kg of wheat grain and Rs 10 in cash for each goat or sheep for a five-month period.

Sherqilla was the first village in which the women organized themselves along the lines of the VO. This happened in June 1983, only four months after the first VOs had been formed in Gilgit. It is important to note that Sherqilla has a 'model school' for girls managed by the Aga Khan Education Services (AKES). Almost all the teachers in the school came from other districts of Pakistan, mostly Karachi, and this might have influenced both men's and women's thinking in Sherqilla. From the very beginning, men and women collaborated in managing income-generating projects; the men being particularly useful in purchasing inputs and marketing in a society where women cannot go to markets outside the village.

In addition to the VOs in Sherqilla, the village had four cooperative societies functioning in 1985 (Hussein 1985). These societies had memberships of 37, 42, 106, and 500 individuals. Together, they had equity and share capital of Rs 600,000; this compared with Rs 111,000 saved by the three VOs by mid-1985. In comparison, the land development loans, provided to the three VOs by AKRSP, totalled Rs 764,000—slightly more than the amount saved by villagers in all their cooperative bank accounts.

Of the four cooperatives in Sherqilla, two appear to be multipurpose societies. The third is for agricultural development and the fourth is a transport society. Many of the investments of these cooperatives have been in non-agricultural activities, particularly consumer shops. Most of these efforts, however, have led to financial loss.

In some cases, these cooperatives have taken out loans from the Federal Bank for Cooperatives at 9 per cent per annum, and reloaned the money to individual members at 12 per cent. The repayment record of the village as a whole is unblemished (Hussein 1985). One way in which the village effects timely repayments is by borrowing from one cooperative to pay off the other's loan. Since cooperative profits are shared by all members, villagers are also particular in repaying their individual obligations to the cooperative. Another incentive for prompt repayment is the significant interest rate of 12 per cent charged by the cooperatives; villagers are well aware that outstanding amounts are subject to this rate of interest.

Thus, Sherqilla shows a considerable variety of institutional and financial mechanisms for income generation and market exchange. It appears to have initiated the transition from a subsistence to commercial economy before AKRSP's arrival. The following points are worth noting.

(1) The villagers had started to apply the spirit of their traditional cooperation to the evolving market economy even before AKRSP arrived on the scene. Most of the cooperative activity, however, seems to have been for the benefit of a minority of the households.

(2) While villagers perceived the benefit of investing in non-agricultural activities, these activities ran at a loss. This would suggest that: (1) although villagers may have the financial assets to invest in non-farm activities, they do not yet have the expertise to be entrepreneurs outside the farm economy or (2) the organizational forms chosen by them (i.e., the cooperatives) to raise capital (through equity and concessional capital) may not be appropriate for the management of non-farm enterprises.

(3) Villagers demonstrated the potential for undertaking new income-generating activities for women by building upon the traditional gender division of tasks. Women's awareness of their collective income-generation potential might have been heightened by their socialization with women from outside the village.

(4) The response to AKRSP's insistence on collective management may have been conditioned by the presence of alternative opportunities for income generation available to the villagers of Sherqilla.

Thingdass

Thingdass is a hamlet-offshoot of Singal Village, the headquarters of the Punyal-Ishkoman Subdivision. It lies at an altitude of about 2,000 masl, some 55 km from Gilgit Town along a dirt jeep-track. It has 42 households belonging to the Ismaili sect. References on Thingdass include: Khan (n.d.), Sakhi (1987), and Semple (1986). Organizations active in Thingdass include the Union Council of the LB&RD System; Aga Khan Education Services; Aga Khan Health Services; the Ismailian Local Council; and the VO and its women's group.

Thingdass was established by a relative of the *Rajah* of Punyal but soon attracted other residents who were given the right to develop the land not given by the *Rajah* to his relative. Whereas the *Rajah's* relative had commissioned the first irrigation channel, subsequent settlers extended the irrigation in the village. Now, irrigation and land are subject to the same rules and conventions that operate on such communal assets in other villages; these resources are no longer considered private property.

Thingdass and Singal, together with a neighbouring village, have access to one of the largest *nullahs* in the district. This *nullah*, however, has been subject to the same kind of overexploitation and depletion that has afflicted other such resources. It is under the control of the Forest Department. Fuelwood and timber collection have become increasingly expensive in the face of longer distances to travel and rising time costs. In response, villagers in Thingdass are planting woodlots within the settled village. Rising time costs and the availability of market substitutes have also led villagers to abandon the cultivation of barley and potatoes in the lower reaches of the *nullah*. Thus, like so many villages with access to non-farm employment and markets, Thingdass is moving away from pastoralism. Virtually none of the men now take their livestock up to the pastures, and there is no longer a rota system to perform that function. The pastures are used by *gujars* who bring their herds from other villages. It is reported that these *gujars* pay toll for the use of the pasture, at the time of the return migration from the pasture, and that this fee is collected by the *numberdars* and divided equally among all households.

Thingdass and its parent village, Singal, are subject to the constant threat of mudflows destroying their channels in mid-season. This represents a substantial risk to agricultural production in both villages. It is not surprising, therefore, that the two villages have an arrangement under which each provides labour to the other in times of emergency. Pooling labour in this manner provides insurance against massive crop failure due to lack of water for irrigation. Villagers report that, in the last 10 years, Thingdass has called upon its neighbours three times and repaid the obligation four times. Villagers also remember a mudflow that required the services of 900 men for three days, meaning that they mobilized more labour than was required for the entire PPI project (an irrigation channel).

The PPI for Thingdass was the extension and widening of an existing irrigation channel. A previous attempt at this, financed by the LB&RD Department, had failed because of poor alignment. The piers left behind from that attempt were used in the AKRSP-sponsored project. Since the completion of the channel, the VO has taken a land development loan from AKRSP. More than half the new land is to be planted with forest trees.

All the VO specialists in Thingdass are active. The plant protection specialist has worked recently in collaboration with the government's Department of Agriculture. Thingdass also is home to 10 of the hybrids introduced as part of the Heifer Project.

CONCLUSIONS AND GENERALIZATIONS

A Model for Sustainable Resource Management: Combining Community Organization with Sustainable Production

The focus of this paper has been on institutions rather than individuals and on changing rather than static institutions. Village and project management systems were described as institutions that may change themselves and the allocation of resources they manage. Such changes are responses to rapid and pervasive change in markets, technology, and the macro-institutional framework of Gilgit. The region is best characterized as being in transition from a traditional, subsistence-oriented, low-income equilibrium to a more modern, commercial, high-income equilibrium. There are signs, however, that the new equilibrium may not be sustainable, at least in terms of the welfare derived from natural resources. These signs have been registered by some of the VOs sponsored by the AKRSP; these VOs have initiated some instructive course-corrections that may enhance the sustainability of the evolving equilibrium.

It is suggested that, in an environment of rapid change, the VO could provide the missing link between income generation from natural resources and their sustainable management—*provided that* the VO can internalize the costs and benefits of resource use. The VO will acquire the capacity to accomplish this if:

- it can devise appropriate rules and conventions governing its members in their use of and investment in various resources; and
- if it can obtain the technology and other inputs required for sustainable resource management at a high level of productivity.

Although institutional and technological innovations appear as complements in the preceding paragraph, AKRSP's experience demonstrates that institution building should precede the delivery of technology. In other words, the investment in technology could be more productive, more equitable, and more sustainable if it is secured by an effective management system. Thus, sustainable and productive resource management is seen to proceed in two broad phases. In the first phase, the community of users adopts the institutional mechanisms needed to internalize the costs and benefits of resource use. In the second phase, the new institution needs to adopt what might be called a sustainable production model. The models of community management and sustainable production together make up the model for sustainable resource management.

The preceding conclusion is analogous to the suggestion made by the World Bank (1987) and adopted by AKRSP (1987b) that AKRSP's successful institutional model needs to be complemented by a well-articulated production model. The emphasis in the present discussion, however, has been on models of community management and sustainable production for natural resource management, neither of which have been operational except on an experimental basis. The remainder of this chapter seeks out operational guidelines for effective interaction between project management and village groups. The next section looks at the subject from the point of view of what could be done by project management, and the final section analyses village responses to project initiatives.

Operational Guidelines for Models of Community Organization and Sustainable Production

Four Broad Concerns for Project Management

When institutions and markets are changing rapidly, how can a development project help villagers respond to change in a productive, equitable, and sustainable manner?

In many Third World communities, market pressures and other changes have led to rapid depletion of resources and the alienation of resource benefits from the host (biological and socioeconomic) environment. Planning and intervention by governments has not kept pace with the pressures of change. In many locations, traditional user groups have responded, sometimes with outside assistance, by devising alternative models to those favoured by governments. There are, indeed, models of community organization, land use, silvicultural practice, pasture management, marketing, etc. For simplicity, the following discussion groups together all aspects, other than community organization, into the category of a model of sustainable production.

The preceding section has argued that community organization is a fundamental component of sustainable resource management. It also argued that a model of sustainable production is a necessary complement to models of community organization. These thoughts will now be re-stated in positive rather than normative terms in order to yield guidelines for development projects. In broad terms, the concern with operational guidelines in the following lines is directed at:

- methods of inquiry and planning;
- resource management systems for the future, particularly models of community organization;

- production possibilities for the future; in particular, approaches to articulating a model of sustainable production; and
- methods of communication.

Methods of Inquiry and Planning

There is now widespread recognition that some of the conventional approaches to research and planning are inefficient at using local knowledge and expertise and may lead to ineffective or counter-productive development interventions. Alternatives include several research and planning approaches that are farmer-oriented and cost-effective. Some of the approaches used by AKRSP are discussed in Husain (1987a). The important elements of these approaches include:

- the recognition that the community of villagers represents a source of knowledge and expertise for action research and planning;
- the use of careful cost-effective methods of data collection, such as semi-structured interviewing and short formal questionnaires (where quantification is essential); and
- the identification of household and village priorities, resources, and opportunities through interactive consultation with villagers.

In essence, these approaches attempt to combine local knowledge and traditional practice with scientific knowledge and modern practice. This has also been the desired goal at AKRSP. There is a need, however, to clearly identify the areas of comparative advantage for villagers and outside experts. For instance, villagers often have an extremely good idea about their priorities and resources and about existing markets (through information on prices) and traditional technology, but, in a changing environment, outside experts may have a better idea about technological options and potential markets. The two sources of knowledge and expertise can be combined in field work through:

(1) Informal methods of inquiry, including site visits and dialogues in the project area as a whole;

(2) Structured long-term monitoring for impact, to observe emerging technologies and management systems, with well-defined indicators and feedback loops, in a small number of villages. This would have two objectives:
   (a) to identify regional trends in order to articulate regional planning needs, so that the project may make informed judgments from time to time on the reallocation of its resources; and
   (b) to identify and help disseminate institutional and technological innovations.

(3) Research to adapt emerging technologies and institutions to the widest possible range of conditions in the project area. The objective is to design replicable models of development, with appropriate institutions and technologies.

Management Systems for the Future

Some important lessons have been learned, from AKRSP's experiences with collective management, regarding the kinds of activity that are suitable for collective management rather than control by individuals. These lessons suggest that:

(1) The VO has the potential to be the **manager of natural resources**. Thus, the VO could play a pivotal role in the transition from feudal authority to open access to common property;

(2) The VO is a versatile **service contractor** since it can:

(a) exploit economies of scale in marketing and input supply; and
(b) facilitate division of labour and specialization by enabling markets to be created in the village where none might have existed before.

(3) **Production units** that are not traditionally common property represent a formidable challenge to collective management. Here, the VO's record is mixed; while there are potential economies of scale to collective management, there is also the distinction between owner and manager that makes it difficult for the VO to manage a unit as efficiently as a single owner-manager might; and

(4) Women are fast emerging as farm managers as men take up off-farm employment opportunities; while this represents a departure from the traditional division of labour, there is little evidence of an increasing role for women in decision-making over common property. Eventually, however, full adjustment to the new circumstances might be consistent with a much greater role for women over natural resources that are traditionally common property.

AKRSP has a studied approach to institutional innovation. In essence, it makes suggestions to villagers based on prior experience in the project area, documents how the villagers respond, and helps disseminate working models that appear to be productive and equitable. Thus, the evolution of management systems for the future is seen to be a learning process for AKRSP and the villagers. There is no blueprint for institutional development.

Articulating a Model of Sustainable Production

Just as there is no blueprint for institutional development, it is difficult to make generalizations concerning terms of a production model suitable for a range of conditions. There are, however, elements of a minimal framework for planning towards a sustainable production model. These elements include:

- developing an **awareness of technological options** available in the project area and elsewhere in similar environments, with particular attention to interactions in the use of various resources;
- **analysing markets**, particularly those subject to change; in the Gilgit context, both output and labour markets are subject to rapid change and reveal the shape of things to come;
- identifying groups of villages or valleys in which particular combinations of resources, technology, and markets can have broadly similar results; in other words, delineating broad **recommendation domains**; and
- maintaining a **balance between** activities that have a **short-term pay-off** (and, thus, can help sustain a community in its collective endeavours) and those with a **long-term pay-off**.

It needs to be emphasized that community action for conservation is seldom forthcoming without the prospect of gain. Thus, the sustainable production model should be able to deliver increases in domestic consumption or market sales within a time frame that is valued by the community.

In the Gilgit context, villages across the district are becoming increasingly differentiated in the way in which they exploit natural resources. There are, however, indications of comparative advantage at a sufficiently disaggregated level to enable AKRSP to develop a menu of production programmes from which VOs can choose the most appropriate options. At the present time, AKRSP is engaged in experimental work on pasture management and valley/watershed planning. Detailed reports by consultants and its own staff are beginning to give direction to the biological component of the sustainable production model. Thus, for instance:

- it is recognized that while new forestry plantations will have to be multipurpose, fuelwood is a priority in upper Hunza and fodder is important in central Hunza;
- it is felt that slow-release nitrogenous fertilizer can increase the effectiveness of fertilizer use on the leaching soils of the area;
- there is discussion about the balance of effort at AKRSP between fodder crop development and pasture development in terms of their contributions to livestock development; and
- there is consideration and testing of a number of options that could make it worth the villager's while to control free grazing and use the land for more productive purposes.

Over time, there is a need to develop many such interventions in response to changes in the region's economy. Each technological intervention demonstrates, too, the challenge to institutional innovation. Addressing the two simultaneously will help AKRSP articulate effective strategies for resource management in the region.

Methods of Communication

Technologies and institutional arrangements passed down to villagers from preceding generations are often described in terms of rules of thumb and conventions. These rules of thumb—many of them expressing quantitative relationships—are simple and clear and can be transmitted cheaply and widely. If innovative practices are to be extended to farmers, they need to be backed up by a strategy of communication that is at least as effective as traditional methods.

Where research funds are small and farmer literacy is low, highly specific recommendations, conditional on continuous variables, may be prohibitively expensive to develop and disseminate (Byerlee 1986). Thus, simplicity in recommendations has particular value in a place like Gilgit and would imply priority to:

- general recommendation, of which a single recommendation is made for all farmers in a recommendation domain, and,
- next in priority, recommendations conditional on discrete rather than continuous variables, e.g., fertilizer level × soil type or crop, rather than irrigation × soil moisture percentage.

If simple and clear recommendations are available, they need to be communicated repeatedly to the farmer, particularly at the 'right' time and in a cost-effective manner. In the past, AKRSP has used the following extension media:

- model farmers and village plant protection specialists trained by AKRSP;
- village meetings and dialogues;
- demonstration plots in the villages;
- Urdu language leaflets; and
- meetings of village representatives held in Gilgit.

In the future, extension efforts might also benefit from the addition of an audio-visual unit and the commissioning of a radio station in Gilgit capable of covering the entire district.

There is a need to focus particularly on communications with village women. One report after another at AKRSP has identified the problems that AKRSP has faced because its field staff are men. AKRSP has been recruiting women to its district-level staff, and these women professionals go on extensive field tours in support of the extension efforts of particular technical sections. Nevertheless, Social Organization Units, based in the villages, remain a male preserve. Since these units are the coordinators of efforts at the field level, and since they are the primary means of two-way communication between AKRSP and the villagers, the absence of women from these teams must be counted as a serious weakness in AKRSP's approach to women in development. While such conclusions have been put forward to AKRSP on a number of occasions, there are delays in formulating appropriate remedial measures. This is unusual for AKRSP and is probably due, in part, to the difficulty of recruiting appropriate female staff.

Questions for the Future

There are parallelisms between AKRSP's response to change and that of the villagers with whom it works. Both AKRSP and the VOs explicitly recognize the need for innovation as a response to change. If markets, technology, and the macro-institutional framework remained static, there would be no payoff to innovation; in particular, there would be no incentive to change traditional patterns of resource allocation and traditional resource management systems. Conversely, where change is greatest, the opportunity for innovation is greatest. The perception of change, and of priority in the reallocation of resources, however, may differ between AKRSP and the VOs, as well as among the VOs.

AKRSP offers a development package that combines collective management with agricultural production and marketing; it does not finance individual enterprise, nor does it sponsor non-agricultural activities. AKRSP is not, therefore, a programme for regional development; it is a small farmer development programme. This raises a fundamental question about future directions for village development: what ways and means will the VO employ in the future to manage its resources in order to respond to all (not only agricultural) opportunities for development? Can one expect equitable and sustainable increases in resource productivity if large portions of economic activity are left outside the planning and coordinating roles of the VO?

REFERENCES

Abidi, S.M. 'Pastures and Livestock Development in Gojal.' Gilgit: AKRSP, Workshop on Women and Resource Management, 1987.

Aga Khan Rural Support Programme. *First Progress Report*. Gilgit: AKRSP, 1983.

Aga Khan Rural Support Programme. *Passu: Securing a Future*. Gilgit: Monitoring, Evaluation and Research Section, AKRSP, 1984.

Aga Khan Rural Support Programme. *A Strategy Paper for the Second Phase*. Gilgit: AKRSP, 1987a.

Aga Khan Rural Support Programme. *Project Area Data Book*. Gilgit: AKRSP, 1987b.

Aga Khan Rural Support Programme. *Fifth Annual Review*. Gilgit: AKRSP, 1987c.

Butz, D.A.O. 'Irrigation Agriculture in High Mountain Communities: The Example of

Hopar Villages, Pakistan.' Unpublished Master's thesis. Canada: Department of Geography, Wilfrid Laurier University, 1987.

Byerlee, D. 'Some Common Sense about Farmer Recommendations and Extension Advice,' mimeo. Islamabad: International Maize and Wheat Improvement Centre (CIMMYT), South Asia Regional Office, 1986.

Caroe, R. 'Women of Gilgit: Time-space Routines in Rural Communities.' Unpublished B.A. Thesis. Massachusetts: University of Cambridge, Pembroke College, 1986.

Conservation for Development Centre. *Sustainable Forestry Development in the Aga Khan Rural Support Programme, Northern Areas, Pakistan.* Gland, Switzerland: CDC, International Union for the Conservation of Nature and Natural Resources (IUCN), 1987.

Conway, G.R., Z. Alam, T. Husain and M.A. Mian. *Agroecosystem Analysis and Development for the Northern Areas of Pakistan.* Report No. 1. Gilgit: AKRSP, Rural Sciences Research Programme, 1985.

Conway, G.R., M.A. Mian, Z. Alam, M.Y. Khan and T. Husain. 'Agroecosystem Zoning of the Hunza Valley: First Iteration' (Draft). London: International Institute for Environment and Development, 1987.

Dani, A., C.J.N. Gibbs and D.W. Bromley. *Institutional Development for Local Management of Rural Resources.* Workshop Report No. 2. Hawaii: East-West Environment and Policy Institute, 1987.

Gohar, A. *Report on Pasture Survey of Chalt-Chaprote Valley.* Gilgit: AKRSP (n.d.).

Gohar, A., S. Khan and S. Rahemtula. *Range Management and Pasture Development: A Series of Case Studies and Tentative Implementation Plans for the Development of Pastures.* Gilgit: AKRSP (n.d.).

Hamid, S. *Village Case Study of Rahimabad I.* Gilgit: AKRSP, 1987.

Hunzai, I.A. *The Political Economy of Forestry: The New Management System for Forests in Chalt-Chaprote.* Gilgit: AKRSP, 1987.

Husain, T. 'Approaches and Techniques for Assessing Mountain Environments.' Kathmandu: ICIMOD, International Workshop on Mountain Agriculture and Crop Genetic Resources, 1987a.

Husain, T. 'Irrigation Development and Water Management.' Gilgit: AKRSP, Workshop on Women and Resource Management in Gilgit, 1987b.

Husain, T. with A. Jan and F. Mahmood. *Village Management Systems and the Role of the Aga Khan Rural Support Programme in Northern Pakistan.* MPE Series No. 10, ICIMOD, 1990.

Husein, M.H. *Case Study of Land Development in Sherqilla.* Gilgit: AKRSP, 1985.

Husein, M.H. and N. Karmali. 'Case Studies in the Development of New Land: Land Use Patterns and Village Management Systems.' Gilgit: AKRSP, Workshop on Women and Resource Management, 1987.

Jan, I. *Comprehensive Village Development: A Case Study of Village Organisation: Rahbat Bala.* Gilgit: AKRSP (n.d.).

Khan, M.Y. *Agroecosystem Analysis of Rahimabad.* Gilgit: AKRSP, 1985.

Khan, M.Y. *Village Agroecosystem Study of Thingdass.* Gilgit: AKRSP (n.d.).

Kreutzmann, H. Background Information on the Pastures of Hunza and Nagar. Interview by Nabeel Malik. Gilgit: AKRSP, 1985.

Magrath, P.A. 'Women's Role in Forest Management and Development in the Gojal

Valley, Gilgit.' Gilgit: AKRSP, Workshop on Women and Resource Management, 1987.

Meghji, Z.M. *Fertiliser Loan Repayment in Rahimabad I and Oshikhandass*. Gilgit: AKRSP, 1984.

Meghji, Z.M. and M. Saleem. *Two Case Studies on the Cooperative Management of Tractors*. Gilgit: AKRSP, 1987.

Meghji, Z.M., K. Tetlay and K. Tejani. *Preliminary Review of Heifer Project Villages in Gilgit District*. Gilgit: AKRSP, 1987.

Neseem, A.A. *Roshanabad-Sherabad Village Organisation: A Case Study of Decentralization for Rural Development*. Lahore: Pakistan Administrative Staff College, 1986.

Sakhi, A.J. *Case Study of Village Thingdass*. Gilgit: AKRSP, 1987.

Saunders, F. *Karakoram Villages: An Agrarian Study of 22 Villages in the Hunza, Ishkoman and Yasin Valleys of Gilgit District*. Rome/Islamabad: Food and Agriculture Organisation of the U.N., Integrated Rural Development Project, Gilgit, PAK/80/009, 1983.

Semple, M. *Village Case Studies on Rahimabad, Broshal, Khaiber and Thingdass*. Gilgit: AKRSP, 1986.

Staley, J. *Words for My Brother: Travels between the Hindu Kush and the Himalaya*. Karachi: Oxford University Press, 1982.

Whiteman, P.T.S. *Mountain Cases: A Technical Report of Agricultural Studies in the Hunza, Ishkoman and Yasin Valleys of Gilgit District*. Rome/Islamabad: Food and Agriculture Organisation of the U.N., Integrated Rural Development Project, Gilgit, PAK/80/009, 1985.

Whiteman, P.T.S. 'Learning about Change in the Mountain Environment: Some Notes from Northern Pakistan.' Ceres 112:42–46, 1986.

World Bank. *The Aga Khan Rural Support Programme in Pakistan: An Interim Evaluation*. Washington, D.C.: Operations Evaluation Department, World Bank, 1987.

30

MOUNTAIN AGRICULTURAL TECHNOLOGY DEVELOPMENT AND DIFFUSION: THE LUMLE MODEL, NEPAL

B. Pound, K. Budathoki and B.R. Joshi

Barry Pound, Kedar Budathoki, and Bhoj Raj Joshi are the Director, Chief Horticulturalist, and Chief Veterinary Officer, respectively, of the Lumle Agricultural Centre, Pokhara, Nepal.

CONTENTS

INTRODUCTION

This paper describes the strategies and methods used by Lumle Agricultural Centre (LAC) to identify, generate, verify, and disseminate agricultural technologies for the Western Hills of Nepal.

Managerial flexibility, continuity of funding, staffing, and strong administrative backup for staff are identified as important institutional requirements for a successful project in a difficult environment.

A multidisciplinary, farming-system, farmer-participatory approach to problem identification and technology generation and verification is used in achieving its goal of providing farmers with relevant technologies in a usable form. Strong linkages with Nepal's Research and Extension Services at national, regional, and district levels are necessary for wide dissemination of technologies in their appropriate recommendation domains.

The History of Lumle Agricultural Centre

The hills of the Western Development Region of Nepal rise from the sub-tropical *Terai*[1] at 300 masl to the Arctic heights of Annapurna and Dhaulagiri Himal which are over 8,000 m. In the low hills, three crops a year are possible. Cultivation of potatoes, barley, buckwheat, and amaranthus is carried out up to 3,000 m, and migratory sheep flocks graze the alpine pastures up to 4,500 m. The people of the area are mostly smallholders, peasant farmers renowned for their industriousness and hardiness. The mid and high hills are a traditional recruiting ground for the world-famous Gurkha soldiers.

In 1968, a resettlement scheme was set up in the foothills of the Annapurna to retrain retiring British Gurkhas in agricultural skills. It was soon evident that training alone was insufficient. Technologies to improve the traditional farming methods, a supply of inputs, and advice at field level were also required to ensure successful re-integration into agriculture. It was as a response to these needs that Lumle Agricultural Centre was born, with the responsibility of providing research, extension, and training services to the farmers of the western hills. Starting with a relatively small command area of 1,000 km^2 (15,000 farming households), Lumle Agricultural Centre expanded steadily and now serves 18,600 km^2 (350,000 farming households). Previously an isolated project, it is now part of the national network of agricultural research stations coordinated by the National Agricultural Research Centre (NARC).

The Objectives of Lumle Agricultural Centre

The major technical objective of LAC is to improve the productivity and income of farmers through the generation, verification, and dissemination of relevant, sustainable technologies. At the same time, LAC is seen as a centre of excellence for the training of national scientists and the pioneering of research and extension methods suitable for hill agriculture. This Centre is now able to demonstrate that mountain farming, including

---

[1] The lowland and southern plains of Nepal, bordering the Gangetic plains of India.

crops, horticulture, livestock, and forestry, can offer a sustainable livelihood for the farmers of the mountain regions of Nepal.

The Context

A more difficult physical, social, and institutional environment in which to carry out development work would be hard to imagine. Paradoxically, this very fact may have contributed to the success of LAC in attracting good quality funding and a team of national scientists dedicated to equalling the challenges set by the environment.

Fig. 30.1 shows the location of LAC, centrally situated in the Western Region of Nepal. The few motorable roads that exist are all marked. The research command area of LAC includes the sub-tropical low hills bordering the *Terai*, the warm and cool temperate middle hill region, and the alpine pastures and Arctic snowfields of the high Himalayas.

Differences in altitude, aspect, slope, and soil multiply the agricultural 'niche' manifold. Annual rainfall varies from 1,500 mm to over 5,000 m. High rainfall, fragile soils, and steep slopes combine to produce landslides, even in well-forested areas.

The area depends upon agriculture, tourism, and military service as the mainstays of the rural economy. Agriculture accounts for 55 per cent of the National GDP and 93 per cent of the population live in rural areas; the majority of these depend upon agriculture for their livelihood. The hills, which contain about 50 per cent of the country's 18 million people, support an average of 5.3 persons per cultivated hectare (World Bank 1989). This is comparable to that of deltaic Asia, a major difference being that in those regions three crops per year are possible, whereas in the hills of Nepal many areas can support only one or two crops per year (Balogun et al. 1988).

The population in the hills continues to rise by about 1.5 per cent per year, despite out-migration to the lowlands and the towns (World Bank 1989). As a result the already small and fragmented farms are further split. Seventy-five per cent of farmers in the hills have less than 1 ha of land to farm and 50 per cent have less than 0.5 ha (World Bank 1989). Thus it is not surprising that the majority of hill districts in Nepal are food deficit areas (i.e., net importers of food) for one or more months of the year.

The resources available to farmers in terms of land, capital, income or underutilized labour are very meagre. GDP per capita is US$ 160 per annum.

In Nepal, and particularly in the hills, farms rely on a complex and complementary mixture of enterprises to spread risk and to provide the bare necessities of existence (Fig. 30.2). The forest provides fuel, timber, fodder, and some food and cash products.

Livestock often depend upon tree fodder and grasses from the forest. While ruminants provide milk products, meat (in the case of buffalo), and draft (in the case of cattle) they are also most important as sources of manure and as an insurance policy for times of need. Manure is the major source of nutrients for crops, inorganic fertilizer use being minimal except for those few areas enjoying good access or unusually intensive development.

In Western Nepal, the majority of farmers live in villages and social cohesion is strong. However, there is social stratification both on the basis of relative wealth and on the basis of caste. Thus, the adoption of any technology depends not only on its relevance to the agro-ecological conditions of the area, but also on the socioeconomic status of individuals within a community.

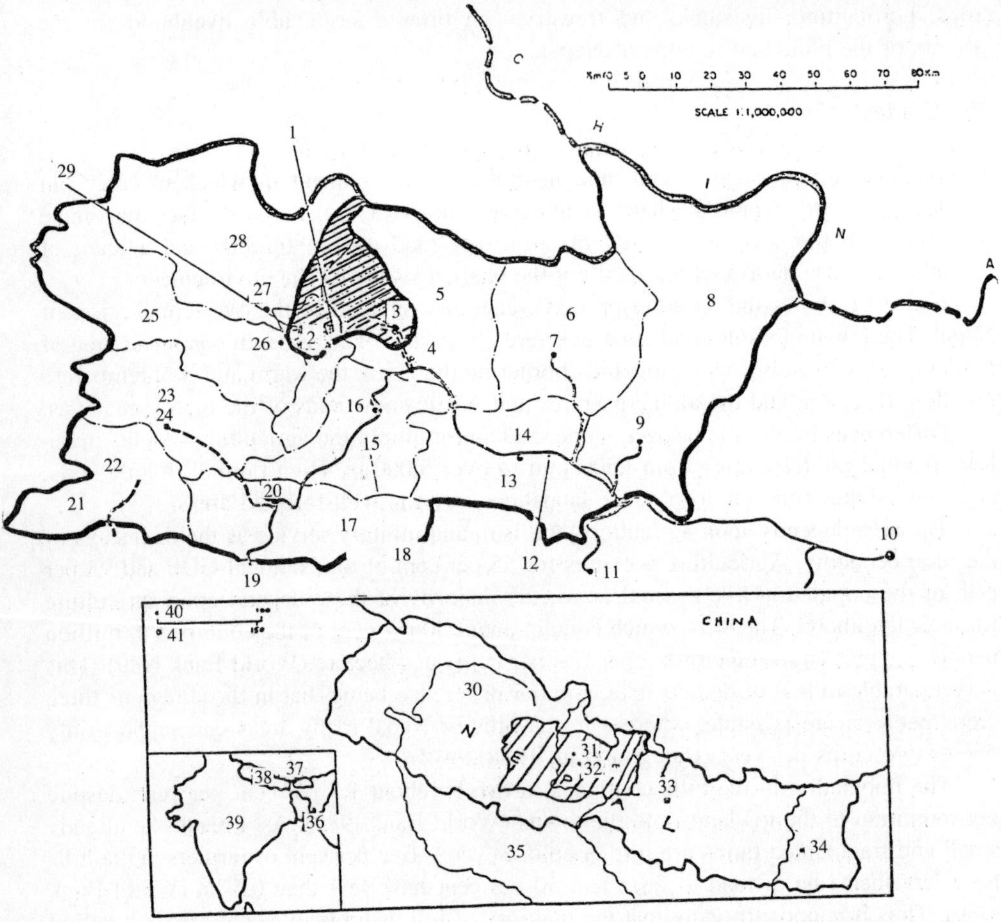

1. Tapu (FSR)
2. Parbat Kusma
3. Lac
4. Pokhara
5. Kaski
6. Lamjung
7. Besisahar
8. Gorkha
9. Gorkha
10. Kathmandu
11. Bharatpur
12. Yampaphant (FSR)
13. Tanahun
14. Damauli
15. Syangja
16. Syangja
17. Palpa
18. Kewre Bhanjyang
19. Butawal
20. Tansen
21. Arghakanchi
22. Sandhikharka
23. Gulami
24. Tamghas
25. Baglung
26. Baglung
27. Beni
28. Myagdi
29. Lopre (FSR)
30. Jumla
31. Pokhara
32. Lac
33. Kathmandu
34. Dhankuta
35. India
36. Bangladesh
37. Bhutan
38. Nepal
39. India
40. Miles
41. Kilometres

Figure 30.1: Map of Lac extension and research command areas
Source: Lumle Agricultural Centre.

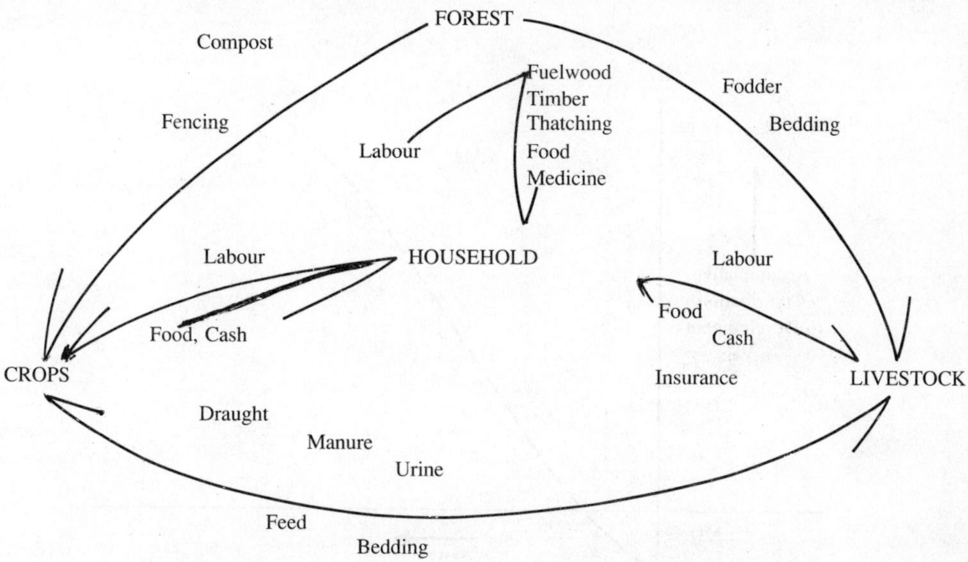

Figure 30.2: A simplistic diagram of the traditional Nepali hill farming system
Source: Lumle Agricultural Centre

Village communities are changing even in remote areas. Most children have access to schooling, and migration and off-farm employment are bringing new ideas and resources from outside. Government services are affecting an increasing proportion of farmers, and 'development', as measured by the quantity of services per head of population, is progressing steadily.

LAC is part of that development process. The approaches it has evolved in order to control the quality of development due to its interventions form the main theme of this paper.

CONSTRAINTS AND FACILITATORS OF DEVELOPMENT IN THE NEPALESE HILLS

It would be wrong to say that LAC has ever had a deliberate policy of minimizing constraints and exploiting the factors that facilitate development. However, as LAC has evolved it has recognized these opposing factors, through close association with its working environment, and modified its approach accordingly (Fig. 30.3).

Constraints

The constraints inherent in the working environment are:

- the increasing population;
- the immense logistical difficulties of working in an area with few roads or services and a highly dissected terrain;
- the meagre resource base, in terms of cultivable land, good quality communal land, soil fertility, capital or liquid assets, professional skills, and available labour;

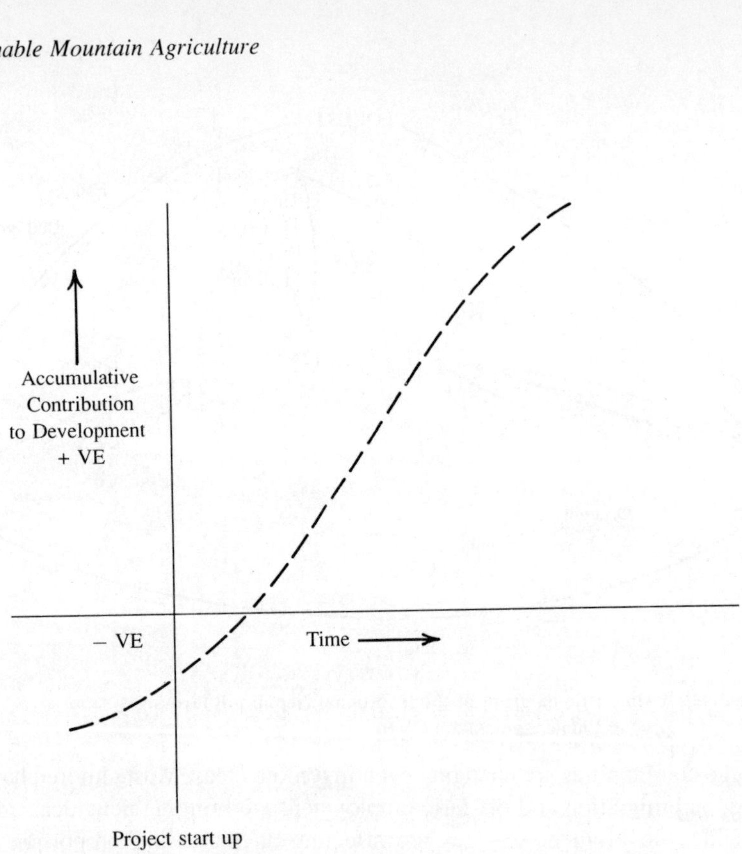

Figure 30.3: A representation of the contribution of donor-supported projects to development with time
Source: Lumle Agricultural Centre (LAC)

- the paucity of local and urban markets, due to a limited population with adequate purchasing power, and the difficulties of international marketing for a landlocked country restricted by its neighbours; and
- the extreme edaphic, climatic, ethnic, and economic diversity of the area.

Facilitators

In contrast to these negative factors, the following must be acknowledged as factors aiding the development process:

- the majority of land is owner-occupied;
- there is good social cohesion and a tradition of self-help;
- labour is cheap by international standards;
- the rural population is receptive to relevant technology;
- there is a considerable inflow of income into rural areas from off-farm sources; and
- the diverse conditions create a wide range of opportunities.

In order to address the problems of the western hills and to exploit their comparative advantages, a favourable institutional environment must be created. Only then can appropriate technical solutions be generated. Partly by chance and partly through a sustained team effort, it is thought that LAC has evolved a combination of institutional characteristics that favour development.

THE INSTITUTIONAL ENVIRONMENT AT LUMLE AGRICULTURAL CENTRE

Location

Initially the siting of LAC, 30 km from the nearest road, (accessible by road now) ensured from the start that there was good contact with the local population and appreciation of the logistical difficulties. Even now all staff spend a considerable amount of time travelling by foot around the command areas of the Centre.

Continuity

Many projects and programmes will have only just emerged from their start-up phase by the time the donor funding runs out. Their contribution is often minimal and could be negative.

Time is required to understand the local conditions, to set up an infrastructure, to initiate working procedures, to build a team of technicians and supporting staff, and to gain the confidence of farmers and local leaders.

While continuity of funding is important, continuity of staff is equally vital. This is true for administrative, technical, and field staff, so that all feel part of a stable but progressive entity.

In order to achieve continuity of staffing, the terms and conditions of service must be adequate and the work professionally satisfying. The first is met by providing good housing, facilities, a good remuneration package, good training opportunities, and a good working environment. In return staff are expected to devote all their energies and abilities to the job in hand.

Administrative Support

A strong administrative section takes much of the day-to-day administrative burden from the technical staff. While the administrative section is on a par with technical sections, it does not dominate, and section heads are responsible for their own budgets, staff management, and programme planning. Administration supports but does not control. Thus, LAC can be described having a professional meritocracy, rather than a bureaucracy.

Funding

- *Continuity*. The agricultural calendar does not include a three-month break at the end of the financial year. Thus, funding must be available throughout the year for programmes to retain their momentum.
- *Long-term commitment*. For any sort of medium or long-term planning and implemen-

tation of programmes there must be security of funding and a declaration of support from both the funding agency and the recipient government.

- *Flexibility*. In-built mechanisms for financial management flexibility at LAC allow the programmes to respond to the dynamic situation with which they are dealing. In this context, it must be recognized that the project is there to respond to the requirements of the farmers, not to mould the farmers to its planned 'model' of development. In fact, the LAC model can be seen to be a response to the requirements of the hill farmers.
- *Direct funding*. The three ideals of continuity, security, and flexibility of financial management are rarely found in development projects. They are made possible because the Centre is directly funded by the British Government. The project director is responsible for the management of the funds, and is fully accountable to the British Government. Within clearly defined limits the director is able to use the funds to the best advantage of the project. It is this degree of on-the-spot control that allows management to be effective and supportive to the development process.

Staff Support

To maintain the vital contact with farmers the staff of LAC must regularly travel on foot to some of the most remote areas of Nepal. These areas are often food deficit areas so food supplies must be taken. The treks must be carried out in the snow during winter and in the torrential monsoon during summer. Staff must be well equipped for such expeditions, they must be supported by a cadre of well-disciplined porters, cooks, and camp managers, and they must be given adequate financial incentive to compensate for the hardship and expenses incurred. These conditions apply equally to field-based extension staff and Centre-based research and training personnel.

All senior (graduate) staff at LAC are issued with a tent, sleeping bag, and sleeping roll, and are expected to use them. Indeed promotion and training opportunities are partly determined by 'trek performance'. Trekking allowances are adequate to cover food, clothing wear and tear, and incidental expenses such as courtesy cups of tea for village leaders.

National Integration

For much of its 22-year history, LAC was independent of the national mainstream development process. Over the last five or six years this has changed so that the Centre is now an integral part of the National Agricultural Research Centre's network of research farms and stations that cover the whole of Nepal, while still maintaining its integrity and character. The Centre bases its overall programme on national policy as embodied in successive Five-Year National Development Plans, and is in constant dialogue with the national, regional, and district HMG/N representatives of the Ministry of Agriculture, Forestry and Soil Conservation, and Land Reform and Management.

This ensures that the efforts of LAC do not duplicate or contradict the efforts of HMG/N, while at the same time ensuring that the work of LAC is understood and approved at all levels and that, where relevant, the recommendations or methodologies developed by LAC can be widely adopted.

TECHNICAL STRATEGIES

The Lumle 'Model' for the Hills of Nepal

In the previous section we have seen that a solid institutional foundation is in place at LAC. This enables the technical staff to carry out their work in a conducive institutional environment.

The next important factor in LAC's model is its comprehensive understanding of the field conditions for which technologies and approaches are being generated for the hills. This is because LAC has been providing extension services for some 15,000 farming households for 15 years. In addition, researchers and trainers carry out much of their work in the rural areas. The roles of researcher, extensionist, and trainer have often overlapped considerably.

The third major structural component in the Lumle model is the participation of farmers in research activities. Of all trials and studies carried out by LAC, between 65 and 75 per cent are carried out on farmers' fields, using farmers' labour. This trend is increasing and we are constantly looking for research approaches that involve the farmers at an ever earlier point in the technology selection process. Two recent examples of this are: Informal Research and Development, in which small packets of advance line seed materials are distributed in remote areas otherwise untouched by research and extension activities, allowing farmers to make their own selection of varieties, and Farmers' Preference Ranking, in which the farmers' preference for varieties is quantified as a complement to the measurement of biological parameters such as yield and time of maturity.

The fourth vital component of the Lumle model is its multidisciplinary approach to research. This has been a fundamental feature in ensuring the relevance of any technologies generated to the overall farming systems of the hills, and reflects the interdependence of all the major components of the farming system at the farm level as shown in a simplified form in Fig. 30.4.

The fifth component is the fact that research, extension, and training are all carried out by the same institution. Each supports the other and there is a synergistic effect resulting in an organisation that is both solid and dynamic.

The Research Process

The research process consists of a number of continuous stages, which are: identification of the problem, technology generation, verification, dissemination, modification, and monitoring (which leads to further modification). At LAC, each stage has been critically examined in the context of the development of approaches thought to be suitable for the conditions in the hills. Several of these approaches are now being adopted at a national level.

Identification of the Problem

Within national policy guidelines, and in consultation with national, regional, and district officials, a broad work programme is drawn up each year, and this is prioritized according to the impact each area of work is expected to have. The prioritization stage requires further strengthening through the setting up of a computerized database.

In order to understand the farmers' problems and to ensure the relevance of any research programme, a special type of Rapid Rural Appraisal has been developed by LAC and its sister organization Pakhribas Agricultural Centre. This is called the *Samuhik Bhraman*. The term means 'travelling together'. Normally, LAC research staff, scientists from the national programmes, and staff of the District Extension Services make up the multidisciplinary team. They work together on a common theme that can be general or specific. In addition to clarifying the issues facing farmers amd learning from them, the *Samuhik Bhraman* is an opportunity to strengthen personal links with individuals of other organizations working in the same area of research (Chand and Gibbon 1990).

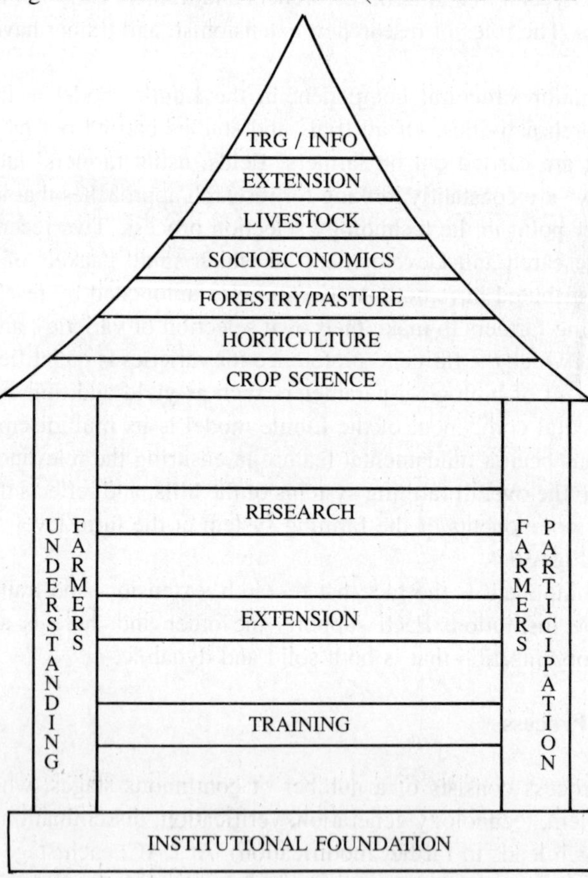

Figure 30.4: The Lumle model for the hills of Nepal
Source: Lumle Agricultural Centre

One consequence of the use of *Samuhik Bhraman* at LAC has been the identification of three major themes (research thrusts) that require special attention in the hills. These are: (1) the maintenance and improvement of soil fertility, (2) the improvement of the availability of fodder, and (3) the identification and encouragement of income-generating activities. These activities are now incorporated into LAC's programme and supplement the 'conventional' single discipline research by providing a mechanism whereby cross-disciplinary research can be carried out. Each thrust has a coordinator and volunteer members from most technical sections. The thrusts have their own programme for research and their own budget, and concentrate on problems that involve expertise from two or more sections such as the use of crop residues as livestock feed or the use of forest litter in the production of compost for crop production (Joshi et al. 1990).

Technology Generation

Many research stations have relied on generating technologies 'within the fence' of the station and then supplying them to the extension services. This approach has long been rejected at LAC. The present approach is to carry out a high proportion of research on farmers' fields. Between 65 and 75 per cent of all trials and studies conducted by LAC are now carried out on farmers' fields using farmers' labour (Fig. 30.5). This has various advantages.

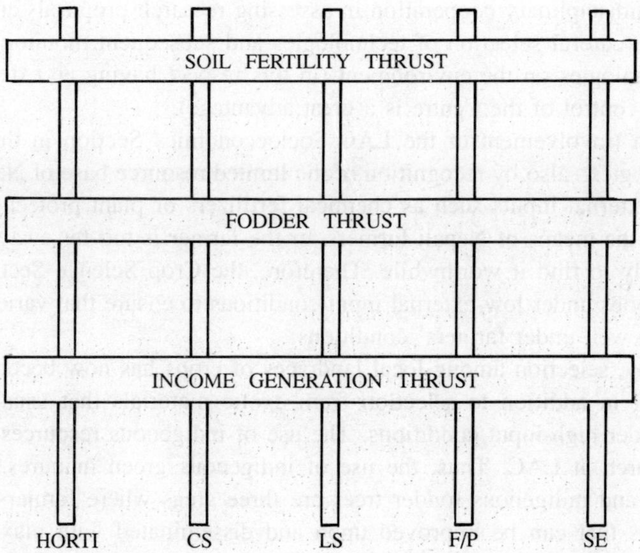

key: HORTI = Horticulture; CS = Crop Science; LS = Livestock; F/P = Forestry and Pasture; SE = Socioeconomics

Figure 30.5: Cross-disciplinary research "thrusts" at LAC
Source: Author

(1) Trials are conducted under farmers' conditions.
(2) Farmers' reactions can be incorporated into technology assessment.
(3) The cost to LAC is reduced.

A three-tier research network, pioneered by LAC, is now being adopted by the National Agricultural Research Centre.

Within the network, trials are carried on 'on Centre' at off-station research sites and at outreach research sites (Table 30.1). Only the first category has a permanent infrastructure, but all have LAC-employed personnel to ensure control of the research programme and good quality research results.

Mechanisms to Ensure the Sustainability of Technologies

It is thought that a technology will be sustainable (i.e., used by a farmer for a considerable period of time without environmental damage or resource degradation) if (1) it fulfills a real need, (2) it is suitable for the physical and social conditions, (3) it does not conflict with other farm enterprises, (4) it does not degrade the environment, (5) it is economically viable, (6) it is significantly better then the farmers' present practice, (7) the benefits outweigh the risks involved, and (8) the necessary inputs are available.

These points are dealt with by LAC in the following ways:

(1) By ensuring accurate identification of problems through *Samuhik Bhraman* and farmer/extensionist feedback.
(2) By carrying out location-specific verification of technologies.
(3) By multidisciplinary cooperation in assessing research proposals and results.
(4) Through careful selection of technologies and subsequent monitoring of the impact of technologies on the environment (in this respect having an extension area under the full control of the Centre is a great advantage).
(5) Through (involvement of the LAC Socioeconomics Section in the) assessment of technologies—also by recognition of the limited resource base of Nepali hill farmers. Many external inputs such as chemical fertilizers or plant protection chemicals are beyond the means of Nepali farmers, or the farmer is too far away from the source of supply to find it worthwhile. Therefore, the Crop Science Section at LAC tests its varieties under low-external input conditions to ensure that varieties selected will perform well under farmers' conditions.
    Similarly, selection among local landraces of crops has now become commonplace at LAC, in addition to selection from exotic materials that usually only perform well under high-input conditions. The use of indigenous resources is given priority in research at LAC. Thus, the use of indigenous green manures, indigenous goat breeds, and indigenous fodder trees are three areas where farmers have traditional practices that can be improved upon and disseminated with maximum speed and minimum disruption of present farming practices.
(6) While statistical analysis can be done on trials that are significantly replicated and controlled, the farmer is the ultimate judge of new technology and must be given ample chance to make his own assessment. This can be done through Pre-Production verification trials (Fig. 30.6), minikits, and informal research and development packages.
(7) As for (6), cost-benefit analysis can give an indication of the economic viability of a technology, but farmer adoption will be the proof of acceptance.
(8) Seed is the major input for agronomic and horticultural crops. LAC produces foundation seeds for a very wide range of crops and uses its own extension command area as a resource area for production of high quality cereals, vegetables, and potato seeds for the benefit of the whole of the hilly area of the Western Region (see Case Study One).

Farmers' variety Farmers' practices	Improved variety Farmers' practices
Farmers' variety Improved practices	Improved variety Improved practices

Figure 30.6: Pre-production verification trial (in this case to verify an improved variety)
Source: Author

Table 30.1. The LAC network of research sites

Facilities	Activities
	On centre
Permanent offices, laboratories, and accommodation. Core scientists Outreach senior staff	Basic research, initial evaluation of new research approaches, collation, analysis and interpretation of results, coordination of off-station and outreach trial programmes. Liaison with national and international research institutions.
	Off-station research sites (up to 8 by 1985)
No permanent infrastructure Graduate Coordinator. Junior technical staff	Basic research for environments not available at the Centre, initial verification, foundation seed multiplication, training, farmers' visits, involvement of AIC, ADB, and district staff.
	Outreach research sites (up to 38 by 1995)
No permanent infrastructure Locally employed recorders supervised by extension services and monitored by LAC-based outreach staff	Verification, involvement of extension services, focus for studies.

Note: AIC = Agricultural Inputs Corporation (Nepal), ADB = Agricultural Development Bank (Nepal).

Verification of Technologies

By the time a technology reaches this stage there has already been a considerable input of farmers' participation in the selection process. We are, therefore, cautiously confident that the technology will at worst be as good as the farmers' present practice and at best be very much superior.

Verification is carried out on farmers' fields at outreach research sites, where good quality data is required, and on farmers' fields away from outreach sites where the farmers' response data give sufficient feedback. The pre-production verification trial is used extensively for verification as it allows two-stage comparison between farmers' conditions and two levels of recommended practices. Verification is a cooperative activity between LAC research staff and the extension services, to the benefit of both parties.

Dissemination of Technology

At LAC, technology transfer occurs at various levels. Technologies come into LAC mainly from farmers and from its own research results, as well as from institutions within Nepal and abroad. At the same time, technologies are disseminated by LAC to extension agencies, farmers, research workers, and other change agencies, including private enterprises (Fig. 30.7).

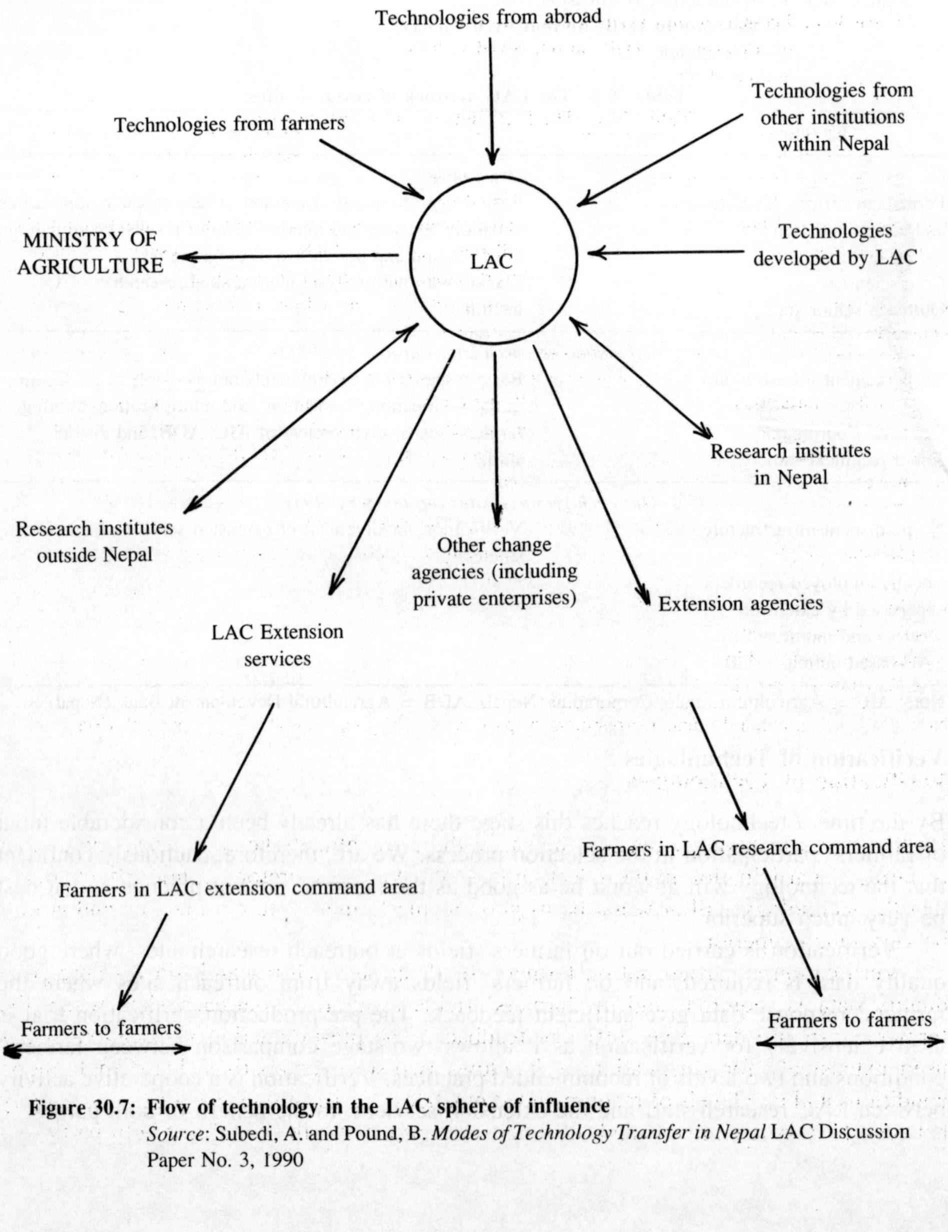

Figure 30.7: Flow of technology in the LAC sphere of influence
Source: Subedi, A. and Pound, B. *Modes of Technology Transfer in Nepal* LAC Discussion Paper No. 3, 1990

LAC has traditionally disseminated its technologies through its own extension staff with whom it has very close links and whose area it knows intimately. However, since 1987, LAC has also been responsible for research support to an increasingly large command area which will eventually include 350,000 farming households. Its clients in this case are the extension services of HMG/N and any other potential users of LAC-generated technologies. So far LAC has relied mainly on six methods of dissemination (Fig. 30.8).

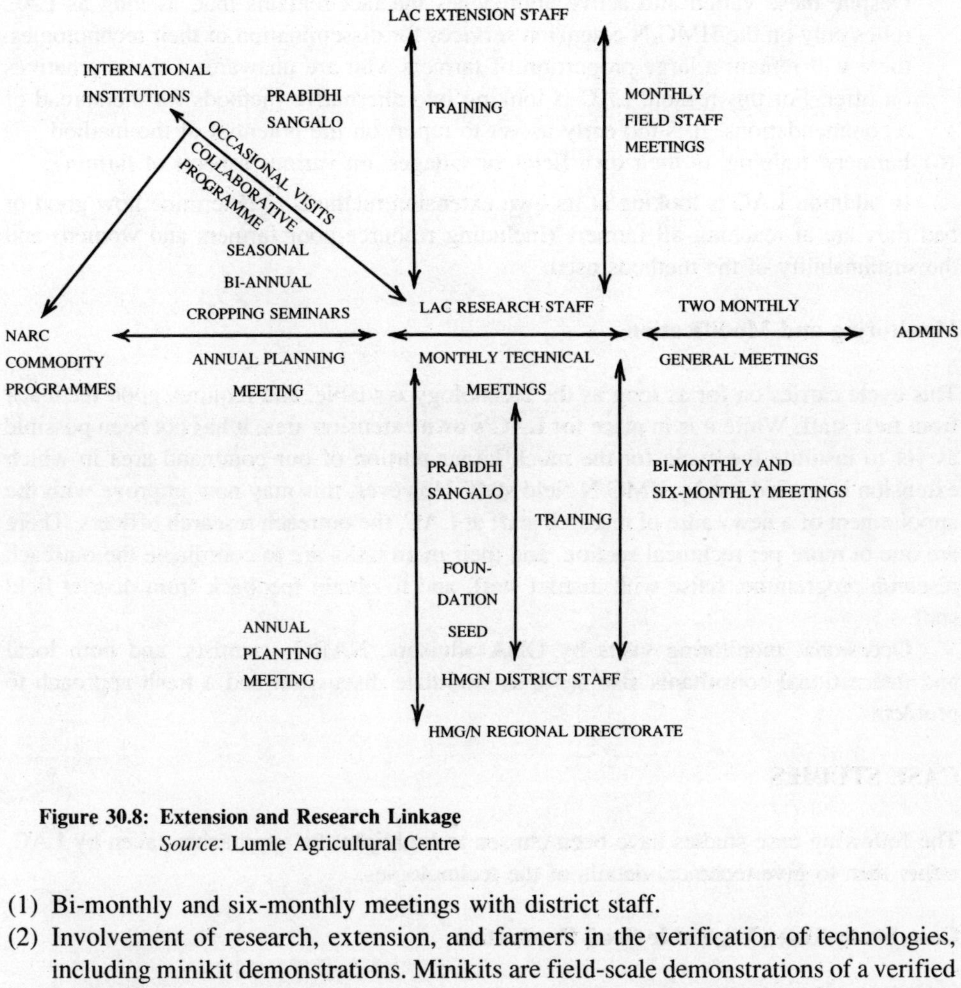

Figure 30.8: Extension and Research Linkage
Source: Lumle Agricultural Centre

(1) Bi-monthly and six-monthly meetings with district staff.
(2) Involvement of research, extension, and farmers in the verification of technologies, including minikit demonstrations. Minikits are field-scale demonstrations of a verified technology put out on farmers' fields. LAC coordinates some 3,000 minikits per year, using pre-released and released varieties from the National Commodity Programmes and from its own varietal improvement programmes.
(3) Published materials, especially *Prabidhi Sangalo* a topical newsletter produced quarterly in Nepali by the LAC technical staff. It covers proven new technologies, developed by LAC, some new ideas, farmer success stories, and hints and tips, as well as letters and poems sent in by readers. It is now in its fourth year of publi-

cation and circulation is about 2,000 copies. Very recently the Royal Nepal Academy for Science and Technology awarded the prize for Science and Technology to LAC, partly in recognition of the contribution of *Prabidhi Sangalo* to the dissemination of technology. It is becoming popular with field staff, district officials, students, and even Ministry staff because it fills a real knowledge gap.

(4) Supply of inputs, particularly foundation seeds of improved varieties of cereal crops, potatoes, and vegetables.

(5) Training of HMG/N extension staff in field experimentation methods.

Despite these varied and active approaches the fact remains that, as long as LAC relies only on the HMG/N extension services for dissemination of their technologies, there will remain a large proportion of farmers who are unaware of the alternatives on offer. For this reason, LAC is looking into alternative methods for the spread of recommendations. It is too early as yet to report on the potential of the method.[2]

(6) Farmers' training, in their own fields or villages, on various aspects of farming.

In addition LAC is looking at its own extension methods to determine how good or bad they are at reaching all farmers (including resource-poor farmers and women) and the sustainability of the methods used.

Monitoring and Modification

This cycle carries on for as long as the technology is viable, and requires good feedback from field staff. While it is in place for LAC's own extension area, it has not been possible as yet to institute the cycle for the much larger portion of our command area in which extension is carried out by HMG/N field staff. However, this may now improve with the appointment of a new cadre of research staff at LAC, the outreach research officers. There are one or more per technical section, and their main tasks are to coordinate the outreach research programme, liaise with district staff, and to obtain feedback from district field staff.

Occasional monitoring visits by ODA advisors, NARC scientists, and both local and international consultants also serve to stimulate discussion and a fresh approach to problems.

CASE STUDIES

The following case studies have been chosen to highlight the approaches taken by LAC, rather than to give technical details of the technologies.

Case Study One: Vegetable Seed Production

Introduction

Those villages fortunate enough to be situated near roads or urban centres have opportunities for comparatively easy marketing of their surplus production. For those

2 This includes use of informal research and development, involvement of NGOs, involvement of schools and health services, increased use of off-station resesearch sites and outreach research sites for extension, and obtaining feedback from the district field staff.

villages one or more days' trek from the nearest roadhead, income generation from agriculture depends upon high-value/low-volume products. This was one of the reasons for introducing vegetable seed production as an income-generating activity to farmers in LAC's own extension command area, where quality control aspects could be well supervised.

Identification of the Problem

The technology also answered a real and growing need for good quality seed throughout the country and reflected the national policy of self-sufficiency in seed production. It also exploited the topography and diversity well. The topography lends itself to isolation of seed-producing pockets, and the diversity allows ideal situations to be selected for a range of activities.

Institutional Consideration

The continuity of the project allowed the technology to develop at a natural pace, and flexibility in financial management allowed payments to farmers to rise from nothing to two million rupees over five years.

Having research, extension, and training under one organization hugely facilitated the coordination of these three vital components in the development, dissemination, and implementation of the technology.

Sustainability

'It doesn't matter now if LAC stays or goes. I know that I must continue producing seed, and I now know how to' (young seed producer—Tapu).

Sustainability is being ensured by three methods. First, and perhaps most important, village seed-producer committees have been set up in the main producing areas. Through these, there exists a degree of internal monitoring of quality as they realize that one poor sample of seed can compromise the reputation of the whole village. The growth in production is shown in Fig. 30.9.

Second, the need for crop rotations that maintain fertility and minimize pest and disease build-up is recognized, and the off-station research sites are being used to evaluate promising 'break' crops.

Third, private enterprises are now very involved in the marketing aspects in open competition with the parastatal Agricultural Inputs' Corporation (AIC). Having initiated the process and helped make it commercially attractive, LAC's future role will be one of quality control and technical backstopping.

Case Study Two: 'Chhomrong Local' Rice

Introduction

Research and extension services neglect farmers living above 1,500 m. These areas tend to be more remote and logistics are more problematic. The major crop research stations (National Commodity Programmes) are all based in the *Terai* or inner *Terai*. It is left to LAC and the Pakharibas Agricultural Centre (PAC) and, more recently, the Hill Crop Improvement Programme to develop technologies for the mid-high hills. Over the

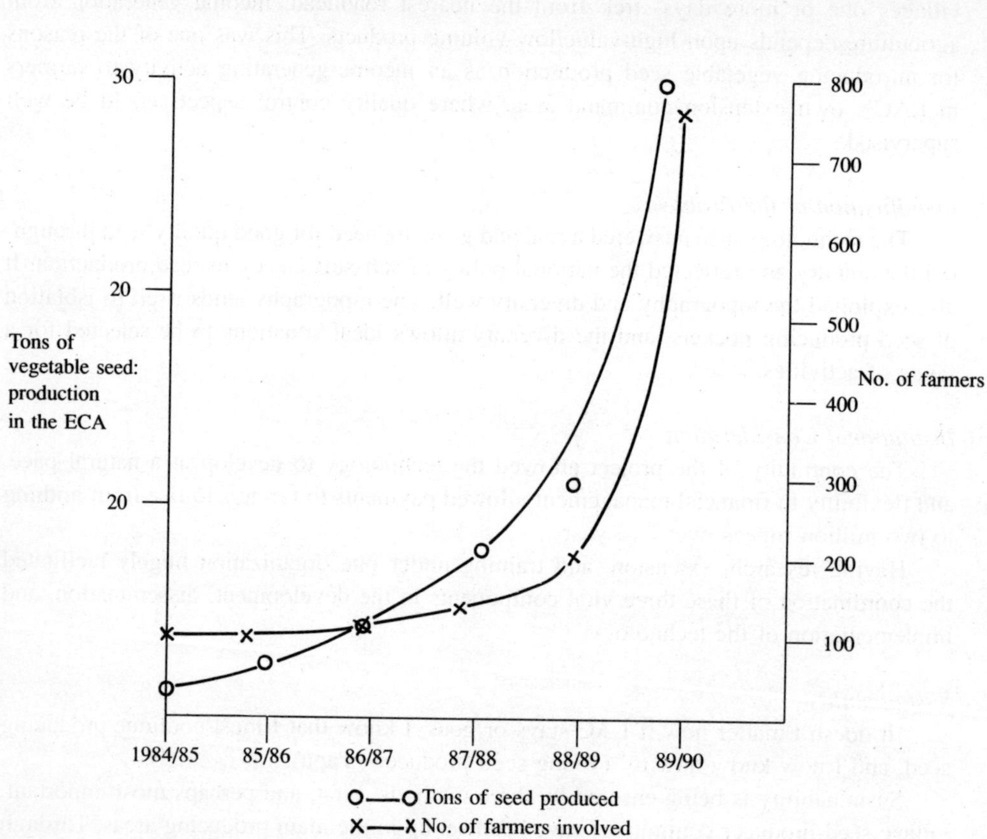

Figure 30.9: The growth in farmers' production of vegetable seeds in the LAC extension command area
Source: Lumle Agricultural Centre

last 15 years, 51 improved agronomic and horticultural crop varieties have been identified by LAC.

Identification of the Problem

Through farmer feedback, it was realized that there were no improved varieties of paddy rice for altitudes above 1,500 m. Due to cold air and water temperatures, most introduced varieties were at least partially sterile at this altitude.

The Research Approach

The approach taken was to select within the most promising local landraces and to improve them through scientific selection for yield components; farmers' selection took place in accordance with other characteristics important to them such as grain colour, cooking qualities, straw type, and length of season.

The selection process has been conducted using low external inputs such as chemical fertilizers and plant protection chemicals to ensure good performance under farm conditions. The work has been carried out at different altitudes from 500 m to 2,000 m using LAC's network of off-station research sites.

The variety to excel in these trials has been Chhomrong Local. Fig. 30.10 shows that, at 1,500 m, it outyielded all other varieties and that at 2,000 m it was outstanding. Chhomrong Local is now about to be released by the National Variety Release Committee and is the only 'Nepalese' variety to be included in the International Rice Research Institute's (IRRI's) International Cold Tolerance Rice Nursery.

This case study demonstrates the need for 'market research'. What are the 'products' of research that will sell? Here a gap in the market was identified and a product produced at a price (in terms of external inputs) that the market (i.e., the farmers) could stand. To embark on such a programme, which takes up to 10 years, requires confidence in the stability of the organization and continuity of staffing. This LAC was able to provide, and the beneficiaries are the highland farmers of Nepal and the international research effort.

Case Study Three: Sustainable Cattle and Buffalo Improvement through Village Livestock Development Committees

The large population of unproductive or under-productive cattle and buffaloes in the hills is not only a cause for concern from an ecological degradation perspective, but also a continuous drain on farmers' labour and resources.

The introduction of high-yielding animals from abroad requires large initial financial resources, a good market outlet for surpluses to justify the high inputs, adequate health services, and quality feed support.

These requirements are not available in most of the hill areas of Nepal. On the other hand, it is also necessary to upgrade the existing animal population so that production can be improved with minimum additional inputs.

Consequently, to improve the productivity from the existing indigenous animal population, LAC introduced a multidimensional approach. This includes genetic improvement by the introduction of genetically superior breeds, feed improvement by the introduction of indigenous fodder and forage species, improved use of crop by-products, health cover by trained field staff, and village-level animal health workers. Most important, these programmes were organized by the farmers' organizations or groups who have some vested interests and goals. These groups are responsible for ensuring the management and continuity of breeding programmes with very little help from LAC, apart from the initial provision of breeding materials (either bulls or artificial insemination services).

The result has been a three- to fourfold increase in milk production from crossbred cows in local environments without any additional inputs and a tremendous increase in the market value of these animals from Rs 200/local cow to Rs 6,000–7,000 per crossbred cow.

Among the present total population of 1,200 jersey-cross cattle and 1,400 cross-bred buffaloes in the ECA villages, 400 cows and 800 female buffaloes are in the productive stage. At present market prices, the total value of these animals is estimated to be about 9–10 million Nepalese rupees. In some areas, cheese factories are now being considered in order to use surplus milk and to provide more cash income to the farmers through fresh milk sales. In order to minimize environmental damage because of the grazing of large ruminants, stall-feeding is being encouraged. This is being rapidly adopted, especially where fodder resources are being increased through the planting of fodder trees and fodder grasses (often made possible only after grazing has stopped, requiring the cooperation of the whole village as organized by the Livestock Development Committee). Hence, with minumum external input, maximum utilization of indigenous resources, and organization of local communities, maximum benefit to the farmers is being provided.

Case Study Four: Fodder Tree Programme

The hill farmers of Nepal are dependent on tree products as an integral part of their farming system. Timber, firewood, compost, animal bedding, and thatching materials, as well as fruits, religious products, and cash products are all obtained from trees. However, it is probably the fodder obtained from trees that is the most important product. For at least part of the year (in some communities throughout the year) the major component of livestock feeds (for buffaloes, cows, goats, and rabbits) is tree fodder. Over 100 species of trees are used for fodder in the Nepalese hills.

Livestock numbers in the hills are declining due to a reduction in fodder resources and an increase in the proportion of children (herders) going to school. This has implications for soil fertility, because it is principally dependent on manure for the maintenance of nutrient status and soil structure.

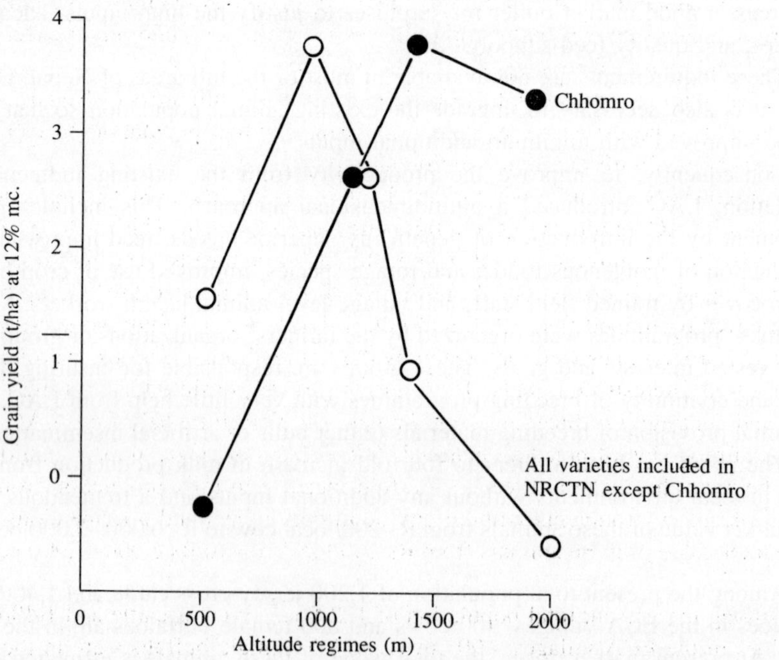

Figure 30.10: Comparative yield performance of chhomrong (●) and mean yields of all other varieties in the national rice cold tolerance (○) across wide altitude regimes, 1987–1988
Source: Sthapit, B.R. Proposal for Release of *Chhomrong* Dhan, LAC, Working Paper 11, 1990

This situation required a radical solution, and has been tackled by LAC and the farmers from its extension command area through the following interrelated programmes.

- Increase in stall-feeding of livestock to reduce grazing pressure on present fodder resources and allow natural regeneration and the establishment of new plantations.
- Identification of suitable fodder species and nursery raising/establishment/management methods through research and observation of farmers' practices.
- Provision of fodder tree saplings through the establishment of LAC-run and privately run nurseries.
- Organization by villages (with the assistance of LAC field staff) of forest user groups and livestock development committees.
- Training of extension field staff (livestock and forestry), nursery staff, and farmers, and visits by farmers and staff to other areas of the country.
- Initiation of special Fodder Saturation Programmes, in those areas with acute need and cattle/buffalo production potential.

These programmes are allowing the transition from a grazing-based to a stall-fed system of livestock raising, with concurrent increase in the fodder resource and a gradual improvement in the environment. During 1988/89 some 350,000 saplings were distributed (approximately 20 per household in the LAC extension command area). Plantations were mainly on private land around the home or on unused, marginal, or degraded land (Shrestha 1989).

Thus, plantations are cared for either by private individuals or by the community. Survival rates are comparatively high at 42 per cent after three years (Balogun and Harrison 1989).

Thus, the effort is a combined one of research, extension, and training. It is a multidisciplinary effort depending upon research by the Forestry and Livestock sections and implementation by forestry and livestock extension staff. It is a long-term programme which is only showing real benefits now after about ten years of sustained effort. The programme is a response to a real need of farmers, and it affects all the major components of their farming system. Its success depends upon using indigenous technical knowledge, as the majority of the fodder species raised in the nurseries are those requested by farmers.

The sustainability of the programme is ensured by the establishment of private nurseries with trained staff who will be able to continue raising seedlings in the absence of LAC. The forestry user-groups and livestock development committees (over 70 by July 1990) ensure that development is in the hands of those who stand to benefit.

HIGHLIGHTS OF INNOVATIONS OF LAC

Lumle Agricultural Centre was founded in 1968. Since 1975 it has been an agricultural research, training, and extension centre for the western hills of Nepal.

The area is characterized by extreme diversity of agro-ecological and socioeconomic conditions. The smallholder farmers have limited resources, which are further threatened by continuing population increases. Farming is mixed, and the different components of the farming systems are interdependent. Markets are poorly developed and relatively inelastic.

LAC has evolved an institutional and technical *modus operandi* to minimize the difficulties posed by its environment and to take best advantage of the opportunities presented. Through long-term, direct funding, LAC has been able to build up a stable, institutional structure capable of giving strong administrative support to its staff. Financial and managerial flexibility have enabled rapid response to hill farmers' problems. Integration into the national agricultural research network has further improved the cost effectiveness of the Centre through facilitating the uptake of new technologies and research methodologies developed at LAC by the national programmes.

The Lumle Model depends upon five major components: a strong institutional foundation, a comprehensive understanding of farmers' conditions, the participation of farmers in all stages of research and dissemination, the interdisciplinary interaction of all sections at LAC, and the synergistic effect of having research, extension, and training under one organization.

For each stage of the research process, LAC has developed or adopted techniques that are appropriate to hill conditions. Problem identification is often the subject for *Samuhik Bhramans* which facilitate a form of multidisciplinary, multi-institutional rapid rural appraisal carried out with a clear set of objectives and a disciplined working procedure. Through one such *Samuhik Bhraman* three major farmer-priority areas of work were identified for the development of agriculture in the western hills: (1) the maintenance and improvement of soil fertility; (2) the provision of fodder; and (3) the identification of income-generating activities. These three areas of work are incorporated into LAC's programme as cross-disciplinary Research Thrusts and support the more conventional single discipline research programmes.

Technology generation is carried out mainly on farmers' fields, and between 65 and 75 per cent of all trials and studies are being conducted on farmers' fields. There is a three-tier research network (on-centre, off-station research sites, and outreach research sites), allowing research to be carried out over a wide range of environments and maximizing farmer participation in the research process.

Technology identification is carried out on farmers' fields together with the extension services. For crops, the pre-production verification trial is the major tool used for verification of new varieties and practices.

The dissemination of technologies depends upon a multi-pronged approach: bi-monthly and six-monthly meetings with extension services, involvement of the extension services and farmers in the verification process, the use of published materials–particularly the *Prabidhi Sangalo* newsletter, the supply of high-quality foundation seed, and the training of extension staff and farmers. Further dissemination methods are being actively sought.

LAC is continuing to seek new extension methods appropriate to the hills and new research approaches that will lead to the development of relevant, sustainable technologies that will enhance the productivity and prosperity of Nepalese hill farmers.

CONCLUSIONS AND RECOMMENDATIONS

The hills of Nepal offer a challenge to agricultural development agencies which is shared by many hill areas of the world. They are characterized by extreme diversity, environmental fragility, and complex smallholder farming systems.

The experiences of Lumle Agricultural Centre can be of use to others, but are unlikely to be replicable in their entirety as situations will be different. With this proviso, the following recommendations are made.

For successful management of the project it is necessary to have:

- continuity of funding for a minimum of 10 years,
- direct funding,
- financial management responsibility,
- executive power,
- autonomy from government control,
- continuity of staffing, requiring a good working environment, and
- good administrative support to allow technical staff to operate.

For a successful programme, in terms of improvement in farmer productivity and income, the following are necessary:

- integration with national policies and programmes,
- understanding of farmers' conditions,
- involvement of farmers in all stages of research and extension,
- multidisciplinary research, to reflect the interdependance of the components of the farming systems,
- research, extension, and training all within the same institution,
- the development of working practices and methods of research and extension that are relevant to the conditions,
- a decentralized research process leading to location-specific recommendations,
- maintenance of a balance between technology generation and technology transfer, and
- sustainability of development through careful selection of technologies, monitoring of impact leading to modification where necessary, and the involvement of village groups in self-determination of development initiatives.

ACKNOWLEDGEMENTS

Lumle Agricultural Centre is funded by the Overseas Development Administration (ODA) of the British Government, and works in cooperation with His Majesty's Government of Nepal. The support of both Governments is gratefully acknowledged. The authors are indebted to Mr Rabindra Kumar Shrestha for his comments on earlier drafts of this paper.

REFERENCES

Balogun, P.A.K., G.B. Gurung and B.R. Sthapit. 'An Approach to Sustainable Agricultural Research in the Hills of Nepal: The Experience of Lumle Agricultural Centre.' Paper presented at the SUAN-IV Regional Research Symposium on 'Sustainable Rural Development,' held at Khon Kaen University, Thailand, 4 July, 1988. LAC Technical Paper No. 88/21. Nepal: Lumle Agricultural Centre, 1988.

Balogun, P.A.K. and A.P. Harrison. *Private Fodder Tree Planting in the LAC Extension Command Area (ECA): A Long-Term Survey of Trees Planted in Monsoon, 1986.* LAC Technical Paper No. 21, Nepal. Lumle Agricultural Centre, 1989.

Chand, S.P. and D. Gibbon. *Samuhik Bhraman: A Rapid and Appropriate Method of*

Prioritising and Replanning Agricultural Research in Nepal. PAC Technical Paper No. 121. Pakhribas Agricultural Centre, 1990.

Joshi, K.D., B.R. Sthapit, B. Pound and J. Gurung. 'Research Thrust: A Multidisciplinary Research Approach to Generate Sustainable Technologies.' Presented at the SUAN-V Symposium on 'Rural-Urban Ecosystems: Interactions in Development,' Bandung, Indonesia, May 21–24, 1990. LAC Seminar Paper No. 25. Nepal: Lumle Agricultural Centre, 1990.

Shrestha, R.K. 'Multipurpose Tree Species Extension in a Multidisciplinary Context: The Experience of Lumle Agricultural Centre, Nepal.' Paper presented at the International Conference on Multipurpose Tree Species Research for Small Farms: Strategies and Methods, in Jakarta, Indonesia. Nov 20–23, 1989.

Sthapit, B.R. *Proposal for Release of Chhomrong Dhan*. LAC Working Paper No. 11. Nepal: Lumle Agricultural Centre, 1990.

Subedi, A. and B. Pound. *Modes of Technology Transfer in Nepal*. LAC Discussion Paper No. 3. Nepal: Lumle Agricultural Centre, 1990.

World Bank. *Nepal Agricultural Sector Review*. Report No. 7693—NEP. Washington, D.C.: World Bank, 1989.

31

MOUNTAIN AGRICULTURAL TECHNOLOGY DEVELOPMENT AND DIFFUSION: THE PAKHRIBAS MODEL, NEPAL

S.P. Chand and B. Thapa

S.P. Chand is the Chief Agronomist and B. Thapa the Chief Forestry Officer of the Pakhribas Agricultural Centre, Nepal

CONTENTS

INTRODUCTION

Pakhribas Agricultural Centre (PAC) was established in 1972 as an agricultural training centre for British ex-Gurkha soldiers. Funding for the Centre comes from the British Government's Overseas Development Administration. The Centre is located centrally within the mid-hills of the Eastern Development Region, about 15 km northwest of Dhankuta Town.

In 1975, at the request of the Government of Nepal, the mandate of PAC was expanded to provide research, extension, and training services for the entire farming community within its target areas. Initially, the Centre's target areas covered two locations: the northern target area which covers eight panchayats of Tehrathum and Taplejung districts, and the local target area, which covers seven panchayats of Dhankuta District. The target areas represent highly heterogeneous environments and contain over 9,000 farm families.

Until 1982, the Centre's agricultural training and extension activities were confined to the northern and local target areas. However, over the years, the emphasis of PAC has expanded, and the Centre's activities are now divided among 10 technical sections: agronomy, forestry, livestock, horticulture, seed technology, socioeconomics, training, extension research, veterinary investigation, and analytical services and information supported by administration. Since 1983, the Centre has been providing technical support to the Koshi Hill Area Rural Development Programme (KHARDEP), another British-funded project. Through this association, PAC has become responsible for agricultural research and development activities in the four Koshi hill districts of Dhankuta, Tehrathum, Sankhuwasabha, and Bhojpur—an area of more than 7,000 sq. km with more than 100,000 farm families.

Following the establishment of the National Agricultural Research Services Centre (NARSC), which coordinates all crop and livestock research in Nepal, PAC has become an important part of the national agricultural research network. From July 1990, the Centre has had a mandate to cover the 11 hill districts in Eastern Nepal—an area of 21,267 sq. km with 252,436 farm families. PAC has also been recognized as one of the important hill centres for forestry research under HMGN's Department of Forestry and Plant Research. PAC's seed testing laboratory has been designated as the official seed testing centre for the eastern hill region by the government's National Seed Testing Programme.

This paper discusses an approach to research and development for hill farming systems, while highlighting the related problems, and reviews recent developments, particularly in light of the experiences of PAC. While the work at PAC is continuing, the review of experiences to date may be valuable for researchers and development workers elsewhere.

AN OVERVIEW OF HILL FARMING SYSTEMS

The region contains diverse farming systems which are primarily subsistence patterns in nature. They change both within and across physical environments depending upon the resources available to the farmers and the needs of the family (Fig. 31.1).

In the High Himal (2500–5000 m), summer grazing of yak is practised on the tundra

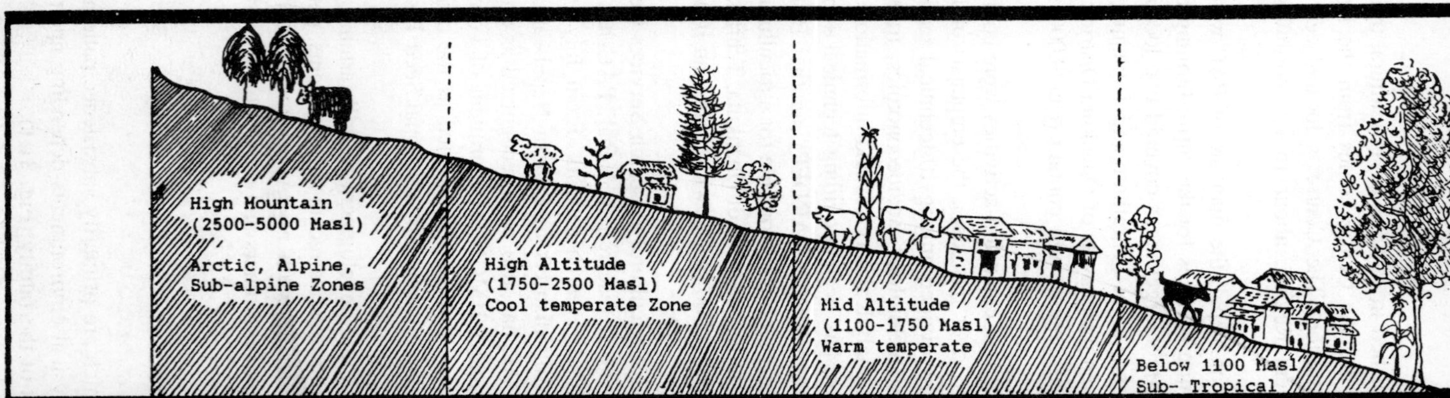

Crops	No arable crop grown, upper limit of summer grazing.	Potato+Maize, Barley, Bari, Wheat, Radish, Beans. Mostly one or two mixed crops/year	Rice-Wheat, Maize-Rice, Maize/Millet Maize+ground nut, cabbage, peas.	Rice-Rice-Wheat, Maize/Ground nuts, Oilseeds, Tomato, Chilli, Brinzal etc. Mostly 2-3 Crops/year.
Livestock	Yak, Sheep, Goat Mainly Livestock enterprises	Goats, Pigs, Chicken, Cattle, Buffaloes, Rabbits	Goats, Pigs, Chicken, Cattle, Buffaloes, Rabbits.	Goats, Chicken, Cattle, Buffaloes, Pigs, Rabbites.
Forestry/ Vegetation/ Fruits.	Conifer trees, Shrubs	Pines, Utis, Gogun, Apple, Pear Plum, Peach, Walnut.	Pines, Utis, Nebharo, Katus, Chilaune, Khanyu, Plum, Peach, Pear.	Sal, Acacia Plants, Khanyu, Banana, Guava, Sub-tropical Fruits, Citrus
Problems	Snow, Access to grazing Pastures.	Erosion, Deforestation, Frost, Hail Stone, Over grazing.	Draught, Soil fertility, Erosion, Hail Stone.	Erosion, Fodder, Desease, Pest.
Opportunity	Pasture management Livestock Development (Marketing, Milk Procucts)	Vegetable Relay/Intercropping. Orchard Stablishment. Livestock Development Forest Plantation.	Forage/Legume intensificaion in rice-fallow pattern, Water Harvesting, Farmers training for management (Livestock, Forest, Crops)	Agroforestry in bari and red soil area. Water Harvesting. Farmer training for management. (Livestock, Forest, Crops)

Figure 31.1: Transect of inter-relationship between land use and altitude in the eastern hills of Nepal

vegetation in open meadows. In these areas, sporadic and unreliable cropping may also occur.

In the high hills (1750–2500 m), the main staple crop is potato, frequently interplanted with maize. Barley and buckwheat are other important crops in this area. The climate is cool temperate, and there are *Quercus* forests.

In the mid-hills (1100–1750 m), a maize and finger millet cropping pattern predominates. Farmers keep a wide range of livestock, namely, cattle, pigs, buffaloes, and chickens. Rice-based cropping patterns are also practised in the mid-hills. However, rice-based cropping patterns predominate at low altitudes (< 1100 m).

In the low hills (< 1100 m), on *Khet* (irrigated lowland), the cropping intensity is very high with a rice-rice-wheat pattern which sometimes reaches 300 per cent. Many grain legumes such as beans, soybeans, cowpeas, and blackgram are interplanted with maize on the *Bari* (rainfed upland) and many of these legumes are also planted on the bunds of paddy terraces. Mustard is grown for its edible oil. Various vegetables, including cucumbers, pumpkins, aubergines, chilli, cauliflowers, cabbages, radishes, and broad-leaved mustard, are grown. Sub-tropical fruits such as citrus fruits, mangoes, bananas, guavas, and papayas are also found. Cash crops include cardamom, tobacco, and groundnuts. Ginger, garlic, and many other spices are also cultivated.

Livestock are an essential and integral part of farming systems. Cattle provide the main form of draft power, and manure is provided by all classes of livestock. The sale of animals and animal by-products is extremely important for small farmers; 55 per cent of the farm income is derived from livestock and small farmers actually earn more cash from this source than other middle and large-scale farmers (Conlin and Falk 1979). Crop by-products are used for animal feed. Recycling of the by-products is thus essential to maintain the integrated farming system.

Forests with a wide variety of vegetation occur in this region. In hill farming systems, the variety of trees and shrubs found on farmlands is of particular interest. These farmland trees and shrubs not only protect the soil from erosion but also provide the much needed fodder for animals, compost materials, timber and poles for construction, and firewood for household use. The intimate associations among the components of the system result in both positive and negative interactions (Fonzen and Oberholzer 1984). The direct interactions involve those among trees and field crops (soil conservation, shading of crops by trees), fodder trees and animals, and cattle manure and crops (Fig. 31.1).

Land holdings are very small and fragmented with 43 per cent of farm families having less than 0.5 ha of land. Moreover, small farmers usually have access to *Bari*, which is less productive than *Khet*. A single farm is rarely one contiguous piece of land, but four or five pieces scattered across more than one agro-ecological zone. This makes management difficult but also satisfies the diverse requirements of farming families and spreads risk.

PROBLEMS OF HILL FARMING SYSTEMS

Most farmers in the area are often isolated both physically and in terms of their knowledge of the opportunities that contact with both governmental and non-governmental agencies might bring. Some of the problems of hill farming systems are briefly described below.

Understanding Complex Farming Systems

In the hill farming system, farmers generally depend on a complex mix of crop, livestock, and forest products for their living. The farming system changes both within and across physical environments depending upon resources available to the farmers and the needs of their families. Stable crop production depends on livestock for manure and draft power, which, in turn, depend on crop residues, trees, and forest land for fodder.

Lack of understanding of complex farming systems, by the agencies involved in agricultural research and development, has been one of the most important drawbacks in generating suitable technologies for the hills. Most researchers and development workers have been brought up dealing with single disciplines. Therefore, an interdisciplinary farmer's complex farming system is not well perceived.

Appropriate Technology

Like most developing countries, Nepal has established a National Crop Development Programme (NCDP) and Livestock and Forestry Research Centres which are centres for generating technologies by the use of formal research methodologies, e.g., breeding and selection. Many of the materials and technologies have been 'brought in' from international research centres. New technologies resulting from research processes are then tested and disseminated to farmers through extension services.

As almost all NCDPs and research centres have been located in the *Terai* and on resource-rich sites, most of the technologies that have been developed to date are not suitable for hill areas where the majority of resource-poor farmers reside. For example, after more than 25 years of research work on the major cereal crops, such as rice, wheat, and maize, the productivity increase of grain crops has been minimal (Annex 1).

As with agricultural research, much of the research work being carried out on livestock is based on the introduction of exotic breeds. There are numerous examples of failure of exotic goat, sheep, cattle, and buffalo breeds in the hills. The main reasons are poor nutrition, limited feed resources, and the inability of exotic breeds to withstand indigenous environments.

Evaluation of suitable fodder trees and grasses is also being initiated. Much research work has been implemented on exotic species such as Ipil-Ipil (*Leucaena* spp.), Eucalyptus, and pines, while emphasis on the evaluation of indigenous grasses and fodder trees is still lacking.

Insufficient Emphasis on Use of Local Resources and Knowledge

At the farm level, especially in the remote hills, farmers have well-established systems of resource management and until recently have been able to provide most of the food requirements (Gibbon and Schultz 1988). With the increasing population, however, these systems are no longer self-sustaining and they now provide a declining proportion of food and other needs of the population. Seasonal off-farm portering or trading is now essential to restore deficits (Schroeder 1985). A further mechanism to adjust to this situation is semi-permanent or permanent out-migration. Nepal has a great wealth of local resources and traditional wisdom that needs to be fully exploited and understood.

Efforts, on the contrary, have been placed on the generation of technologies based on exotic and non-renewable materials. For example, the national strategies for crop research are based on the high input/output concept that came with exotic materials from the international research centres. New crop varieties are selected or bred using high levels of imported chemical fertilizers and pesticides. As Nepal is totally dependent upon imported agricultural chemicals, such technologies have been adopted by only a few resource-rich farmers in the *Terai* who have economic and political access, whereas the majority of the resource-poor hill farmers have not benefited (Chand et al. 1990). Due to uncertainty in the procurement and supply of imported chemicals, resource-rich farmers are also at risk. The present research system provides little emphasis on the comparison of local resources and knowledge in research strategies.

Depletion of Forest Resources and Soil Degradation

Hill areas are vulnerable to landslides and soil erosion that result in the loss of cultivable or grazing land. Such hazards hit poorer groups more severely as they usually have access to limited areas of land. The slips and erosion may be partly due to the removal of forest cover for fodder, fuelwood, timber, and leaf-litter, as well as to the over-grazing of upland vegetation, and partly due to the feature of general instability in the fragile hill ecosystem. This is considered to be one of the many hazards faced by farmers in the hill areas.

There is much evidence of soil degradation due to very intensive cultivation and failure to add an adequate amount of soil nutrients. Sherchand et al. (1990) have reported that a significant amount of soil is being lost annually due to the farmers' traditional method of maize cultivation.

One of the reasons for the farmers' lack of interest in the sustained management of forest resources in the past is said to have been the government's legislation; this has not been very sympathetic towards the farmers, particularly in relation to the use of forest resources. Although the legislation is now more sensitive to farmers' needs, there is a legacy of mistrust of the forestry staff and a lack of knowledge among the forestry staff that perpetuates this attitude (Gronow 1987).

PAC'S APPROACH TO ADDRESSING SOME OF THE PROBLEMS OF HILL FARMING SYSTEMS

Interdisciplinary Approach

PAC now believes that individual disciplinary research cannot successfully gain a holistic understanding of farmers' problems. Considering the subsistence or semi-subsistence nature of complex hill farming systems described above, development decisions are more complex than those encountered in the *Terai* systems, which more closely resemble commercial farming.

From the beginning, and with the establishment of various disciplinary sections at PAC, the Centre developed a farming systems approach to agricultural research and extension. Several sections developed techniques and approaches that indicate a sensitivity to the needs and environment of farmers, farmers' knowledge, and the need for multidisciplinary interaction in the identification and planning of research processes. As a result, programmes also underwent significant modification over time (Chand 1987, Khadka and Gibbon 1988, Thapa 1985).

The sections developed multidisciplinary work at different rates, depending upon the strength of individuals and available manpower support. However, there was inevitably a tendency to develop programmes in isolation from each other and to focus on those parts of farming systems that were perceived to be of high priority by individual disciplines. Although many important linkages were identified, and resulted in adjustments to programmes and recommendations, there was no mechanism for broadening a programme to examine these linkages in greater depth.

In order to address this rigidity in research programmes and to deal with complex farming systems in greater depth, in 1988 PAC developed a number of interdisciplinary working groups (Fig. 31.2).

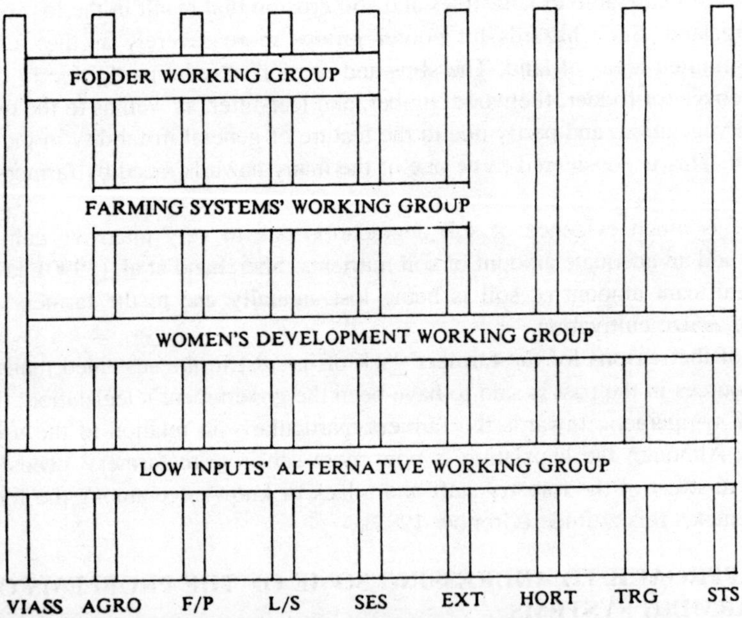

Figure 31.2: PAC's interdisciplinary working groups

Key: VIASS: Veterinary Investigation and Analytical Services Section; AGRO: Agronomy; F/P: Forestry and Pasture; L/S: Livestock; SES: Socioeconomic; EXT: Extension; HORT: Horticulture; TRG: Training; STS: Seed Technology Section.

Source: Pakhribas Agricultural Centre.

Interdisciplinary Working Groups

A working group is an interdisciplinary group of agricultural scientists, livestock experts, socioeconomists, and foresters. It tackles interdisciplinary research problems identified in the field through coordinated programmes. Each group has one coordinator and seven to eight core members representing the various disciplines concerned. There are currently four working groups at PAC dealing with major interdisciplinary problem areas. Their rationale, approach, and implications are outlined below.

Working Group on Low Inputs Alternative Technology

The work of this group evolved from an examination of the implications of high external input technologies that are recommended for farmers in the hills of Nepal. The group collects information regarding the knowledge of traditional farmers on the use of indigenous resources. It also conducts research on various methods of livestock and crop protection that rely on local materials, as well as research on the further development of composting and manuring technology and systems of resource management that concentrate on enhancing long-term stability rather than short-term production payoffs.

Working Group on Women

In common with other research systems, it took PAC many years to recognize the important role that women have in hill farming systems in Nepal. However, this group has been set up also because of the fact that a significant number of household heads are women (Thapa 1989). This group aims to collect detailed information about the involvement of women farmers in different aspects of farming, e.g., fodder and fuelwood collection, crop cultivation, livestock rearing, and decision-making processes. It is quite probable that this activity will have a useful linkage to the experience of community forestry development programmes and other works with groups who have common interests, and a linkage to recognized nutrition, food sufficiency, and cropping strategies (Cassels et al. 1987). The setting up of this group has highlighted the fact that the Centre has far too few female staff to work confidently in this field, considering the problems related to the further development of this work.

Working Group on Fodder

Livestock are an important link between the agricultural and forestry systems of hill farming. The organic manuring of fields, associated with animal husbandry, is an essential aspect of subsistence agriculture. In order to sustain this system, fodder supply, particularly at critical periods of the dry season, is most important. The group, therefore, provides a forum for coordinating work that until recently has taken place in three sections—agronomy, forestry, and livestock. It has a mandate to examine the important interactions in the flow of fodder and nutrients through the system. The work is focussed on the provision of fodder quality and quantity throughout the season and examines the implications of evidence for a decline in fodder availability and investigates potential for improvements.

Working Group on Farming Systems

The main objectives of this group are to collect and analyse information from secondary and primary sources on resource distribution and use in the hills, to characterize the major farming systems, to quantify labour and distinguish gender issues in labour use, to understand changes in the systems and how intra- and inter-household decisions are made, and to provide the understanding of farming systems by all PAC staff including HMGN staff in the eastern hills. The work of this group relies on the participation of all scientific and extension staff and the cooperation and support of farmer client groups. The approach is aimed at generating a more complete, dynamic picture of farming systems so as to provide more relevant information for future research programmes. The group encourages innovative approaches to the development of alternative technologies that could make much greater use of social science inputs and farmer knowledge in the planning, implementation, and evaluation of research programmes.

Samuhik Bhraman

Another interdisciplinary approach that has been found most effective for understanding farmers' needs and prioritizing research programmes is *Samuhik Bhraman*. *Samuhik Bhraman* (the Nepali name means 'group trek by a multidisciplinary team') is a method of initiating, prioritizing, and replanning research that has been developed in response to the particular environment of the hills of Nepal (Chand and Gibbon 1989). It is also considered to be an appropriate, rapid, informal interdisciplinary procedure that can be used to understand the farmers' environment (e.g., farmers' traditional knowledge, resources, and development capacity). *Samuhik Bhraman* enables multidisciplinary groups to interact with farmers in their own fields in order to better understand their problems and priorities. Details on the methodology of the *Samuhik Bhraman* and its role in eliciting farmers' interests and skills can be found in Chand et al. (1990).

Farmer-oriented Research Approach

PAC has been designated as one of the agricultural research and resource centres in the Eastern Region of Nepal. It conducts both on-station and on-farm research in crops, horticulture, forestry, and livestock. Most of the on-station research is coordinated with various NCDPs, horticultural centres, forestry research divisions, and livestock farms. PAC also designs and conducts its own on-station trials to meet its regional needs. Promising technologies are further tested in farm trials which are also a source of ideas for future research. On-station research at the Centre also takes due interest in farmers. For example, the agronomy section of PAC has been involving male and female farmers for the evaluation of crop breeding and varietal selection. The on-farm research of PAC emphasizes the fulfillment of the needs of farmers in the hills. The overall objective of the on-farm research programme is to produce recommendations for improved technologies which are of relevance to the majority of farmers in the hills. Detailed steps, adopted by PAC for conducting farmer-oriented research, are described below.

Site Selection

To implement on-farm, farmer-oriented research, the Centre initially worked together with the KHARDEP project to select representative sites, within the new research command area, using the cropping systems programme methodology. In 1983, three agricultural service centres were selected in each of the four Koshi Hill Districts. Each centre has a command area of four to seven *panchayats*, and in each area one representative *panchayat* was selected for on-farm farmer-oriented research (Fig. 31.3). Those representative *panchayats* were then subjected to site description surveys, using secondary data and key informant farmers' surveys to provide background information on available resources, infrastructural development, institutional involvement, existing common cropping patterns, and livestock practices.

In 1985, PAC adopted the concept of recommendation domains, and this helped to further define and establish research priorities at the on-farm, farmer-oriented research sites, e.g., some domains are more important than others (Kayastha et al. 1989). On-farm research trials are now designed for the three agro-ecological zones (high, middle, and low altitudes) and three land types (rain-fed upland, irrigated lowland, and rain-fed lowland). As each on-farm, farmer-oriented research site consists of a number of different recommendation domains, research staff have focussed on producing recommendations for the farmers within those particular domains.

As more information and experience is gathered from on-farm research sites, the recommendation domains are being constantly re-evaluated and refined. Once the sites were selected on the basis mentioned above, PAC followed certain procedures for setting research priorities. As the research needs identified by the KHARDEP survey could not be accommodated at one time, PAC started to prioritize the research needs of farmers based on wider information than the site description survey described above. The following procedures were followed for setting research priorities.

Setting Research Priorities

Land-use Mapping Data

These are derived from air photo interpretation and ground surveys, leading to the allocation of broad priorities to cultivated areas of particular crops and cropping patterns as well as the forest distribution type within a particular area. Annex 2 shows that 78 per cent of all land in the eastern hills grows at least one maize crop per year, compared to 33 per cent of the land which grows at least one rice crop per year. Of the rain-fed *Bari*, the most important cropping pattern is the maize/finger millet system (67% of all *Bari*). Therefore, the research priority is geared to the maize and maize/millet cropping system.

Feedback from the District-based Staff

The district-based staff of PAC, who normally spend 50 per cent of their time in the field, have been able to fine-tune the data from site description surveys. For example, field agronomists were able to clarify that among the reasons for farmers' adoption of cropping patterns, factors such as irrigation reliability, grazing problems, labour shortages, and availability of fertilizer were important. Similarly, livestock staff were able to identify the problems of fodder shortages, farmers' preferences, and prevalence of animal diseases.

Samuhik Bhraman

PAC believes that researchers cannot successfully gain and understand the farmers' perspectives until and unless they develop ways of learning from farmers and colleagues in the field. *Samuhik Bhraman*, as described above, is one of the strong methods of prioritizing research activities.

On-farm Research Methodology

PAC conducts various on-farm trials on livestock, e.g., animal breed comparison

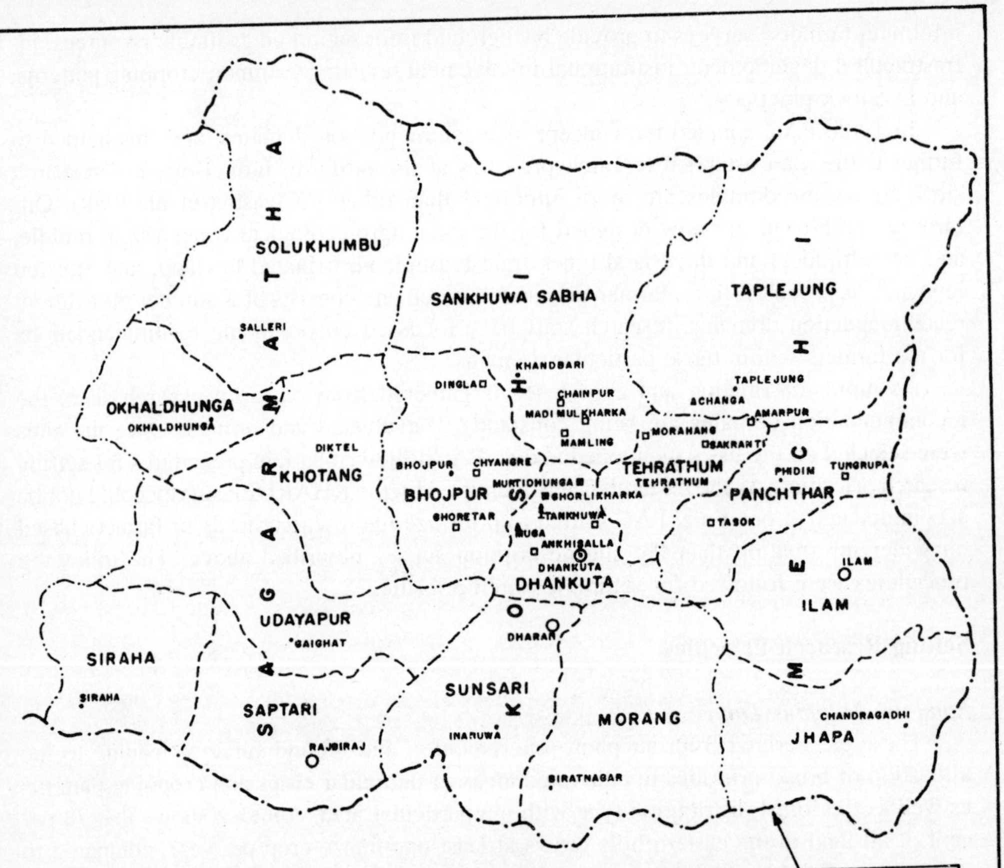

Figure 31.3: PAC's research sites

Source: Pakhribas Agricultural Centre

study and forage evaluation study. Similarly, the Forestry Section does various silvicultural and management studies in the farmers' fields. Crops are tested on small plots in farmers' field trials, in the beginning, and then better technologies are verified on larger plots, in pre-production verification trials. Most of the trials carried out on livestock, forestry, and crops are farmer-managed with guidance from the respective scientists. Details of on-farm methodology can be found in Oli (1987) and Chand (1987).

Group Approach

Over the years, based on its own experience, PAC has learned that there are some innovations that are adopted and disseminated quickly if the community or groups of farmers are made integral partners in the research and development process. Women farmers in kitchen garden programmes, the community approach to intensify rice-fallow systems, and the involvement of farmers' groups in community and private forestry programmes are some of the approaches instigated by PAC.

Women Farmers Group

In 1986, PAC initiated a women's development programme, initially on a limited scale, with local women farmers. Later on, as the programme expanded, home-based women motivators were recruited to run the programme. The women motivators assist in organizing women's groups and in conducting training. So far, women's groups have been formed in three *panchayats* of the local target areas. The committees are formed at the ward level where members are selected representing 7 to 10 neighbouring households. Among these members, one is chosen to represent the ward on the main *panchayat*-level committee. Officials for the *panchayat*-level committee, such as chairperson, vice-chairperson, secretary, and treasurer, are chosen from among the ward representatives. PAC does not intervene in the process, but leaves it entirely to the women farmers.

PAC has an indirect role in the committee of assisting in developing linkages with other offices or institutions within the districts. At the request of the committee, PAC provides the necessary inputs and expertise for training. However, training needs and other logistics are identified and managed by the group. The committee meets every month to discuss their problems and future programmes. The kitchen garden programme, which was implemented in seven *panchayats* of Dhankuta District, was the first thrust. Gradually, other skills such as fruit and vegetable preservation, seed selection and storage, small livestock keeping, and fodder planting were introduced. Women's participation in the programme has increased dramatically since the inception of the programme. More details can be found in Thapa et al. (1990).

Community Approach for Crop Intensification

The rice-fallow cropping system is one of the important cropping patterns in the eastern hills. About 10 per cent of the total cultivated rain-fed lowland comes under this pattern. Usually, after the rice crop the field is left fallow because of lack of sufficient moisture to grow a subsequent crop. It has been found that even some of the drought-tolerant crops, legumes, and forage crops cannot be grown successfully due to grazing pressure. The Agronomy and Livestock Sections have demonstrated that drought-tolerant legumes, such as chickpeas, and some of the forage crops, such as vetch, can be grown successfully with the involvement of a group of farmers. Grazing is avoided as the farmers involved feel responsible for saving the crop. In this way, farmers can obtain an extra crop and quality forage out of their fallow land after the rice harvest.

Approach for Group Involvement in Community and Private Forestry

The Forestry Section of PAC has successfully demonstrated that farmers' groups can be mobilized effectively for the establishment, protection, and management of community and private forests. The procedures followed for such an approach are: studying in detail the physical and social environment of the area; identifying local interest groups; understanding their needs and opportunities; and involving them right from the programme formulation stage, where decisions are taken by the farmers themselves. The innovative approach adopted by the Centre in planning, designing, and implementing community and private forestry programmes through community self-help groups (user groups) is considered to be one of its most successful programmes (more details can be found in Thapa et al. 1990). PAC has demonstrated that, given suitable encouragement and guidance, possibilities exist for the most successful protection and management of forests over large areas of Nepal through local cooperation. Forests in the hills now attract attention through approaches that involve the village people and communities and there is a growing optimism about eventual results.

Linkages with Different National and International Institutions

Although PAC is autonomous in its organization and management, it is technically integrated within the national government research network of agricultural and livestock farms and stations under NARSC and is recognized as a regional research centre by the NCDPs and the Forestry Research Division, Department of Forestry and Plant Research. PAC's location in the mid-hills and its sub-centres at high and low altitudes offer excellent hill testing sites for agronomical and horticultural crop varieties and husbandry trials organized by NCDPs and central-level horticultural farms. An excellent example of this cooperation with national programmes is the release of a new white maize variety, Manakamana-1, which was jointly developed by the research staff at PAC and the National Maize Development Programme. Similar work is in progress for wheat and rice.

PAC's technical section has established a close link with the district-based line agencies such as the Agricultural Development Office (ADO), the District Livestock Development Office (DLDO), and the District Forest Office (DFO). PAC'S Livestock Officers and Agronomists are posted in each of the Koshi Hill Districts. These technical staff assist ADOs and DLDOs on technical matters such as reporting and providing advice on the latest information on livestock and crop development and providing training support to their staff. The Forestry Section of PAC has already started a joint programme with the DFO on community and private forests. Similarly, the Horticultural Section of PAC has started a pilot vegetable production project jointly with the ADO near the district headquarters. The Livestock Section of PAC has taken the responsibility of supplying quality inputs to support the DLDO's programme. From 1985, the Agronomy Section, with the consent of the NCDPs, has taken the responsibility for the timely supply of well-proven crop minikit sets to meet the ADO's targets in the four Koshi Hill Districts.

PAC has a strong link with the National Farming Systems Research and Development Division Programme. It actively participates in the FSR working group meetings, and both the institutions send their staff to the *Samuhik Bhraman* organized by both institutions. PAC also maintains strong links with the work at the Lumle Agricultural Centre through exchange of information and frequent staff visits. PAC's on-farm research team has cooperated with the Mechi Programme (Netherlands Development Organization—Nepal). The Centre has been helping the Mechi Programme by supplying trial sets and minikits, and organizing field staff training (Chand et al. 1987). In turn, PAC is receiving feedback on crop performance. PAC also provides crop extension support to other institutions, both in the area and outside, such as Uttarpani Technical School, Small Farmers' Development Programme, British Welfare Centre, and Action-AID Nepal.

PAC has established good links for the exchange of research materials with international research centres such as International Rice Research Institute, International Wheat and Maize Improvement Centre, International Institute for Tropical Agriculture (Nigeria), International Potato Centre, Centre International de Agricultura Tropical (Colombia), and International Crop Research Institute for Semi-Arid Tropics through concerned NCDPs. The Centre has also established a working relationship with Centre for Tropical Veterinary Medicine (University of Edingburgh), Overseas Development Group (University of East Anglia), and Oxford Forestry Institute (University of Oxford).

HIGHLIGHTS OF SOME ACHIEVEMENTS

PAC, through its multidisciplinary farmer-oriented research activities, has been able to achieve considerable success in the resource management and diffusion of technologies to the farmers and has also been successful in bringing about changes in the attitude of the majority of farmers towards improved and sustainable farming.

Indigenous Resource Use and Management

One of the aims of PAC has been to encourage and promote the use of local resources for sustainable farming. The working groups of PAC started numerous research activities that emphasize the identification and use of local plant materials for soil nutrients and the control of crop and livestock diseases.

To find viable alternatives to the use of external inputs, PAC has conducted many on-station and on-farm research trials to examine the effect of locally available green manure on crop yield. It has also conducted studies on the possibilities of extending and, in some cases, re-establishing some of the traditional green manure materials. This has been particularly true in areas where heavy promotion and use of chemicals has resulted in a 'loss of knowledge' regarding local materials traditionally used as green manure (Chand et al. 1990). Preliminary research shows that the use of local green manuring plants has beneficial effect on rice yield. Some local green manures, such as *Adhatoda vesica, Albizzia lebbek, Hularrhwa antidysentrica,* and *Jatropa curcas*, produced 23, 17, 14, and 11 per cent more yield over the control (no green manure) and in some cases produced higher yields than the application of 60:30:0 NPK/ha (Sherchand et al. 1990).

Similarly, studies on the effect of Azolla on rice yield showed interesting results. The highest rice yield (4.5 t/ha) was obtained from 30 kg N/ha with Azolla incorporated (20 t/ha) as a basic application, and this gave 31 per cent increase over control.

Results from trials conducted at PAC to test different locally available plant materials to control *Sitophilus orazai* in wheat, indicated that local plant materials such as *Azadirachta indica* and *Zanthoxylum alatum* are as effective as 5 per cent Malathion dust.

Scientists at PAC have developed an oral drug from a wild shrub, locally called *Sihundi* (*Euphorbia royleana*). The drug has been found to be very effective in controlling *Ascaris summs* of pigs (Mahato and Rai 1988). Details on farmers' traditional knowledge of plant and animal disease control can be found in Chand et al. (1990).

Adoption of Technologies

PAC believes that any improved technology given to the farmers should perform well under their management. This is why PAC modifies the conventional research design into a form that accommodates farmers' practices or management. For example, the pre-production verification trial which was introduced by the cropping system programme, was designed originally to test new improved varieties to be used as a package with local varieties. PAC modified this into a 'diamond trial design' that splits improved packages into two components. The pre-production verification trial can now provide information on local versus improved packages, local versus improved varieties, local versus improved husbandry practices, and the effect of the interaction between variety and husbandry.

PAC has so far recommended 13 wheat, 11 maize, 12 rice, 3 potato, 5 barley, 2 millet, and about 17 different winter and summer legume crops for general production. Thapa (1985) reported that about 99 per cent of the farmers in the local target area cultivate improved wheat and more than 60 per cent grow improved maize. Over the last 15 years, PAC has been producing seeds of many crops and vegetables in its local and northern target areas. After the inception of the PAC/Koshi Hills Agricultural Programme seed programme in 1988, it started multiplying seeds of many improved cereal, legume, and vegetable crops with the involvement of private growers and local cooperatives in four Koshi Hill Districts (Annex 3).

While recommending any crop varieties and livestock breeds, PAC takes into greater consideration the socioeconomic conditions of farmers. For example, coarse, low-quality rice varieties (which give a feeling of fullness in the stomach) are mainly liked by resource-poor farmers who often have to do hard labour. On the other hand, resource-rich farmers prefer soft and long grain rice. PAC takes into account these factors while recommending any crop variety and also considers the importance of crop by-products for animal feed. Almost all rice varieties recommended by PAC produce higher or equal straw yields than the local varieties (PAC 1988). Seed rates of maize in on-farm maize trials are increased by 50 per cent over the national recommendation, since farmers use thinned plants to feed their animals and a higher plant population in the beginning provides insurance against poor germination and disease pest damage.

Due to the failure of most of the improved breeds of livestock, PAC started research work on the evaluation of cross-bred and indigenous animals. The PAC pig, a cross between an improved breed and the local black pig, produces 32 per cent more meat in a year than the local breed. Similarly, cross-bred poultry produce significantly more eggs and meat than the local breed under a semi-intensive system. After evaluating promising breeds of native and exotic goats, PAC came to the conclusion that the indigenous goats are best; they produce more meat than the exotic goats and are more tolerant to many diseases and stresses.

Development and Promotion of Community and Private Forestry

In Nepal, PAC is one of the organizations that has approached private tree planting in a structured manner. The approach chosen by the Centre to promote this programme involves more than just the distribution of seedlings. It includes raising trees and crops together, planting fodder and fuelwood blocks on land that would otherwise be unproductive, and the continuous monitoring of the programme. Over the years, more than 1,000,000 seedlings of fodder, fuelwood, and timber species have been distributed for private planting, and survival rates as high as 60 per cent have been achieved. As a result, more than 1,000 fodder and fuelwood blocks have been established. Thirty-four farmers have already registered their private forests and more applications have been received by the District Forest Office.

Various species of fodder grasses and legumes have been screened, evaluated, and introduced in the villages for planting on terrace risers (which constitute about 25% of the total cultivated land area), intercropping in fodder and fuelwood blocks, and in community forests. Yields as high as 1,700 kg of dry matter per hectare per year of *Setaria anceps* have been achieved by the farmers. This has not only increased farmers' fodder resources but has also helped to stabilize terraces, reduce soil erosion, and restore land fertility.

Since 1983, PAC has been providing support to the District Forest Office for the initiation of community and private forestry programmes through the formation of user groups. Major focusses of the programme are development and promotion of appropriate methodologies for implementing community and private forestry, through the user group approach, in collaboration with concerned agencies, and the long-term study of these user groups, in terms of their participation in forest protection, management, benefit-sharing, and sustainability.

Soil and Nutrient Loss Assessment

It has been observed that excessive amounts of soil and nutrients are lost annually from cultivated land. PAC, for the first time, initiated a soil and nutrient loss assessment programme. The programme involves comparing different cultivation practices to find out the most appropriate method of crop cultivation.

Preliminary results of the study on maize cultivation showed that soil loss can be reduced significantly (at 17 t/ha) if maize is sole-cropped and sown on slopes with minimum tillage and mulch, in comparison to the normal practice (maize as the sole crop), which caused high soil loss at 36 t/ha. Similarly, nutrient losses, especially nitrogen and organic matter content, were found to be highest under the local cultivation method (Sherchand et al. 1990).

CONCLUSIONS AND RECOMMENDATIONS

PAC, since its establishment as a Centre for providing agricultural training and extension support to a limited number of British ex-servicemen, has expanded over quite a large area, accommodating a larger number of farm families residing in the eastern hills of Nepal. Over the years, due to its continuous effort to improve the farming systems of hill farmers, PAC has been recognized by the Government of Nepal as an important hill research centre for the eastern hills. Its link with national and international research centres has enabled PAC to develop technologies that can be transferred to other areas. Lack of understanding of complex farming systems, by many research organizations, has led to the development of technologies that have not been fully adopted by the majority of hill farmers. PAC, through its multidisciplinary farmer-oriented research programmes, has been able to incorporate farmers' needs, resources, and knowledge in the research programme in order to develop long-term sustainable farming systems. PAC has also been able to employ a group approach in forestry, crop, and women's development programmes. In order to bring about more desirable changes in the hill farming systems, the following recommendations have been made.

- There is an urgent need to deal with the fertility and soil conservation issue. The first step for this would be to quantify actual fertility and soil erosion problems in the farmers' fields. For this, there is a need to conduct area surveys; understand farmers' interests, constraints and opportunities; and document the methods that farmers are traditionally following to minimize erosion and soil degradation problems.
- The conventional high input/output concepts presently employed in Nepal, based on external inputs, should only be concentrated in the *Terai* and other accessible areas and should not be employed in the hills where land holdings are small (average 0.5 ha) and most households live at subsistence level. Separate research programmes and strategies should be developed for the hills which take into consideration the local needs, tap traditional knowledge bases, and maximize the exploration and use of local resources for increased and sustainable food production.
- The success of any forestry programme in Nepal depends largely on effective farmer participation at all stages of programme implementation. Recently, much has been written on the importance of working with the farmers, but little has been written on how this can be achieved within the context of community and private forestry programmes in Nepal. PAC experience suggests that, if such schemes are to be successful, local groups with common interests must be clearly identified. Considerable time must be spent in the field with farmers exploring their needs and opportunities so that their requirements are met. Identifying such groups is a difficult task. However, experience has shown that there are existing resource user groups and other social groups which can provide an ideal point of entry into the local community. The way in which such groups are approached is of immense importance. To get the most out of farmers' participation requires a great deal of skill, and many of those who are involved in forestry development may not have such skills. New skills and field procedures will have to be developed to enable this type of scheme to succeed. This is an important area in which a greater concentration of inputs and resources is required.
- At present, there are many forest regulations that affect the use of trees on both gov-

ernment and private land. Many of the regulations are difficult to interpret due to frequent amendments. These regulations should be simplified and adequate information disseminated to the farmers in order to encourage community and private forestry on a wider scale.

- It is observed that technology developed or identified in one part of the Hindu-Kush area may be transferred successfully to another part where the environment is similar. There is evidence that some crop varieties developed in the northern hills of India can be grown successfully in the eastern hills of Nepal. Similarly, some livestock species brought from Himachal Pradesh, India, have adapted well to the eastern hills. International institutions such as ICIMOD can play a vital role in exchanging research materials for and organizing exchange visits with researchers.

There is a need for research scientists to build up a wider scope rather than being confined to and trained in single disciplines. This is an area in which ICIMOD can play an important role through organizing training for the researchers.

- It is recommended that ICIMOD establish formal links with hill agricultural research centres such as PAC and LAC in order to facilitate the further development of hill farming systems.

REFERENCES

Basnet, S.R. *The 1988/89 Seed Multiplication and Production Programme in the Koshi Hills*. PAC Working Paper No. 6, 1990.

Cassels, C, A. Wijga, M. Pant and D. Nabarro. *Monitoring and Evaluation of the* KHARDEP, *Nepal*. Results of the KHARDEP Impact Studies (1980–85), 1987.

Chand, S.P. *On-Farm Crop Research in the Hills of East Nepal: The Experience of Pakhribas Agricultural Centre*. Proceedings of the 18th Asian Rice Farming System Working Group Meeting. Pakistan, Aug; 30–Sept. 4, 1987.

Chand, S.P. and D. Gibbon. '*Samuhik Bhraman*: A Rapid and Appropriate Method of Prioritising and Replanning Agricultural Research in Nepal.' Paper presented at the Ninth Annual Farming Systems Research/Extension Symposium, Arkansas, USA, Oct. 8–11, 1989.

Chand, S.P., B.D. Gurung and P.G. Rood. 'Farmer's Traditional Wisdom: Where Does It Stand Within the Present Agricultural Research System of Nepal.' Paper presented at the International Workshop on 'Sustainability Through Farmers' Involvement in Technology Generation, New Delhi, Feb. 6–10 1990.

Chand, S.P., R.K. Shrestha and R.B. Katwal. *From Farmer's Field to Farmer's Field: A Possibility of Extrapolating Koshi Hill Crop Recommendation to the Mechi Hills*. A report written for the Mechi Programme (Netherlands Development Organization—Nepal) to develop Research Strategies and PAC's involvement, 1987.

Conlin, S. and A. Falk. *A Survey of the Socio-economy of the KHARDEP Guidelines for Planning an Integrated Rural Development Programme*. KHARDEP Report No. 3., 1979. Department of Food, Agriculture and Marketing Service, HMG Nepal, Annual Report, 1987.

Fonzen, P.F. and E. Oberholzer. 'Use of Multipurpose Trees in Hill Farming Systems in Western Nepal' *Agro-Forestry System*, 2:187–197, 1984.

Gibbon, D. and M. Schultz. *Agricultural System in the Eastern Hills of Nepal: Present*

Situation and Opportunities for Innovative Research and Extension. PAC Technical Paper No. 108, 1988.

Gronow, C.J.V. 'Developing Forest Management in Nepal.' *Banko Jankari*, 1(4):37–44 FRIC, Kathmandu, 1987.

Hildreth, G. Land Resources Mapping Project. Summary Report of Kenting Earth Sciences, Ltd. Ottawa, Canada: KES, 1986.

Kayastha, B.N., S.B. Mathema and P.G. Rood. *Organisation and Management of On-Farm Research in the National Agricultural System*. ISNAR OFCOR Case Study No. 4, 1989.

Khadka, R.J. and D. Gibbon. 'The Contribution of Pakhribas Agricultural Centre to the Development of Farming Systems in the Eastern Hills of Nepal.' Paper presented at the 8th Annual Farming Systems Research/Extension Symposium. Arkansas, USA, Oct. 9–12, 1988.

Mahato, S.N. and K. Rai. The Efficacy of *Euphorbia Royleana* as an Anthelminthic against *Ascaris summs* Infection in Pigs. *PAC Veterinary Newsletter*, Vol. 1, 1988.

Oli, K.P. On-Farm Research Methodologies for Livestock Development at Pakhribas Agricultural Centre. Revised version of a paper presented at the Second Farming System Working Group Meeting. Kathmandu, June, 14–17, 1987.

PAC. *Review of Achievements* (1972–87). Kathmandu: PAC, 1988.

Schroeder, R.F. 'Himalayan Subsistence Systems: Indigenous Agriculture in Rural Nepal,' *Mountain Research and Development*, 5(1):31–44, 1985.

Sherchand, D.P., S.P. Chand, Y.B. Thapa, T.P. Tiwari and G.B. Gurung. 'Soil and Nutrient Losses in Runoff on Selected Crop Husbandry Practices on Hill Slope Soils of Eastern Nepal.' Unpublished paper prepared for presentation at the International Symposium on Water Erosion Sedimentation and Conservation, Dehra Dun, 1990.

Thapa, B., L. Joshi and S.L. Sherpa. *A Community Scheme to Encourage Private Tree Planting in the Hills of Nepal*. PAC Technical Paper No. 134, 1990.

Thapa, H.B. 'Agriculture Extension and Training in PAC Target Areas.' Paper presented at the Workshop on Agriculture in the Hills of Nepal, Nov. 26–28 1985.

Thapa, H.B. 'Women in Extension—A Group Approach to the Extension of Pakhribas Agricultural Centre.' Paper presented at Institute for Agriculture and Animal Sciences (Rampur, Nepal)/U.S. Agency for International Development Second Annual Workshop on Women in Farming, Oct. 27–29 1989.

ANNEXURES

ANNEX 1: Area, Production, and Productivity of Main Cereal Crops

Table 31.1. Area production, and productivity of rice from 1968 to 1989

Fiscal year, year in BS	Year in AD	Area (000 ha)	Production (000 mt)	Productivity (kg/ha)
2025/26	1968	1,162	2,178	1,874
2026/27	1969	1,173	2,241	1,910
2027/28	1970	1,183	2,304	1,945
2028/29	1971	1,201	2,344	1,952
2029/30	1972	1,140	2,010	1,763
2030/31	1973	1,227	2,416	1,969
2031/32	1974	1,240	2,452	1,978
2032/33	1975	1,256	2,605	2,074
2033/34	1976	1,262	2,386	1,891
2034/35	1977	1,264	2,283	1,806
2035/36	1978	1,263	2,339	1,853
2036/37	1979	1,254	2,060	1,642
2037/38	1980	1,276	2,464	1,932
2038/39	1981	1,297	2,560	1,975
2039/40	1982	1,265	1,833	1,449
2040/41	1983	1,334	2,757	2,066
2041/42	1984	1,377	2,709	1,968
2042/43	1985	1,391	2,805	2,016
2043/44	1986	1,333	2,372	1,779
2044/45	1987	1,423	2,982	2,096
2045/46	1988	1,423	3,200	2,250

Source: Department of Food, Agriculture and Marketing Service, HMG Nepal. Annual Report, 1987.

Table 31.2. Area production, and productivity of wheat from 1968 to 1986

Fiscal year, year in BS	Year in AD	Area (000 ha)	Production (000 mt)	Productivity (kg/ha)
2025/26	1968	208	233	1,119
2026/27	1969	226	265	1,173
2027/28	1970	228	193	846
2028/29	1971	239	223	933
2029/30	1972	259	312	1,204
2030/31	1973	274	308	1,126
2031/32	1974	291	331	1,137
2032/33	1975	329	387	1,178
2033/34	1976	348	362	1,039
2034/35	1977	366	411	1,123
2035/36	1978	356	415	1,166
2036/37	1979	367	440	1,199
2037/38	1980	392	477	1,218
2038/39	1981	400	526	1,315
2039/40	1982	484	656	1,358
2040/41	1983	471	634	1,343
2041/42	1984	451	535	1,181
2042/43	1985	483	598	1,239
2043/44	1986	535	701	1,310

Source: Department of Food, Agriculture and Marketing Service, HMG Nepal, Annual Report, 1987.

Table 31.3. Area production, and productivity of maize from 1968 to 1986

Fiscal year, year in BS	Year in AD	Area (000 ha)	Production (000 mt)	Productivity (kg/ha)
2025/26	1968	422	765	1,814
2026/27	1969	433	795	1,836
2027/28	1970	446	833	1,869
2028/29	1971	439	759	1,730
2029/30	1972	446	822	1,845
2030/31	1973	453	814	1,795
2031/32	1974	458	827	1,805
2032/33	1975	453	748	1,652
2033/34	1976	445	797	1,790
2034/35	1977	445	740	1,664
2035/36	1978	454	743	1,635
2036/37	1979	432	554	1,281
2037/38	1980	457	743	1,629
2038/39	1981	475	752	1,581
2039/40	1982	511	718	1,405
2040/41	1983	504	761	1,511
2041/42	1984	579	820	1,417
2042/43	1985	614	874	1,421
2043/44	1986	627	868	1,384

Source: Department of Food, Agriculture and Marketing Service, HMG Nepal, Annual Report, 1987.

ANNEX 2
Estimates of the Area (ha) of Net Cultivated Land under Different Cropping Patterns in the Eastern Hills

Cropping pattern	Zone: Sagarmatha Hills	Zone: Koshi Hills	Zone: Mechi Hills	Total Eastern Hills: ha	Total Eastern Hills: % of total
Bari					
maizefallow	162	290	1,226	1,680	1%
potato + maize	4,611	9,132	10,245	23,988	9%
maize + soya	3,219	3,984	2,690	9,894	4%
maize-cereal	4,619	1,753	4,932	11,304	4%
maize/millet]					
maize-millet]	39,204	44,521	30,980	114,704	45%
maize-mustard	802	3,046	1,555	5,403	2%
Total *Bari*	53,014	65,631	51,700	170,346	67%
Khet					
rice-fallow	2,128	16,983	5,677	24,788	10%
rice-wheat	5,311	13,034	8,395	26,740	11%
maize-rice	8,226	7,066	11,490	26,783	11%
rice-rice-fallow	626	1,884	1,154	3,664	1%
rice-rice-wheat	387	724	571	1,681	1%
maize-rice-wheat	346	252	309	906	0.4%
Total *Khet*	17,024	40,046	27,493	84,563	33%
Total *Bari* + *Khet*	70,038	105,677	79,193	254,908	100%

Source: Hildreth, *Land Resources Mapping Project Sciences Ltd. Summary Report of Kenting Earth* Ottawa, Canada: KES, 1986.

ANNEX 3

Targets and Achievements of the PAC/KHAP Assisted Seed Programme during the Fiscal Year 1988/89

Districts	Crops	Varieties	Total area (rop)	No.of farmers involved	Collection target(kg) (kg)	Collection achievement	Remarks
	Maize	*Manakamana*-1	33	9	600	361	
	Wheat	RR-210	240	46	12,000	8,200	
	Soyabean	Lumle-1	6	4	140	76	
	Millet	Okhle-1	6	2	150	75	
DHAN-	Rayo	KBL/MBL	8	10	188	14	
KUTA	Cress	Nepali	18	21	450	122	
	Peas	S/Local	29	41	1,140	148	
	F/Bean	K/Wonder	7	10	290	22	
	Broccoli	G/Sprouting	2	4	60	27	
	Radish	JWN	30	23	900	296	
	Maize	Manakamana-1	33	7	975	100	AIC target
	Maize	H/Composite	30	2	900	251	is not
	Wheat	RR-21	80	19	4,000	3,737	included.
BHOJ-	Soyabean	Lumle-1	3	1	75	51	
PUR	Radish	Mino Early	100	40	2,500	410	
	Rayo	KBL	21	20	300	75	
	Peas	S/Local	5	5	200	76	
	F/Bean	K/Wonder	2	2	100	16	
	Maize	Manakamana-1	22	3	550	715	
S/SABHA							
	Maize	Arun-1	8	1	200	NA	
	Wheat	RR-21	320	48	16,000	14,100	
	Peas	S/Local	40	20	2,000	300	
	Maize	Mana-kamana-1	9	3	450	NA	
	Maize	H/Composite	15	4	550	NA	
	Wheat	RR-21	176	27	11,000	7,070	AIC target
	Soyabean	Lumle-1	8	13	206	24	is not
	Rice	P/Masino	8	4	400	106	included.
	Radish	Mino Early	64	43	1,750	368	
TEHRA-	Radish	White Neck	—	—	—	64	
THUM	Rayo	KBL	16	18	485	323	
	Rayo	MBL	2	1	40	7	
	Peas	S/Local	6	9	300	32	
	F/Bean	K/Wonder	2	5	100	12	
	Cress	Nepali	12	12	300	86	
	Broccoli	G/Sprouting	1	1	50	57	

Source: Basnet 1990.

32

AGRICULTURAL RESEARCH EXPERIENCES OF THE ARID ZONE RESEARCH INSTITUTE (AZRI) IN BALUCHISTAN, PAKISTAN

J.D.H. Keatinge and B. Roider Khan

J.D.H. Keatinge is the Principal Scientist, International Centre for Agricultural Research in the Dry Areas (ICARDA). Quetta, Pakistan, and B. Roider Khan is the Director of Pakistan Agricultural Research Council's (PARC) Arid Zone Research Institute (AZRI), Quetta, Pakistan.

CONTENTS

INTRODUCTION

This paper deals with problems specifically associated with, but not necessarily limited to, the operation of a research programme to aid agricultural management in mountainous and highland areas. These problems include the requirement for quantified primary information in the design of a successful research programme, the advantage of adopting a partnership arrangement in providing the necessary long-term continuity of a research programme, the need for timely reappraisal of strategic thinking after sufficient results become available, and the importance of ensuring that the impact of the research process, particularly in socioeconomic matters, is evaluated in a timely manner and takes the known climatic variability fully into account.

The problems associated with improved agricultural production in the arid and semiarid mountainous areas (> 1000 m) of West Asia have received little or no effective attention from research workers in the past. In raising small ruminants, traditional farming systems have depended on either a nomadic or a transhumant mode with some sedentary subsistence agriculture based on winter wheat production. However, with the rapid increase in human population, experienced recently in West Asia, and in anticipation of the continuation of this trend into the foreseeable future, the governments of countries such as Pakistan have become oriented towards change. They are now relying upon increased and sustained agricultural production from less developed agricultural areas, such as the arid highlands, to help meet expected food production deficits in the future (GOP 1988).

AGRICULTURAL RESEARCH PROBLEMS SPECIFIC TO DRY MOUNTAIN AREAS

The arid mountainous areas of West Asia share the common ecological characteristics of cold, heat, and drought, which pose a set of specific problems to the researchers' goal of increasing and sustaining agricultural output. At the core of these problems is the interaction between moisture supply and temperature. In order to grow crops with consistent success in such an environment, it is necessary to attune maximum crop development and moisture use with the brief period in which the influence of temperature constraints is minimized. Furthermore, research designed to monitor independently the direct effects of environmental variables, such as heat, cold, and aridity, will often result in a gross oversimplification of their influence. Rather, it is vital to consider the combined interactive process of the physical environment on crop growth.

One major environmental factor, which is characteristic of arid highland areas and has a substantial influence on both crop and animal productivity, is the extreme inter- and intra-year climatic variability which can be experienced (e.g., Figure 32.1). It may be difficult to design an agricultural system that can be functional in a dry but consistent environment, but it is much more demanding to design a flexible system for a predominantly dry environment that is also capable of responding to considerable variability in annual rainfall and its distribution, as well as the timing, duration, and intensity of frost, heat, and other climatic events. Unfortunately, arid highland areas generally have environments that fall into the latter category. If research programmes are to be effective in such areas, environmental uncertainty and risk factors have to be fully taken into account.

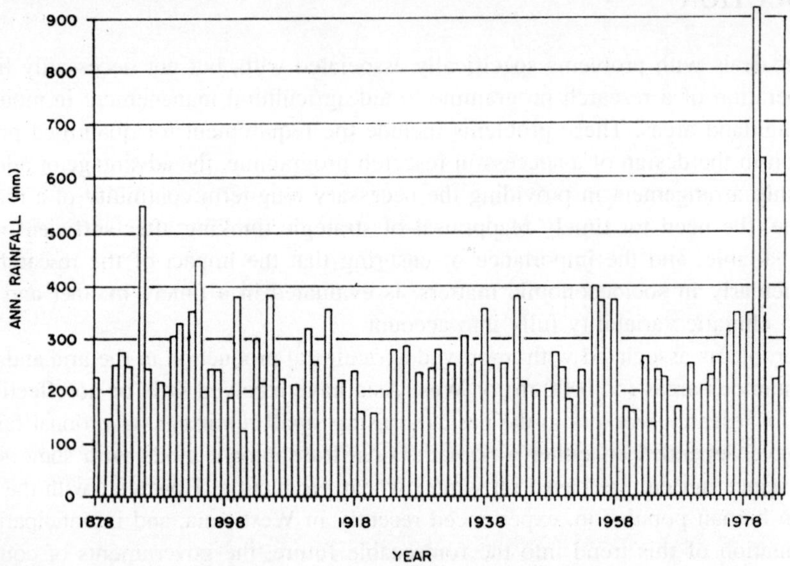

Figure 32.1: Annual rainfall(mm) at Quetta in 1878–1986

Source : Keatinge and Ress, 1988

OPERATIONAL IMPLICATIONS: THE EXAMPLE OF HIGHLAND BALUCHISTAN

In order to operate an effective research programme in a particular dry mountainous area, it is necessary to appreciate that, although there may be climatic, social, and infrastructural problems that are common to many such areas, the actual research operation occurs in a very specific physical and human environment. AZRI's experience suggests that research programmes need to be initially focussed on specific problems with the expectation of generalizing at a later date, rather than dealing with general issues and hoping for specific local spin-off effects. Therefore, an intimate knowledge of the system for which the research is targeted is a primary requirement before research plans can be properly formulated.

In this paper we discuss the operation and impact issues arising from AZRI's research programme in highland Baluchistan. We seek to demonstrate that the research strategy adopted by AZRI is beginning to have appreciable results that may have a substantial impact by increasing productivity and reducing the annual uncertainty of agricultural output in the highlands. In addition, we hope to show that the applied research partnership adopted by ICARDA and AZRI may be useful examples which could be effectively copied in other areas.

THE AGRICULTURE OF HIGHLAND BALUCHISTAN

In the arid highlands of the Baluchistan Province of Pakistan, the production of sheep and goats on approximately 20,000,000 ha of rangeland and subsistence production of winter wheat on a further 125,000 ha of land are the predominant dryland agricultural activities. Irrigated fruit production is also important in economic terms but this is only carried out on a very limited area. Annual productivity of both crops and ruminants is very irregular and generally low, because of the harsh and agriculturally marginal climatic conditions.

Most of the people in the area are almost completely dependent on agriculture for their livelihood, and the population has approximately quadrupled in size in the last 40 years. This has led to a corresponding increase in the population of sheep and goats from around 1 million head in 1950 to more than 18 million head in 1986. These numbers are well above the economically efficient and ecologically sustainable carrying capacity of the rangelands (ICARDA 1990), and, thus, severe overgrazing, made worse by shrub gathering for fuel, has occurred in a widespread and sustained manner. As a result, the productivity of the now highly degraded rangeland vegetation is very low (Figure 32.2), and, in consequence, annual off-take from small ruminants is sub-optimal. Thus, the standard of living of dryland agricultural communities is inevitably low and, compared with the population of Pakistan as a whole, the farmers of highland Baluchistan are at a great disadvantage.

This deteriorating and distressing situation is probably common to other West Asian countries with similar ecological conditions, such as Iran and Afghanistan. However, ICARDA, in partnership with the Pakistan Agricultural Research Council's Arid Zone Research Institute, implemented a joint research programme in the highlands of Baluchistan in 1985 which seeks to address the problems of low and unsustained dryland agricultural productivity.

DESIGN, CONTINUITY, DIRECTION, AND IMPACT OF AGRICULTURAL RESEARCH PROGRAMMES

Design: The Need for Quantitative Primary Information

In the development of a research strategy for dryland agriculture, a shortage of reliable quantitative information is a major drawback in ensuring the effective investment of scarce human and financial research resources. Decisions regarding which research problems should receive priority are critical and often require extensive survey work before a realistic long-term strategic plan can be formulated (Keatinge et al. 1988).

For example, in highland Baluchistan, winter wheat is the principal dryland crop grown, while barley covers only 5 per cent of the cereal area. On the surface, this decision by farmers appears not to be very rational in economic and biological terms, as barley, traditionally a more efficient and reliable crop in areas of less than 300 mm annual rainfall, should be more consistently productive in the highlands of Baluchistan as the average annual rainfall there is closer to 200 mm (Kidd et al. 1988). Results from the AZRI agronomy trials over the last five years have supported the supposition that barley would be a more productive crop to grow than wheat.

Therefore, should AZRI be concentrating its resources on developing increased barley production? In light of the chronic and increasing deficit in animal feed in Baluchistan, the answer to this question is probably yes. However, AZRI's survey work with cereal farmers and ruminant herders has highlighted the complexity of this issue. Farmers have

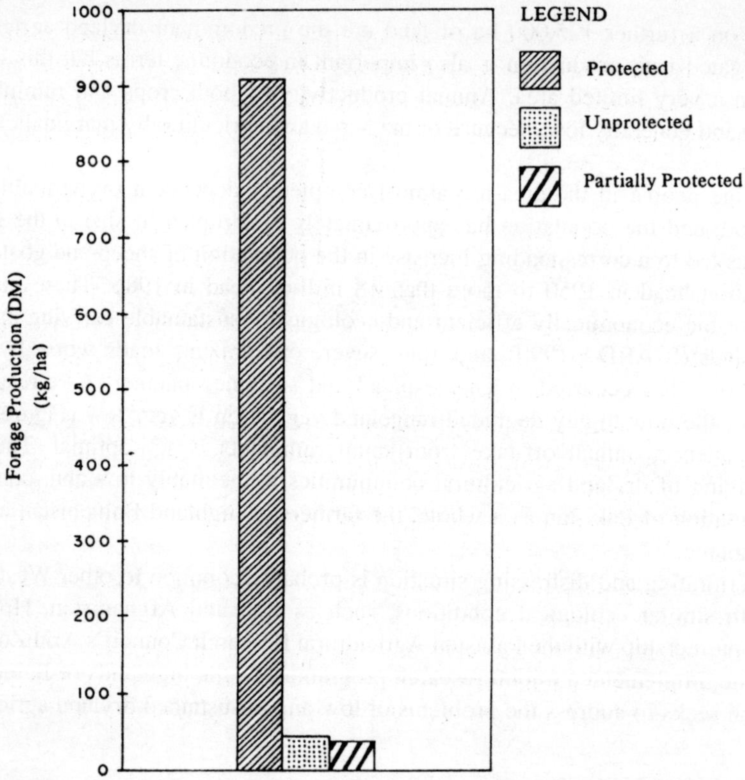

Figure 32.2: Productivity of native range vegetation (kg/ha/yr) at Tomagh, Pakistan

Note: protected treatments are means of six exclosures which have been ungrazed for three years. Partially protected treatments have been ungrazed for two six-month periods in the last three years under a rotational grazing scheme.
Unprotected treatments experience normal grazing pressures from local flocks.

Source: AZRI. 1990

indicated that, without a yield premium of at least 50 per cent, they would be unwilling to shift from wheat to barley production. Their reasons include a continuing desire for personal food security and the small and uncertain market for barley (Table 32.1). This, in turn, is linked to the need for farmers and herders to appreciate the value of small amounts of supplemental feeding, as demonstrated by AZRI, to improve considerably potential ruminant off-take (Table 32.1). Unless this occurs, the market demand for barley, which would be an ideal, locally produced source of animal feed supplement, cannot create a sufficient incentive to persuade farmers to change their traditional cropping pattern.

Continuity: The Advantage of a Partnership Arrangement

Once a series of research experiments has been identified as having high priority for a research institute's programme, it is vital that they be continued for a sufficient time period to yield meaningful results. This time period is, of course, variable depending on the type of experiment; but it is safe to assume that in uncertain and marginal environments,

Table 32.1. Ranking of reasons why farmers do not grow more barley in Highland Baluchistan[a]

Farmers' reasons for not growing more barley	Top Three Rankings[b]		
	First reason	Second reason	Third reason
	% farmers responding		
Grow wheat for food security	97	3	0
No secure barley market	0	47	21
Land shortage	2	16	38
Prefer wheat to barley straw	0	13	9
Wheat price higher than barley	0	13	6
Orchards are more profitable	0	5	20
Barley yields less than wheat	1	2	2
Labour shortage	0	1	2
Not profitable to feed barley	0	0	2
Total	100%	100%	100%

[a] Khuzdar, Kalat, Kachhi, Quetta, Pishin, Loralai, and Zhob Districts.

[b] Farmers were asked to choose and rank the three most important reasons as to why they do not grow more barley.

Source: Nagy et al. 1989.

Table 32.2. Reproductive performance of Harnai ewes at Tomagh

1.	Date of breeding	September 15, 1987
2.	Number of ewes at breeding	81
3.	Number of ewes at lambing	81
4.	Number of ewes having lambed	76
5.	Number of lambs aborted	0
6.	Number of lambs born	76
7.	Number of lambs dead (0–90 days)	1
8.	Number of lambs weaned (90 days)	75
9.	Fertility rate (4)+(5)/(3)	94%
10.	Prolificacy (6)/(4)	100%
11.	Weaning rate (8)/(6)	99%
12.	Number of lambs weaned per ewe put to breeding	0.93
13.	Number of oestrous availed by the ewes to conceive One—75, Two—6, Three—0, Four—0	
14.	Services/conception	1.14
15.	Average gestation period (days)	155
16.	Sex ratio (male:female)	1:1.2

Source: Atiq-ur-Rehman et al. 1989.

such as those of the arid highlands, more experimental years will be needed to ensure meaningful results than, for example, irrigated agricultural systems. A long-term breeding or selection programme, for example, with an expected duration of at least 5 to 10 years of very dry or extremely cold years in which crop failure is certain, can have a negative impact on progress. As these years of harsh environmental conditions are commonly experienced in the arid highlands, how are replacement seed reserves of selected material to be preserved to ensure continuity of the selection programme from every growth year? A disrupted research programme, in experience, is often either abandoned or in future accorded much lower priority. This has important implications for the attainment of an institute's research strategy.

In AZRI's case, its research partnership with ICARDA helps to ensure vital programme continuity and to reduce research costs in Pakistan. For example, AZRI uses ICARDA's international nurseries, which are specifically targeted at highland areas, as the initial basis for its selection nurseries and yield trials for a range of crops. Evaluation of material selected in the first year of testing is then carried out at multiple locations over several years. Seed supply is always a critical determinant of the number of test locations which can be employed in any given year, particularly when lines are being promoted from single-rowed observation nurseries to multi-rowed replicated yield trials. If a very harsh year occurs, as in highland Baluchistan in 1987/88, crop failure at all locations ensues due to extreme drought. Should AZRI then lose all its experimental selected seeds, it would negate three years' progress in selection. However, one way in which ICARDA helps the AZRI programme is by acting as a seed supply reserve from its headquarters in Aleppo, Syria, so that more seeds from selected material are almost always available in experimental amounts for the purpose of programme continuity in the harsh seasons. The alternative of AZRI's duplicating each nursery under irrigation not only would be very expensive but also would not fully guard against the risk of crop loss if there was a sudden cold spell.

Direction: How Results Should Influence Strategic Thinking

Once sufficient information is collected and a long-term strategy is devised in which research priorities are clearly defined, it is important to realize that change in the relative weighing of priorities can, and probably will, occur. This may well be in response to the perceived timing and size of the predicted impact of an intervention. This will obviously alter as the research process generates results and it becomes evident that one research approach is proving to be fruitful more immediately (or more certainly in the long run) than another.

For example, in 1986 AZRI initiated a series of rangeland management and rehabilitation trials at its sub-stations in Tomagh and Zarchi. These included large-scale management trials of native ranges, with and without complete or partial protection (deferred grazing), and small-scale rehabilitation areas in which new species of fodder bushes were introduced. As late as 1988, in AZRI's research plans, the grazing studies were accorded a higher priority than the range rehabilitation experiments (AZRI and MART/AZR 1988).

However, this situation was reversed by late 1989 as it transpired, from our experimental results, that the regrowth rate of native range vegetation was extremely slow even with full protection, deferred grazing as a management technique appeared to have little promise, and social grazing-protection techniques were less effective than had been previously contemplated (Figure 32.2). In direct contrast, range rehabilitation techniques, through the creation of small areas of fodder bushes (*Atriplex canescens*), indicated that the introduced species was well adapted to the environment and could be extremely productive up to 6 t/ha/yr of dry matter (Aro et al. 1988). Furthermore, in subsequent growth and digestibility trials with local sheep, the introduced fodder was found to be suitable for use as a winter supplement in maintaining body weight and to have potential for flushing ewes before mating (Atiq-ur-Rehman et al. 1990). In addition, social control of grazing on these planted forage reserves near farmers' houses and other well-defined sites seems to be much more effective than simply making improvements on the general condition of the range.

Thus, the creation of fodder reserves as a range rehabilitation intervention seems likely to have a much greater potential impact and be more readily accepted than the potential interventions in range grazing management proposed in 1990. As a result, the AZRI's research programme in the planned period from 1990 to 1993 is now heavily biased in the direction of range rehabilitation, without necessarily cancelling the grazing management experiments. These have been reduced in scope, as well as in time requirement, in order to accommodate these trials by scientists (often the limiting factor at AZRI). In the setting of priorities and investment of resources, a judicious balance between long- and short-term experimental return must be maintained, but, within this balance, change can and should occur at appropriate intervals when the importance of additional research results has been digested. This need for a shifting order of priority should not, however, be regarded as a licence to make wholesale changes in an experimental programme on an annual basis. These changes should occur only when there is sufficient data to fully warrant amendment of strategic thinking, perhaps at two- to four-year intervals.

Impact: The Need for Socioeconomic Determination over an Adequate Time Period

In ensuring that a research programme will have sufficient impact to maintain or, better still, increase research support levels for the arid highlands, care must be taken to evaluate the social and economic impact of an intervention over an adequate time period. We have already stressed the importance of inter-year climatic variability on the research process. This is also of considerable importance in the evaluation of farmer acceptability and impact in both social and economic terms.

For example, since 1986, AZRI has been testing improved catchment basin water harvesting systems for improved crop agronomic management (Figure 32.3). In the period between 1986 and 1989, both wet and dry years have occurred, and the economic benefit of taking some land out of production for water harvesting will obviously vary in the amount and distribution of grain in any one year. However, the crop year 1987/88 was very dry throughout most of Baluchistan and a large majority of farmers failed to get grain from their wheat fields. This will probably recur in two to four years out of ten, depending upon the specific location (Kidd et al. 1988).

In contrast, in AZRI experimental treatments of water harvesting techniques, some plots yielded enough grain (< 100 kg/ha) to return sufficient seed for farmers to plant in the following year as well as a good straw-yield for animal feed. Such experimental results in a very dry year will have a big impact on the collaborating farming community, as consistency of yield across the years is almost a more important parameter than extra yield in a favourable season, because domestic food security considerations for family and livestock are critical (Table 32.1). However, in strictly economic terms, averaging experimental results across the years may de-emphasize the results of a specific year such as 1987/88 and, as a result, underestimate potential research impact. The need, therefore, for continuous assessment of impact over a number of years and the inclusion of the known element of climatic variability are essential in the socioeconomic as well

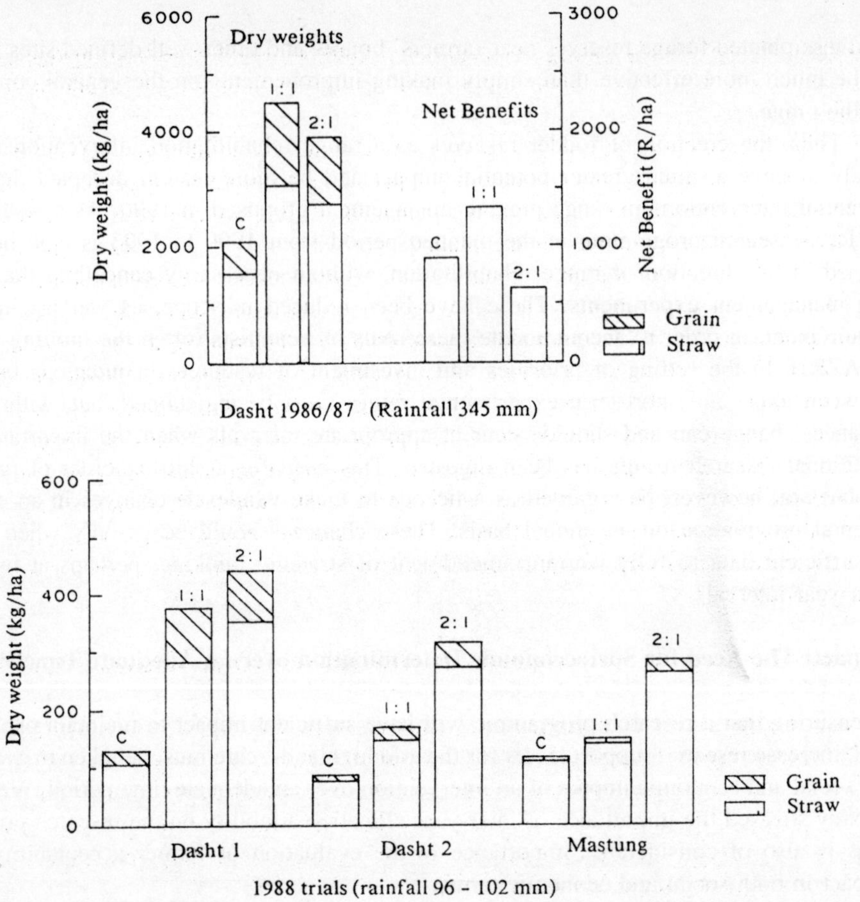

Figure 32.3: Productivity and net benefits of water harvesting treatment with wheat at Dasht and Mastung, Pakistan, in the 1986/87, 1987/88 and 1987/88 crop growth seasons

Source: Rees et al. 1989.
Note: 1US$ equalled approximately 18 Pakistani Rupees.
Treatments
1:1 = Half treated catchment:half cropped area
2:1 = 2/3 treated catchment:1/3 cropped area.

as the impact calculations, in order to avoid premature recommendations and injudicious investment in widespread adoption.

CONCLUSIONS

We have attempted to demonstrate in this paper that, in the design, conduct, and impact of a research programme, for the dry, mountainous environment special consideration needs to be given. An attempt to transplant into the highlands a research programme, derived from experiences of a less harsh lowland environment, is likely to be ineffective. The problems of mountainous environments are both complex and highly specific. In consequence, agricultural research, although urgently in need of further investment, may be too daunting a task for national governments to fully accept the challenge of providing the required level of human and financial resources. Initially, partnership arrangements between national and international agencies, such as ICARDA and ICIMOD, can be a fruitful way to demonstrate research impact on a small scale, in order to convince governments to adequately support the research required for the development of the disadvantaged agricultural communities in the arid highlands.

REFERENCES

Arid Zone Research Institute. *Annual Report for 1989*. Quetta: AZRI, 1990.

Aro, R.S., M.I. Sultani and M. Asghar. *Introduction of Fourwing Saltbush* (Atriplex canescens) *into Degraded Rangelands in Upland Baluchistan*. MART/AZR Res. Rep. 22. Quetta: ICARDA, 1988.

Atiq-ur-Rehman, R.S. Aro, M. Asghar and M.I. Sultani. *Effect of Flushing and Late-gestation Supplementary Feeding on Fertility and Productivity of Harnai Ewes*. MART/AZR Res. Rep. 27. Quetta: ICARDA, 1989.

Atiq-ur-Rehman. S. Rafique, A. Ali and M. Munir. *Nutritive Evaluation of Fourwing Saltbush in Growth and Digestibility Trials with Harnai Lambs in Highland Baluchistan*. MART/AZR Res. Rep. 58. Quetta: ICARDA, 1990.

AZRI and MART/AZR. *Research Plans for 1988-89*. MART/AZR Res. Rep. 14. Quetta: ICARDA, 1988.

Government of Pakistan. *Report of the National Commission on Agriculture*. Islamabad: GOP, 1988.

International Centre for Agricultural Research in the Dry Areas (ICARDA). *High-Elevation Research in Pakistan*: The MART/AZR project annual report for 1989. ICARDA Res. Rep. 158EM. Aleppo: ICARDA, 1990.

Keatinge, J.D.H. and D.J. Rees. *An Analysis of Precipitation and Air Temperature Records in the Quetta Valley, Pakistan : The Implications for Potential Improvement in Agricultural Productivity*. MART/AZR Res. Rep. 9. Quetta: ICARDA, 1988.

Keatinge, J.D.H., B. Roidar Khan, D.J. Rees, R.S. Aro and C. Tallug. *A Strategic Plan for the PARC Arid Zone Research Institute 1990–2000*. MART/AZR Res. Rep. 23. Quetta: ICARDA, 1988.

Kidd, C.H.R., D.J. Rees, J.D.H. Keatinge, F. Rehman, A. Samiullah and S.H. Raza. *Meteorological Data Analysis of Baluchistan*. MART/AZR Res. Rep. 19. Quetta: ICARDA, 1988.

Nagy, J.G., G. Farid Sabir, N.A. Shah, M. Afzal, D.J. Rees and J.D.H. Keatinge. *Barley Production and Its Scope for Improvement in the High Elevation Rain-fed Farming Systems of Baluchistan*. MART/AZR Res. Rep 26. Quetta: ICARDA, 1989.

Rees, D.J., M. Islam, A. Samiullah, Z. Qureshi, S. Mehmood, F. Rehman, J.D.H. Keatinge, H. Raza and B.R. Khan. *Water Harvesting and Nitrogen Fertilizer Applications as a Means of Increasing Crop Water Use Efficiencies in Sub-optimal Conditions in Upland Baluchistan*. MART/AZR Res. Rep. 31, Quetta: ICARDA, 1989.

Rees, D.J., Z.A. Qureshi, S. Mehmood, F. Rehman and S.H. Raza. *Catchment Basin Water Harvesting as a Means of Improving the Productivity of Rain-fed Land in Upland Baluchistan*. MART/AZR Res. Rep. 40, Quetta: ICARDA, 1989.

33

AGROFORESTRY AS AN OPTION FOR MOUNTAIN AGRICULTURAL DEVELOPMENT

J. Denholm and N.S. Jodha

Jeannette Denholm is a professional staff member of the Mountain Farming Systems Division of ICIMOD and N. S. Jodha is the Head of that Division.

CONTENTS

INTRODUCTION

Mountain Agriculture: Weakening Regeneration Processes

Despite increased public investment and interventions, mountain agriculture in most parts of the Hindu-Kush-Himalaya region is showing symptoms of decline. This decline is manifested by both stagnant or falling productivity levels and the diminished health of the production base. One of the basic reasons for the decline is the rapid erosion of the natural resource base of agriculture. In the past, mountain agriculture, broadly defined to include different land-based activities such as crop production, horticulture, forestry, and animal husbandry, was sustained largely through effective linkages among the above components. The latter helped in food and other biomass supplies, nutrient cycling, and regulation of moisture flows, essential for the stability of agriculture in mountain habitats. The organic integrity of mountain agriculture, based on the linkages and interactions of the above-mentioned components, has suffered due to increased population growth, market forces, and public interventions. The mechanisms (as well as immediate manifestations) of organic disintegration include reduced area under forests and other perennials, reduced diversity of agricultural production systems, and increased runoff as well as soil erosion. In various ways, these changes obstruct natural regenerative processes that are essential for the health and productivity of the natural resource base of mountain agriculture. The new agricultural technologies based on high yielding varieties (HYVS), in spite of their ability to raise crop productivity under favourable circumstances, offer no solution to the above problem. The resource-centred technologies do generate some hopes in this context (Jodha 1990c). Based on an understanding of the potential of resource-centred technologies to meet the twin goals of productivity and conservation, agroforestry is suggested as one means to contribute to the sustainable development of mountain agriculture.

What is Agroforestry?

A technical practice that is commonly found in many traditional farming systems throughout the world, including mountain regions, is that of cultivating annual and perennial crops on the same landholding. This strategy, as an intentional manipulation of the land base to satisfy subsistence needs for food, fodder, fuelwood, and/or soil protection, is today called an agroforestry system. A definition of the term that is widely accepted is that of Lundgren (1982):

> . . . an approach to land use in which woody plants are deliberately combined on the same land management unit with herbaceous crops and/or annuals either in some form of spatial arrangement or in sequence. The concept of an agroforestry system implies both ecological and economic interaction among the components of the system.

Agroforestry encompasses a wide variety of land-use systems to realise increased productivity and more dependable economic returns while promoting the conservation of soil, plant, and water resources. A classification system proposed by Nair (1985) is based on four criteria: structure, function, agro-ecological setting and socioeconomic scale, and management levels of the system. From these factors, systems can be grouped as agrisilvicultural, silvopastoral, and agrisilvopastoral.

Shifting cultivation, home gardens, *taungya* afforestation, alley cropping, and wind breaks and live fences are sometimes considered as additional categories of land-use systems (ICRAF 1983).

The agrisilvicultural system is defined by multi-storey, multipurpose cultivation of agricultural, horticultural, and tree crops on a single field. Intercropping (alley cropping), shelterbelts, production during fallow stages on shifting cultivation sites, shade trees planted on commercial plantations, and fruit or fuelwood trees grown on field boundaries are examples of this system.

When tree species are grown in combination with forage crops or pasture for livestock feed production, the system is termed 'silvopastoral'. Examples are animal production systems in which multipurpose woody perennials provide fodder (and fuelwood) or act as living fences around grazing or croplands. Trees retained for shade, fruit, timber, or fuelwood on pasture lands are also examples of this system.

Most commonly encountered in the HKH region, however, is the agrisilvopastoral system which combines elements of perennial and annual crops with animals or pasture. The crop/tree/livestock mix typically found around homesteads presents an example of this. The practice of grazing animals on fallow croplands, establishment and retention of fodder trees around fields, and use of woody shrubs and hedges for mulch, browse, and green manure point to the multipurpose uses of this diverse system.

INDIGENOUS AGROFORESTRY

As a public intervention, agroforestry may be of recent origin, but, historically, the system has evolved over a much longer time. Planting of annual crops within forest areas (as under *taungya* cultivation) and rotating crops with forests (as under shifting cultivation) are two well-known examples of spatially and temporally focussed agroforestry systems. However, the most important form of agroforestry observed in mountain areas is the planting or retention of trees and shrubs with annual crops.

Two key factors that have contributed to the evolution of this last form of agroforestry system are: (1) the crucial role of biomass from perennials in farming systems and in the sustenance strategies of farmers, and (2) the changing patterns of availability and access to the tree products, following increasing pressure on lands producing perennials (e.g., common property lands).

Conditioned by these two factors, rural communities in mountain areas (and elsewhere too) have evolved different arrangements to ensure the availability of perennial plant-based biomass to complement their supplies from annual cropping to meet their production and consumption needs (Jodha 1988). These arrangements are ultimately reflected in the relative allocation of land to perennials and annuals in an overall village (or community) context and individual farm context. The spatial dimension of these arrangements (e.g., a block of community forest land or maintaining trees on private field borders), proportions of space for perennials and annuals and relative composition of naturally growing or planted species vary according to changing availability and productivity of land, availability of alternative options (e.g., chemical fuels or fertilizer as against their biomass-based counterparts), and economic and technological changes influencing land use.

Agroforestry (planting perennials along with annual crops) in a way represents a compromise between extensive land use (possible under a low population pressure situation) and intensive land use (forced by the current demographic situation). Besides, this helps to ensure the availability of and access to biomass in the absence of vast uncultivated lands.

Farmers' Strategies: Focus on Biomass

To understand the fundamental importance of perennials (from forests), which necessitated their incorporation in private farming through agroforestry, a brief discussion on farmers' resource management and production strategies will be useful.

Fertility Maintenance

An essential component of traditional farming systems in mountain areas has been the emphasis on maintaining soil fertility through internal resource regeneration (Whiteman 1988). Without the input of chemical fertilizers, the fertility levels of the light and shallow mountain soils are heavily dependent on the addition of farmyard manures and forest litter.

Farmers have devised numerous ways to replenish soil nutrients and protect fields from the loss of soil through erosion. Terracing is the most visible modification that farmers of the HKH have made on hill slopes for soil conservation and cultivation. A crop rotation system with a fallow period allows the regeneration of soil health by ploughing in crop residues and allowing for a resting period. The high altitude farmers of Bhutan burn straw that has been placed on croplands to provide potash, destroy weeds, and warm the soil before ploughing and sowing. Intercropping legumes with crops is another technique used to increase fertility; soil is conserved through more complete plant coverage and root activities.

However, by far the most widespread and significant strategy for maintaining soil fertility is through the transfer of nutrients from forests and grazing lands via livestock. The fodder element is the linchpin in the farming systems of the hills and mountains. In the form of grass and fodder leaves, biomass is recycled through cattle, sheep, goats, buffaloes, and yaks to produce the valuable manure on which crop productivity depends. Crop residues also contribute to feed resources, and forest litter is incorporated into the compost as well, but fodder from trees and grasses constitutes well over 50 per cent of the diet of livestock on an annual basis. Seasonal fluctuations in supplies result in livestock being fed mostly green feed in the summer months and crop residues after the harvest season.

Provision of Products

In addition to the provision of tree and grass fodder, forests are often the sole suppliers of timber and fuel. Far from the urban centres, where kerosene, charcoal, and electricity are available, mountain families depend on another free resource from the forest—fuelwood. Even in the sparsely vegetated arid highlands of Pakistan, most families must struggle to collect fuelwood from far-off 'forests' consisting of scattered trees. Grass peat is the fuel used by West Sichuan mountain households.

Bhutanese farmers supplement their otherwise simple diet of rice, millet, and chillies with mushrooms, tubers, and fruit from the forests. Many mountain residents depend on herbal remedies from forest plants to cure illness. Fencing, thatching, and construction materials all come from communal forest lands (even land legally owned by the government is here termed community land, based on its de facto use status).

Hedging against Risk through Diversification

With the primary objective of hedging risks and securing the provision of basic food supplies, a basic strategy employed by farmers is to diversify production. Crop/livestock-based, mixed farming, backed up by heavy use of forest and pasture resources, facilitates the flow of varied products throughout the seasons. Besides supplying organic fertilizer and draft power, livestock are used for milk, meat, eggs, butter, cheese, and hides. A very important advantage of livestock-raising for the small farmer is the fact that animals and products can be sold for cash at any time. Vegetables and fruits may also present cash-earning opportunities, but the transport of these to markets is more problematic for farmers without access to roads. Moreover, the mobility of livestock means that a farmer's means of subsistence are not entirely tied to local conditions with their accompanying hazards of drought, hailstorms, or cold temperatures.

The operation and success of farmers' above-mentioned strategies are closely linked with the complementary use of private property resources (e.g., for private farming) and common property resources (CPRs) (e.g., CPRs such as community forests and grazing lands which ensure a supply of biomass not available on one's own farm). Hence, the decline of the commons as a source of biomass may adversely affect farmers' resource management and production strategies (Jodha 1990d). This is happening at a rapid rate as discussed below.

PROCESS OF CHANGE

The biomass supplies and services crucial to the operation of farmers' above-mentioned strategies are dependent upon the availability of forest and grazing lands as common property resources. The changes in the status of the latter have significantly influenced the farmers' afore-mentioned strategies.

Transformations Affecting Farmers' Strategies

Strategies and practices developed and used by mountain farmers under conditions of low demand for resources are becoming increasingly unfeasible and ineffective. Population growth, market forces, and public interventions have placed new pressures on the fragile resources of mountains and their natural environment. This, in turn, is leading to the unsustainability of the whole mountain farming system (Jodha 1990b).

Population Growth

The most important component of the process of change is population growth and its consequences for farmers' traditional resource management strategies. Population growth has strained the farmers' traditional strategies in several ways. Most importantly, in various direct and indirect ways, it has contributed to a reduction in the area of forests and pastures. Crop cultivation has been slowly extended to a large proportion of these areas. Even through over-grazing and over-extraction of supplies from the remaining CPRs, people are not able to meet the increased demands of the increased population. Thus, from the biomass supply side, farmers' strategies are adversely affected by population growth.

From the demand side, increased population has forced farmers to employ more intensive use of their small landholdings. This calls for yield-raising inputs. Yield-raising chemical fertilizers are costly and non-available in most cases; the other option is to increase manure production. This needs additional animals, additional labour to feed and graze them, and, most importantly, additional feed material or space to graze the animal. This amounts to asking the impossible in the face of the already stated decline of area and productivity of CPRs and reduced fallow periods of croplands. In the context of the biomass situation, farmers are faced with the difficult choice of whether to select HYVs, which can provide an increased amount of food for human beings, or traditional varieties with high stalk-grain ratios that are therefore valuable also as sources of fodder.

Backlash of Development Interventions

Farmers' biomass-centred strategies have further suffered due to the backlash of modern development processes which are quite indifferent to the utility and value of biomass in farming systems. For instance, due to the strong focus of public support systems on limited crops and limited crop attributes (e.g., grain production mainly), the diversity of agriculture has been reduced. Emphasis on commercialization and cash crops has also reduced the importance of crop by-products and biomass in general.

The cash nexus has further induced farmers to over-extract the remaining sources of biomass (forests and pasturelands) for commercial purposes. This delinks the forest from the resource regenerative processes crucial for mountain farming systems. Evidence of over-extraction is visible everywhere. Sparse natural lands in the North West Frontier Province (NWFP) of Pakistan have been stripped bare through such extraction. Similarly, the trekking industry in Nepal has provided a ready market for fuelwood. Trekking tourists consume twice the amount of wood used by their porters (ERL 1989).

The introduction of modern crop technologies involves components (e.g., hybrid seed, fertilizers, pesticides, even tools and equipment) that are usually external resource-centred. Hence, they have little sensitivity to local resource regeneration and biomass stability. The importation of mechanical technologies designed for lowland conditions can prove harmful to fragile environments. Tractor cultivation, introduced in Bhutan (and other parts of the HKH Region as well) for cash crop production, for instance, increases erosion and discourages the planting of trees and bushes in fields and on field borders (Gupta 1986).

Transformations Affecting Social Arrangements

Historically, the stability of biomass supplies was also a result of the proper management and regulated use of CPRs. The traditional institutional mechanisms to enforce the same have slackened with the process of development.

Under conditions of overcrowding and overuse of resources, traditional management systems to regulate forest cutting and pasture grazing have broken down in many areas, notably in India and Pakistan (Jodha 1990d and Mulk 1990). The inability of community leaders to enforce regulations and obligations on users to contribute taxes and labour has resulted in uncontrolled exploitation and poor maintenance of these resources.

As the hunger for land grows, it is almost impossible for the boundaries of communal forests, pastures, and marginal lands to remain unaffected. Encroachment on these lands has transformed their use patterns to those of crop cultivation. Privatization of sources of natural biomass for crop cultivation has been encouraged by public land policies (Jodha 1990d). The consequence of privatization is the loss of access for others to fuel and fodder supplies. This affects the rural poor more than any other group.

Apart from the large-scale privatization of CPR sources of biomass, public policies have also led to disruption of their management. In several cases, public interventions have been directed to replacing local village authority by formal administration from above. For instance, community forestry legislation in Nepal, by handing over plots of State-owned forest land to panchayats, has disrupted the existing forest management systems formed by smaller groups of local users. By providing a cash salary to forest guards, the government has superseded the authority of the user-group which had previously paid him in-kind. The guard becomes answerable to government forestry officials instead of to local users.

Through cooperation among all households and the feudal rulers (*mirs*) in Pakistan's northern areas, local institutional arrangements were evolved for the management of pasture and irrigation systems. The enormous labour investments required to maintain these systems, particularly hill irrigation systems, were possible only through sanctions and their enforcement by the *mirs*. The decline and abolition of their authority over the last four decades have led to a parallel decline in the effectiveness of village-level institutions (World Bank 1990). A similar situation has been observed in parts of India (Jodha 1990d).

These examples are indicative of the alienation that the State has created among villagers from their common property resources. Nationalization of forest and pasturelands has converted CPRs into government property resources. Unless communities can be assured that the products of such lands will be available to them to meet their biomass needs on an equitable basis, government initiatives to encourage community management will be subverted (Agarwal and Narain 1990).

AGROFORESTRY AS AN OPTION

The above discussions show that the integrative role of biomass was the key element in the farmers' traditional strategies. Under a situation of demand on mountain resources, these strategies helped to sustain agriculture.

However, the recent forces of change have disrupted the institutional (and technological) arrangements conducive to uninterrupted supplies of biomass for effective operation of farmers' strategies.

The traditional systems, involving land-extensive and subsistence-oriented activities, are less feasible. The objective circumstances require increased use intensity of land resources. New crop technologies can help in this. But, because of the very nature of fragile land resources, increased use intensity without resource conservation/protection (possible through biomass, nutrient cycling) may prove counter-productive. Hence, the need for an approach that can blend land-intensive and land-extensive uses. In other words, a need for options that can ensure the 'biomass-generating functions' of forests and pastures without having sufficient land under these use categories. It is in this context that agroforestry emerges as one of the potential options to reverse the emerging unsustainability trends in mountain agriculture.

In the context of mountain areas, agroforestry can not only help in performing some functions of the traditional CPRs, but it can also fit well into the circumstances characterizing mountain habitats. In fact, it is the latter attribute of agroforestry which makes it the more appropriate option for mountains in today's context of land scarcity, conservation needs, and pressure for higher production. To elaborate on this we may briefly discuss specific characteristics of mountain areas vis-a-vis the attributes of agroforestry as a system of land use and cropping.

Mountain Specificities and Imperatives for Agroforestry

Farmers in mountain areas occupy a habitat that imposes a set of constraints and opportunities which differ widely from those of the lowland areas. Physical isolation, distance, transportation difficulties and costs, climatological and environmental hazards, limited production opportunities, and diverse agro-ecological conditions are some of the factors affecting mountain farming systems. These can be grouped under the following categories of mountain specificities as outlined by Jodha (1990a): inaccessibility, fragility, marginality, diversity, 'niche', and human adaptation mechanisms.

These mountain characteristics, to be elaborated upon later, generate a number of objective circumstances. Any intervention that helps in handling these circumstances or has the potential to benefit from these circumstances is likely to be more relevant and effective in the mountain areas. Table 33.1 illustrates this point with reference to agroforestry as an intervention for the development of mountain agriculture. In column one, the table lists the important operational implications of different mountain specificities. In column two, relevant attributes of agro-forestry as an activity are listed. The information helps to see the degree of convergence between requirements imposed by mountain conditions and production-cum-development possibilities associated with agroforestry. Table 33.1 also lists a few imperatives of the above-mentioned degree of convergence for agroforestry interventions. The latter can be used (as attempted towards the end of the paper) for assessing the relevance of recent agroforestry programmes in mountain areas. The issues mentioned above can be elaborated with reference to individual mountain specificities.

Agroforestry, based on the complementarity of annuals and perennials, has potential for local resource regeneration and recycling. It reduces dependence on external inputs also. Accordingly, it greatly matches with the requirements of relatively inaccessible habitats.

However, to benefit from the potential of agroforestry in inaccessible areas, the focus on choice of species and programme management (e.g., through a decentralised approach) will have to change. Such imperatives can be worked out in detail for specific locations.

Inaccessibility

The most obvious feature of mountain areas, inaccessibility, is a result of altitude, slope, terrain conditions, and periodical seasonal hazards, such as snow, storms, and landslides, that impede mobility. Isolation, distance, poor communication logistics, relative closedness of the system, with a tendency towards limited dependence on external inputs, and focus on local regeneration and recycling of inputs and products are important manifestations of this mountain specificity.

Agroforestry, as a system of land use, potentially facilitates the harnessing of complementarity between annual and perennial plant-based activities and helps in local resource regeneration processes with little external dependence. This potential can be more effectively materialized, provided both technological and institutional measures under agroforestry interventions are focussed accordingly. Table 33.1 indicates the broad directions for this purpose.

Table 33.1. Areas of convergence between implications of mountain specificities and attributes of agroforestry

Mountain specificities/ implications	Relevant attributes of agroforestry	Imperatives for agroforestry interventions
INACCESSIBILITY		
Closed system, poor transport logistics; limited external input dependence, focus on local resource use recycling.	Complementarity of annuals and perennials; local resource regeneration; nutrient/moisture cycling.	R & D focus on species with maximum complementarity; least dependence on external input/ market. Decentralized operational approach to agroforestry programmes.
FRAGILITY		
Rapid resource degradation with use intensification; limited production options, low resource productivity.	Higher resource use intensity with conservation through annual-perennial (tree, shrub) inter-cropping, balancing extensive-intensive usages.	Focus on tree species with high soil-binding/soil-building potential; and positive (nutrient-, moisture-, shade-related) impacts on crops. Location specific differences in programmes.
MARGINALITY		
Problem soils (due to low fertility, unmanageable topography, etc.), low productivity; people's poverty, risk aversion, low resource base, subsistence orientation	Usability of micro-level (plot) variability of resource base; gradual and upgrading resource/productivity; multiplication and diversification of production options for risk reduction, with limited out-of-pocket cost; contribution to input product recycling.	Focus on species with wider adaptability; multiple uses, varied maturity/harvest cycles; potential for soil building, resistance to physical hazards. 'Target group' orientation of programmes.
DIVERSITY		
Biophysical heterogeneity as basis of diversified interlinked activities, source of stability and resilience of farming systems.	Potential for activities/ products/land uses with non-covariate flows of products and input needs; diverse and multiple products and interlinked second-level activities; fuller use of micro-environment, seasonality, etc.	Focus on species with wider adaptation, multiple uses, variable characteristics (maturity, input needs, rationing, tolerance to stresses, etc.); decentralisation, area/group specific changes in programme contents,

Table 33.1 Contd.

'NICHE'		
Mountain areas' greater suitability for specific activities/ products as source of comparative advantage.	Scope for harnessing environmental/resource-related uniqueness of mountains through use of specific perennials by incorporation in inter cropping.	Search for and use of species unique to mountains, focus on quality and performance improvement, their adaptation to inter-cropping situation. Programme focus on high value, processing, and marketing potential of agroforestry components.
ADAPTATION MECHANISMS		
People's resource management and sustenance strategies directed to stability of food and biomass through interlinked, diversified land-based activities, collective management recycling; and adjustments to collapse of traditional arrangements (e.g., common biomass).	Conducive to diversification of interlinked, land-based activities with multiple products and services; neutral to scale of operation; group action no precondition for success.	Focus on species, productivity, etc. to be guided by both subsistence and market needs; farmers' adaptation to wider market economy requires element for market orientation of agroforestry programme.

Source: Adapted from N.S. Jodha, 'Economic Development in the Himalaya: The Missing Mountain Perspective' (ICIMOD Internal Document, 1990).

Fragility

Fragility is a condition that results from the geologic, edaphic, and biotic factors associated with steep slopes and high altitudes. It makes the natural resources or ecological system vulnerable to rapid or irreversible degradation even with slight disturbances.

By implication, this makes land suitable for only low intensity uses (e.g., forestry as against intensive annual cropping). This restricts the range of production options and limits the input absorption capacity as well as the productivity of land.

Agroforestry, owing to the involvement of a mixture of annuals and perennials, offers a compromise solution for such lands, where use intensity of fragile lands could be enhanced without exposing them to rapid erosion as under pure annual cropping. However, as indicated in Table 33.1 effective balancing of intensive and extensive uses of land through agroforestry can be achieved, provided the interventions in this area are fairly location-specific and have an explicit focus on the attributes of species (both crops and trees) that are conducive to soil building and soil binding, besides providing sufficient economic returns.

Marginality

The feature of marginality often infers being outside of the mainstream processes that occur in the larger socioeconomic, and political area. In this sense, marginality is something that counts least in decision-making processes.

The term may also refer to the condition of biophysical resources. Any resources that are too fragile, too unproductive, and too unattractive to be worthwhile for any use could be described as marginal. Both natural and man-made circumstances contribute to the marginal status of any resource or any community. Such circumstances are many in the mountain context.

Most of the operational implications of marginality (as far as they relate to biophysical resources) are similar to those of fragility. When related to socioeconomic marginality, the implications are manifested by poverty, neglect, and exploitation by the mainstream decision makers, and little participation in decisions affecting the self.

While highlighting the relevant attributes of agroforestry one can again refer to the biophysical and socioeconomic dimensions of marginality characterizing mountains. In discussing biophysical marginality, the attributes of agroforestry that match with the requirements of fragility may be mentioned. Besides, agroforestry, both because of resource regenerative attributes and the potential for high pay-off options, can help in upgrading the marginal resource base. Regarding socioeconomic marginality, some features such as scale-neutrality of the system, diversification potential for hedging against risk, contribution to resource/product recycling, and limited out of pocket cost may be mentioned. For these features, agroforestry can suit the needs and means of poor people. However, to achieve such goals, the choice of species with enough variability in specific plant characteristics and a 'target group' orientation of agroforestry programmes will be essential (Table 33.1, col. 3, row 3).

Diversity

This characteristic refers to the immense variation relating to different mountain specificities and their components, within and among the ecozones of mountain regions. This contributes to the extreme heterogeneity of production and of living environments, animal and plant life, as well as socioeconomic groups and their survival strategies.

Diversity is the source of a complex of constraints and opportunities for interventions in mountain areas. The response to diversity takes the form of diversified and interlinked activities that are a source of the stability and resilience of farming systems in the mountains. However, the conventional development interventions, with their bias for uniform and standardized contexts, often treat diversity as a major constraint to development through replicating options evolved in non-mountain situations.

Agroforestry, depending upon the choice of components (i.e., species with varied characteristics) offers the opportunity to fully respond to the diversity of land resources and production environments. It can encourage land uses and production activities with non-covariate flows of products and input needs (e.g., in the case of trees and annual crops), and can provide a basis for diversified, interlinked second-level (i.e., agro-processing) activities.

The major imperatives of the above discussions for agroforestry development interventions would call for a focus on species that meet diverse requirements and group or location-specific variations in the programme contexts (Table 33.1, col. 3, row 4).

'Niche'

Mountain areas are endowed with resources and an environment that make some activities and products extremely suited to them. In these activities (e.g., potential for irrigation and hydropower, timber production, and horticulture), mountains potentially have comparative advantages over the plains. If properly harnessed they may help in the sustainable development of mountain areas. If over-extracted (as usually happens) they may lead to resource degradation.

Some of the minor 'niches', based on perennial and even annual plants, offer unique opportunities for harnessing through agroforestry. This potential can be harnessed through specific focus on unique and high-value species which could fit into the intercropping of annuals and perennials. Their advantages could be further enhanced if agroforestry initiatives are supplemented by programmes in the field of processing and marketing (Table 33.1, col. 3, row 5).

Adaptation Mechanisms

Mountain people have devolved their own resource management and sustenance strategies. They evolve measures to amend mountain conditions (e.g., by terracing and irrigation) or adapt to the conditions (e.g., by choice of diversified, interlinked activities). The strategies involve both technological measures (e.g., ethno-engineering and folk agronomy) and institutional arrangements (e.g., provision of collective risk-sharing and common property resources).

Agroforestry, as indicated above, offers opportunities for the above two-way adaptation processes. Diversification and the potential choice of multipurpose options are conducive to such adaptations. Thus, there is a wide range of convergence between attributes of adaptation strategies and agroforestry systems. In fact, indigenous agroforestry is based on such convergence.

The degree of convergence can be increased through focussed public interventions. In fact, with the right focus, agroforestry can help meet subsistence needs and can encourage farmers' participation in the wider market economy. Viewed in this way, agroforestry could be a useful intervention for the transformation of mountain farming from a subsistence system into a semi-commercial system (Table 33.1, col. 3, row 6).

Some Agroforestry Interventions with Reference to Implications for Mountain Area Development

Much of the work done on agroforestry systems in the HKH region is limited to research on species' trials and descriptions of indigenous practices. Literature outlining the strategies and impacts of government and private interventions in the field of agroforestry is scanty and diffused.

Government-sponsored agroforestry programmes are apparently far more common in lowland than hill or mountain areas, according to a preliminary survey of existing projects in India, China, Nepal, Pakistan, and Bhutan. A further investigation of programmes in the mountain regions of these countries is ongoing (ICIMOD, in press). Samples of a few agroforestry interventions initiated by governments, NGOs, donors, and, indirectly, by market forces are described here to examine their relevance to mountain conditions.

Multipurpose Tree and Fodder Production by Village Organisations in Remote Areas

The Aga Khan Rural Support Programme (AKRSP) in Chitral, Gilgit, and Baltistan, Pakistan, offers technical and managerial assistance for farmers living high in Pakistan's mountainous northern areas[1]. AKRSP has recently initiated a programme to encourage

---

1 Derived from The Aga Khan Rural Support Program: Second Interim Evaluation. Washington, D.C.: World Bank, March 1990.

multipurpose tree planting on both public and private lands. A variety of species, including poplar, willow, and an exotic leguminous species, are produced in decentralized nurseries operated by village organizations (VOs). Fruit tree planting, long a tradition on the irrigated lands of this region, is done in a system of intercropping with alfalfa and vegetables to increase soil fertility and to generate an income source. The fruit trees themselves provide fodder and fuel from loppings and prunings.

The project provides seeds to encourage the traditional practice of planting alfalfa on newly developed (irrigated) lands. The AKRSP staff and farmers are conducting trials on new fodder legumes and appropriate management systems. Fodder cultivation is encouraged through the introduction of alternative crops and tree species, and free grazing of livestock is discouraged, by the VOs. The quality and digestibility of cereal straw feed is being improved through urea supplements and a new practice of chopping the residue into smaller pieces.

The success of AKRSP's initiatives in the area of agrisilvopastoral agroforestry, as for its other activities, may rest largely on the strength of the VOs and reliance on them for technology transfer. The programme has been responsible for a great increase in the availability of improved seeds and fertilisers through provision of credit and supplies. The construction of the feeder roads linking the area to the Karakoram Highway and the input supply system encouraged by the AKRSP have significantly reduced the negative impacts of inaccessibility. The project's emphasis on the development and maintenance of irrigation systems has allowed subsequent investment in the production of complementary annual and perennial crops for food and fodder.

Virtually all agroforestry production in the area is dependent on irrigation facilities. This focus has reduced the constraints posed by the inherent marginality of this remote region, strengthening the resource base and permitting more intensive land use.

Against this backdrop of improved accessibility to markets, technological inputs, and water resources, AKRSP's agroforestry activities aim to diversify production, build soil fertility, provide sources of income, and improve local timber, fuelwood, and fodder resources all through the actions of VOs and based on locale-specific needs and opportunities (World Bank 1990).

Viewed in the context of agroforestry development imperatives, the AKRSP programmes, described above match fairly well with the requirements of mountain specificities. In other words, they have a strong mountain perspective.

Pastureland Improvement with Low Levels of Input[2]

In Bhutan, a country with two-thirds of its land under forest cover and a sparse population, the need for agroforestry interventions *per se* has not arisen (although the indigenous method of shifting cultivation is still widely practised). Most farm households in the mountainous region are within a close distance from forests which are used for the collection of timber and fuelwood and as grazing lands for livestock. However, the overuse and deterioration of pasturelands is becoming increasingly obvious, particularly in the south where migrating herds of cattle are concentrated in the winter months. These

2 Information from Country Programme: Swiss Cooperation with Bhutan Helvetas. Bern: Swiss Development Cooperation, 1989, and Helvetas, *A Study Tour to Bhutan*. Kathmandu: Integrated Hill Development Project, 1990.

lands, unsuitable for cropping due to their steepness and thin soils, are particularly fragile and easily damaged by trampling.

The Royal Government of Bhutan (RGOB), with assistance from a Swiss donor, Helvetas, has initiated a livestock and Fodder Development programme, under the Animal Husbandry Department, to improve the productivity of livestock through an integrated approach. The programme focusses on 'scientific breeding', better feed resources, improved management, and effective animal health care. The secondary objectives are to establish a marketing system for dairy products and to reduce cattle migration. The government recognizes that activities to upgrade livestock conditions must also include measures to prevent ecological damage to grazing lands.

Given the shortage of manpower in Bhutan, the programme has encouraged the adoption of low labour input practices to cultivate food and forage crops together. Buckwheat and clover seeds (plus fertilizers) are sown simultaneously on communal pasturelands that have been apportioned off to private farmers. After harvesting the buckwheat crop, which has benefited from the nitrogen-fixing abilities of the clover, the plot is already established as productive pastureland. Clover will provide soil fertility and soil protection for several years, requiring no additional input of labour. This intercropping system, however, has no explicit place for perennials and is therefore not a true example of an agroforestry system. In many areas, forests surround large areas of grazing land, and bamboo is a common pastureland species.

The RGOB/Helvetas programme promotes the production of forage grasses in recognition of the 'niche' provided by these steep, sloping highland areas for the support of livestock production and to provide a marketing potential for dairy products. The fragility of such lands is well understood, so that species are selected for their ability to improve soil stability and soil nutrients. The constraints posed by labour shortages are also attended to.

However, despite RGOB's commitment to the guiding principles of decentralization, people's participation, and self-reliance, this programme depends heavily on the plans, technical advice, and inputs of seed and fertilizer from experts situated far from the site of the improved pasturelands. There is no evidence of an attempt to incorporate the traditional adaptation mechanisms of farmers, such as knowledge of indigenous grass species or existing social arrangements, to manage pasturelands (Helvetas 1990).

The above example indicates the sensitivity of public interventions to at least some of the important mountain specificities such as fragility, marginality, and 'niche'.

A Market-oriented Agrisilvopastoral System: Cardamom Plantations in the Eastern Himalaya[3]

Large cardamom is a plant native to Sikkim, where it grows under shade trees and is used as a spice and medicine. Originally, it was collected from natural forests, but, with the decline in area and density of trees in Sikkim's forests, farmers have begun to cultivate it on gentle sloping and steep-sloping lands. Today, 27 per cent of the land of Sikkim under agricultural production is used for cardamom cultivation. Most plantations are less than one hectare in size.

---

[3] Derived from K.A. Singh, R.N. Rai, Patiram, D.T. Bhutia. 'Large Cardamom (Amomum subulatum Roxb.) Plantation—An Age-old Agroforestry System in Eastern Himalaya'. In Agroforestry Systems 9:241–257, 1989.

Because cardamom requires an over-storey for shade, farmers plant various species of fuelwood, timber, and fodder species. Alder (Alnus) is the most common associated species; it is fast-growing, deciduous, and nitrogen-fixing. Therefore, it quickly provides shade, enriches the organic matter fertilisers for cardamom, and also yields fuelwood that is useful for the curing process.

In the middle hills of Sikkim, where milk production is the aim of the farming system, the planting of fodder trees instead of Alnus is increasing. Fallen leaves increase the soil's organic content, rendering it unnecessary to apply manure or fertilizer. The trees are lopped after the cardamom harvest at a time when there is a shortage of other green fodder material.

Cardamom is thus a low input crop with no need for tilling, manure, fertilizer, or pesticides. The Government Spice Board gives extension services and improved planting materials plus subsidies for fencing and replanting. The farmer's gross financial returns from cardamom cultivation are equivalent to those of rice production, but the profit margin is higher on account of lower input into the system.

This agrisilvopastoral system is well suited to the fragile lands of the steep-sloped Eastern Himalaya. With no ploughing of the soil and little removal of vegetation or woody biomass, the system resembles an undisturbed natural forest ecosystem. Tree species are selected for their complementarity to the cardamom system, to provide protection from the impact of sun and raindrops, to improve the soil base through leaf litter and nitrogen fixation, and to increase moisture retention through the infiltration and recharging of rainwater.

As the access to State-owned forests is curtailed by legislation, the secondary products of fuelwood, fodder, and timber have gained in importance.

The cardamom system demonstrates an appropriate use of the comparative advantage Sikkim has to produce a high-value, easily transportable product with a high market demand. Cultivation of the crop is ideal in that it provides for both subsistence and market-oriented systems in an ecologically sustainable way on marginal lands.

Although not an intervention of the government as such, the popularity of cardamom production is a direct response by farmers to market opportunities and other support offered by the government. As its production system is well known to Sikkimese farmers, little intervention is necessary to popularize cultivation or adapt the system to local conditions (Singh et al. 1989).

Promoting Agroforestry among Marginal Upland Farmers

Action Aid Nepal (AAN) has been implementing an integrated rural development programme in a mid-hill and high hill region of Central Nepal since 1982.[4] The organisation has identified food deficiency and non-sustainable agricultural production as the primary causes of poverty in the area, and has thus developed an approach to minimize dependency on external inputs, boost productivity, and create sustainable agricultural systems. Programme interventions include farmer-based research, farmer awareness-building, encouragement of local seed producers, improving quality and availability of compost manure and organic alternatives to chemical fertilizers, improving quality and quantity

---

4 Described in *Action Aid Nepal Progress Report (July 1988–June 1989) and Annual Management Plan.* Kathmandu: Action Aid Nepal, July 1989–1990.

of fodder, increasing livestock productivity, and providing alternative cash crop opportunities.

Farmers are motivated through literacy classes and meetings of grassroots groups to plant fodder trees on marginal land and terrace risers, and to stall-feed their animals. Improved grasses, including legumes, are introduced and propagated in grass nurseries for distribution of seeds and vegetative cuttings. These activities are closely linked to those of the livestock programme; farmers receiving improved breeds of cattle and goats are required to grow a specified quantity of fodder trees and/or grasses. Fodder seedlings of local species, according to those preferred by the farmers in the area, are raised in community nurseries managed by groups of trained and interested farmers. Women, as the target group of AAN's agricultural programme, often manage these nurseries.

Farmers of one area adjacent to an extensive tract of government forest have become avid tree planters on private lands after strict closure of the forest by the army denied them access to their sole source of timber and fuelwood materials.

Fruit tree and cardamom production are two activities aimed at increasing cash crop opportunities for the area's poorest farmers. Fruit trees are planted near homesteads and on nearby field borders, while cardamom is cultivated under alder tree groves in gullies and other marginal lands. AAN's strategy to reduce dependence on external inputs solves the problems posed by inaccessibility and makes its programmes relevant to the cash-poor marginal farmers who form the main target group. The focus on fodder production is to strengthen the weakest link in the farming system while improving the ecological base. Soil erosion losses from fields can be reduced by planting trees and grasses on otherwise barren terrace risers; stall-feeding reduces trampling and grazing on marginal lands, already prone to high erosion rates, and allows the natural regeneration of such lands. The promotion of local species taps the farmers' extensive knowledge of fodder trees and provides them with diverse options for planting.

The programme's reliance on community groups for motivation and extension, particularly those formed around the common intent to learn literacy skills, is its strength. AAN possesses the skills in community organization to form and work with true users' groups, and to plan and implement activities in a decentralized fashion with high levels of farmer participation (Action Aid Nepal 1990).

REFERENCES

Action Aid Nepal. Progress Report (July 1988–June 1989) and Annual Management Plan. Kathmandu: Action Aid Nepal, 1990.

Agarwal, A. and Narain. Village Ecosystem Planning. Dryland Networks Programme Issues Paper No. 16. London: International Institute for Environment and Development, March 1990.

Environmental Resources Limited (ERL). Natural Resource Management for Sustainable Development. Kathmandu: ERL, 1989.

Gupta, A.K. *Socio-Ecology of Natural Stress: 'Technological Change and Human Response in Bhutan.'* India: Indian Institute of Management, 1986.

Helvetas. Helvetas Country Programme: Swiss Cooperation with Bhutan. Bern: Swiss Development Cooperation, 1989.

Helvetas. A. A Study Tour to Bhutan. Kathmandu: Integrated Hill Development Project, 1990.

International Council for Research in Agroforestry (ICRAF). *An Account of the Activities of ICRAF*, Nairobi; 1983.

Jodha, N.S. 'Institutional Aspects of Range Resource Management in the Arid Zone of India.' Paper presented at the *Third International Rangeland Congress*, New Delhi, 7–11 November, 1988.

Jodha, N.S. A Framework for Integrated Mountain Development Mountain Farming Systems. Discussion. Paper No. 1 Kathmandu: ICIMOD, 1990a.

Jodha, N.S. 'Mountain Agriculture: Search for Sustainability.' *Journal of Farming Systems Research*—Extension, 1(1) 1990b .

Jodha, N.S. Sustainable Agriculture in Fragile Resource Zones: Technological Perspectives. Mountain Farming Systems. Discussion Paper No. 3. Kathmandu: ICIMOD, 1990c.

Jodha, N.S. 'Rural Common Property Resources: Contributions and Crisis.' *Economic and Political Weekly*, Quarterly Review of Agriculture. 25 (No. 26) 1990d .

Lundgren, B. Introduction. *Agroforestry Systems* 1 (1):3–6 1982.

Mulk, Masud ul. Farmers' Strategies and Their Implication for Sustainability of Mountain Agriculture in Chitral Pakistan. International Centre for Integrated Mountain Development and the Aga Khan Rural Support Programme. Kathmandu: ICIMOD, 1990.

Nair, P.K.R. 'Classification of Agroforestry Systems.' *Agroforestry Systems* 3:97–158, 1985.

Singh, K.A., R.N. Rai, Patriam and D.T. Bhutia. 'Large Cardamom (Amomum Subulatum Roxb). Plantation—An Age-old Agroforestry System in Eastern Himalayas.' *Agroforestry Systems* 9:241–257, 1989.

Wenner, Carl G. 1983. 'Trees in Erosion and Soil Conservation.' In *Agroforestry Systems for Small Scale Farmers*. Proceedings from ICRAF/BAT Workshops Nairobi: ICRAF/BAT Ltd.

Whiteman, P.T.S. 'Mountain Agronomy in Ethiopia, Nepal and Pakistan.' In N.J.R. Allan, G.W. Knapp and C. Stadel, (eds.), *Human Impacts on Mountains*. New Jersey: Rowman & Littlefield, 1988.

World Bank. The Aga Khan Rural Support Programme Second Interim Evaluation. Washington D.C.: World Bank, March 1990.

INDEX